102 # Topics in Current Chemistry

Fortschritte der Chemischen Forschung

Managing Editor: F. L. Boschke

Inorganic Ring Systems

With Contributions by
T. Chivers, R. Keat, J.-F. Labarre,
R. Laitinen, V. I. Lavrent'yev,
R. T. Oakley, R. Steudel, M. G. Voronkov

With 111 Figures and 39 Tables

Springer-Verlag
Berlin Heidelberg New York 1982

This series presents critical reviews of the present position and future trends in modern chemical research. It is addressed to all research and industrial chemists who wish to keep abreast of advances in their subject.

As a rule, contributions are specially commissioned. The editors and publishers will, however, always be pleased to receive suggestions and supplementary information. Papers are accepted for "Topics in Current Chemistry" in English.

ISBN 3-540-11345-2 Springer-Verlag Berlin Heidelberg New York
ISBN 0-387-11345-2 Springer-Verlag New York Heidelberg Berlin

Library of Congress Cataloging in Publication Data. Main entry under title: Inorganic ring systems.
(Topics in current chemistry; 102 = Fortschritte der chemischen Forschung; 102)
Bibliography: p. Includes index.
1. Cyclic compounds — Addresses, essays, lectures. I. Chivers, T. (Tristram), 1940 —.
II. Series: Topics in current chemistry; 102.
QD1.F58 vol. 102 [QD197] 540s [546] 81-24064 AACR2

2152/3020-543210

Table of Contents

Up-to-date Improvements in Inorganic Ring Systems as Anticancer Agents[1]

Jean-Francois Labarre

Laboratoire Structure et vie, Université Paul Sabatier, 118 Route de Narbonne, 31062 Toulouse Cedex (France)

Tables of Contents

---

[1] Plenary lecture delivered at the 3rd IRIS Meeting held in GRAZ (Austria), August 17–22, 1981

1 Introduction

Cancer does not exist. There exist actually 104 different types of cancers — or more precisely of *human tumors* — which may attack any part of the body and which nearly always enter by furtive and mysterious routes, depending on so many intangible factors, that mankind has up to now been completely powerless to prevent cancerisation in humans.

Thus, the only possible way to escape from the terrifying corners into which cancers are every day driving more and more humans is, of course, to detect them as early as possible and to start immediately the appropriate treatments, if any.

Unfortunately, the reality of the situation is not so simple. However large the arsenal of weapons that clinicians may have at their disposal, there are more than 40 kinds of tumors that cannot be cured at all nowadays by any medical approach. Thus, for such tumors, diagnosis means unstoppable slides to death. The situation is scarcely better for about 40 other tumors, mainly when they are so spread out over the whole body that surgery and radiotherapy cannot be used, so that chemotherapy and immunotherapy have to be employed.

In other words, in spite of the efforts performed and the relative successes obtained since the end of the 2nd world war within the field of this struggle for life, i.e., against cancers, we are still in the Stone Age. Thus, any new idea, approach or concept, even apparently crazy or fully outside the scope of the well-established dogmas, must be examined, criticized and tried out.

2 The Specific Role of Chemotherapy Amongst the Anticancer Weapons

Sword (surgery), artillery (radiotherapy), asphyxiating gases[2] (chemotherapy) and jiu-jitsu or aïki-do (immunotherapy) are the four weapons against cancers. They must be used in concert in any clinical treatment of localized tumors which can be concomitantly excised, irradiated and/or size-reduced by a drug.

However, as we mentioned above, chemotherapy and immunotherapy are still the only weapons applicable either when solid tumors are delocalized over a large area of the body or when the tumors are liquid (ascites tumors like leukemias).

Thus, chemotherapy must be considered as a curative technique which may be of vital help for any kind of tumor. Such a privileged role appears a bit surprising if we remember that about 60% of human cancers are generally assumed, according to world Health Office Statistics, to be induced by our chemical environment (tobacco, cosmetics, food dyes, nitrosamines, . . .). Consequently, treating cancers by chemistry looks a priori paradoxical. However, one may understand that some chemists have to be under the obligation to repair damage created by other chemists, so much the more that the percentage of chemists who become cancerous is about 25% larger than the one which is observed for non chemists.

---

[2] The first really efficient anticancer drugs were indeed "nitrogen mustards", derived from yperite, one of the notorious poison gases of the 1st World War.

3 How to Design an Anticancer Drug?

When surveying the literature on this subject over the last thirty years, we may notice that pathways for such a discovery were essentially as follows.

(i) About 80% of the drugs used to-day at the clinical human level were detected through the huge systematic screening developed by NCI (National Cancer Institute) in the States during the 1950's. Within the last three decades NCI has tested more than 800,000 chemicals on probably billions of tumor-bearing mice and rats. This "blindeyes" investigation has yielded about 30 effective drugs against several types of human tumors. The incredibly small ratio of the two previous figures may stupefy beotians. However, this somewhat desperate method has provided many powerful anticancer drugs which would have never been detected by any other, pseudo-logical, approach.

(ii) The discovery by Barnett Rosenberg of many very active platinum drugs started from a shrewd observation of an unexpected experiment: mitosis of cells in NH_4Cl buffer solution appeared to be deeply inhibited when subjected to an electric field produced by two "inert" (!) platinum electrodes ... Rosenberg could have concluded that the electric field was the inhibiting factor. But he did not fall into this trap; rather he demonstrated excellently that such an inhibition was actually due to some $(NH_3)_2PtCl_2$ entity produced by chemical reaction between NH_4Cl and the so-called inert Pt electrodes. The first cis-platinum drug was discovered in this way and everybody knows how fruitful this lucky find has been for further treatment of many cancers.

(iii) The antitumoral properties of some plants (roots, stems, leaves) and fungi were intuitively known by many tribes or peoples, sometimes for several centuries. For example, in 1609 a Dutch clinician reported at the end of a medical trip to Moluccas Islands that natives were successfully treating cancer of the nostrils by repeated applications of ground roots of a local plant called *Elliptica*. More than three centuries later, some Australian researchers prepared some anticancer alkaloïds derived from a molecule they called Ellipticine, owing to the fact that this chemical had been demonstrated as being actually the active principle of the Moluccan elliptica. Incidentally, ellipticine was also extracted from some other plants, i.e., apocynaceae (ochrosia moorei and excavatia coccinea).

In other words, several anticancer drugs are present in nature and they may be isolated by appropriate chemical techniques. The yield of such extractions, however, remains obviously very low and chemists are normally required to prepare synthetically large quantities of these natural products.

Other examples of natural drugs may be pointed out: streptozotocin (from streptomyces achromogenes), bleomycin A_2 (from streptomyces verticillus), adriamycin and daunomycin (from streptomyces pencetius), mitomycin C (from streptomyces caesipitosus), vincristine, vinblastine and vindoline (from catharanthus roseus or vinca rosea L.).

(iv) In contrast with the previous approaches, one may envisage a more logical route based on some structural molecular peculiarities which would be common to several individual drugs or series of drugs.

Let us consider for example the geometrical structure of Rosenberg's active platinum drugs. Up to now, the most efficient anticancer derivatives of the series have a square-planar Pt (II) uncharged cis-structure (Fig. 1). The activity of such drugs was demonstrated as occurring through a *dialkylation* of DNA on N7 and O6 sites of guanine (Fig. 2) by the mean of the two labile Cl atoms of the molecule. Rosenberg claimed that the distance (3.4 Å) [1] between these chlorines may be considered as a "magic number" for suitable strong dialkylation of DNA.

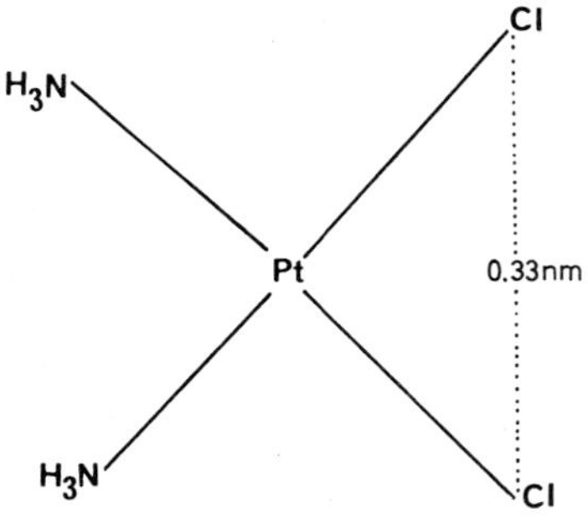

Fig. 1. Geometry of Rosenberg's cis-platinum

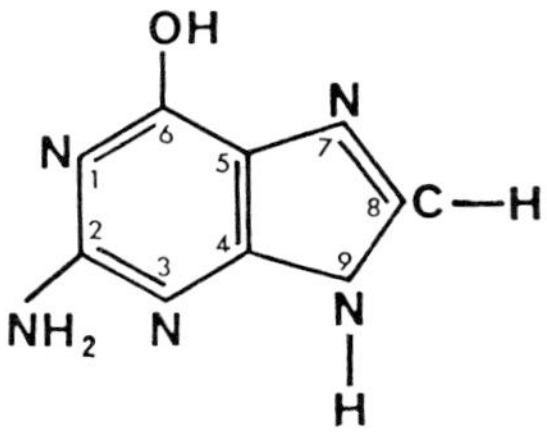

Fig. 2. Numbering of atoms in Guanine

Let us now consider some other anticancer drugs such as ellipticines or adriamycin, which prevent the replication of vicious DNA through a process of *intercalation* between plates (A ... T and G ... C) of bases. The effectiveness of these drugs seems related to the plus or minus chemical stability of such an intercalation and this stability is generally determined by strong hydrogen bonds between endocyclic nitrogen atoms or molecular oxygen atoms with sites on the ribose backbone of DNA. This binding mechanism obviously depends on the basicity of the N and O atoms in question: the larger the basicity, the stronger the stability of interaction and, consequently, the higher the effectiveness.

Thus, if a molecular structure contains both (i) pairs of chlorine atoms in a "square-planar-like" 3.4 Å situation and (ii) a planar ring with highly basic endocyclic N or O atoms, we may expect that the coexistence of these two structural peculiarities will confer a potential antitumor activity on the molecules in question.

This is actually the case with hexachlorocyclophosphazene (Fig. 3) and relatives and this is the starting idea of the investigations, which we began in 1976, on the application of cyclophosphazenes as anticancer agents.

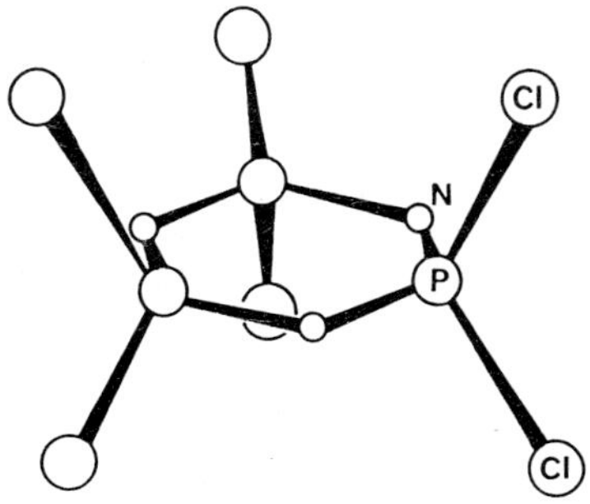

Fig. 3. Geometrical structure of $N_3P_3Cl_6$

4 Cyclophosphazenes as Anticancer Agents

The first investigations on the antitumor activity in vivo of cyclophosphazenes were started late in 1976 on murine L1210 and P388 leukemias (DBA/2 female mice) and on sub-cutaneous (s.c.) B16 melanoma (C57 black female mice).

For the first step, we chose 8 cyclophosphazenic systems containing either chlorine pairs (Rosenberg magic number) or highly basic endocyclic nitrogen atoms (intercalation process), namely $N_3P_3Cl_6$ (I), $N_4P_4Cl_8$ (II), $N_3P_3Az_6$ (III), $N_4P_4Az_8$ (IV), $N_3P_3Pyrro_6$ (V), $N_4P_4Pyrro_8$ (VI), $N_3P_3Morph_6$ (VII) and $N_4P_4Morph_8$ (VIII) (Az = Aziridinyl; Pyrro = Pyrrolidinyl; Morph = Morpholinyl).

4.1 Synthesis and Purity

(I) was obtained at that time from Fine Chemicals (purity > 93%), (II) was obtained from R. A. Shaw (Birkbeck College, London) to whom we are indebted for this generosity. These materials were recrystallized at least 6 times from acetonitrile for (I) and from petroleum ether (60–80 °C) for (II). n—Hexane should be avoided as crystallizing solvent since (I) exhibited arrangement therein, leading to (II), $N_5P_5Cl_{10}$ and higher isologs, as demonstrated by ^{31}P NMR spectroscopy ($\delta^{31}P$ = −19.5, +8.0, +17.0 and +19.5 ppm respectively for (I), (II), $N_5P_5Cl_{10}$ and higher isologs with 85% H_3PO_4 as a standard).

The three trimers, (III), (V) and (VII), were prepared using Rätz's procedure [2] in the presence of ammonia; the tetramers, (IV), (VI) and (VIII), were obtained by the same process but in the presence of triethylamine.

4.2 Solutions

Since all the derivatives studied, except (III), had a very poor solubility in water (<2 g/l), they were inoculated as suspensions in 4‰ hydroxypropylcellulose (Klucel J. F., Hercules Co) solutions in water. Such Klucel solutions were shown to be non-toxic after i.p. (intra-peritoneal) inoculation. Moreover, the size of the molecular aggregates in such Klucel solution was much less than 1 µm, as demonstrated by electron microscopy.

(III) on the contrary is highly soluble (without any hydrolysis) in water, about 100 g/l, and could consequently be used in 0.9% aqueous NaCl solution (saline) as for both the i.p. and the i.v. (intraveinous) route.

4.3 Toxicity Measurements

The toxicity of the drugs was determined either on female Swiss or on female DBA/2 mice. The lethality — which happens systematically 5–6 days after administration — was recorded as a function of the dose inoculated and allowed to deduce the LD_0 values which correspond to the highest non-lethal doses. The results do not depend on the species of mice we used.

LD_0 values are gathered in Table 1. Two points may be emphasized.

(i) For a given substituent X the trimer $N_3P_3X_6$ and the corresponding tetramer $N_4P_4X_8$ have practically the same LD_0 value, except in the case X = Cl where the trimer (I) appears to be about 15 times more toxic than its tetrameric isolog (II).

(ii) For (III) the i.p. and i.v. LD_0 are quite similar, about 40 mg/kg.

Table 1. Highest non-lethal doses LD_0 for the 8 cyclophosphazenes studied. (in mg/kg)

$N_3P_3Az_6$	(III)	40
$N_4P_4Az_8$	(IV)	75
$N_3P_3Pyrro_6$	(V)	10
$N_4P_4Pyrro_8$	(VI)	20
$N_3P_3Morph_6$	(VII)	150
$N_4P_4Morph_8$	(VIII)	>600
$N_3P_3Cl_6$	(I)	20
$N_4P_4Cl_8$	(II)	300

4.4 Antitumor Tests

The L1210 and P388 cells were maintained by weekly passages (i.p. inoculation) of ascites cells in female DBA/2 mice (Centre de Selection et d'Elevage des Animaux de Laboratoires du CNRS, Orléans—La Source, France). The B16 cells were maintained by 10–14 days passages (s.c. inoculation) of solid tumor cells in female C57 black mice (same origin). Experiments were conducted on mice weighing (20 ± 2) g which were about 2 months old. Fifteen mice were used per group and the deaths were recorded daily at the same hour.

The mean survival times of the treated mice (T) and of the control (C) were used to calculate the percentage increase in median life time over control

$$\%\,\text{ILS} = \frac{T - C}{C} \cdot 100$$

which is significant for antitumor activity only when higher than 25%.

For B16 melanoma, the median diameter of the tumors on the 12th and on the 14th day after the tumor graft were measured for the control and treated sets of mice: this blind procedure does indeed enable us to appreciate the way in which a drug delays and/or inhibits the growing of a solid tumor.

The antitumor tests were performed using the standard NCI protocols [3]: the leukemia or the melanoma was transplanted on the day D, and the drug was inoculated by i.p. route (except for (III) where i.p. and i.v. routes could be used) on the day D + 1 (monoinjection protocol) or within a QnD schedule (the same dose being injected at intervals of n days from the day D + 1).

4.5 Antitumor Activity of Compounds (I) to (VIII) [4)]

4.5.1 Effects on the P388 Leukemia

In Table 2 are shown the activities of (III), (IV) and (VI) under various conditions. The five other cyclophosphazenes, including the chlorine derivatives (I) and (II), were indeed found to be just at the limit of a significant activity (i.e., % ILS ~25) within heavy Q4D (1, 5, 9, 13, 17) schedules. In other words, it is clear that the magic number assumption does not work as well as we could have expected; however, $N_3P_3Cl_6$ and $N_4P_4Cl_8$ must be considered as antitumor agents on P388, even if a repeated long polyinjection schedule is required to make their effectiveness conspicuous.

Table 2. Antitumor activity of some cyclophosphazenes against P388 leukemia

Compound	Schedule	Dose (mg/kg/day)	%ILS
	Once, day 1 (i.p.)	2.5	21
		10	51
		20	101
$N_3P_3Az_6$	Q4D; days 1, 5, 9	10	100
(LD_0 = 40 mg/kg)	Once, day 1 (i.v.)	10	17
		20	47
		30	69
	Once, day 1 (i.p.)	10	18
		20	47
$N_4P_4Az_8$		30	49
(LD_0 = 75 mg/kg)		40	54
		50	57
	Q4D; days 1, 5, 9	40	101
$N_4P_4Pyrro_8$	Once, day 1 (i.p.)	5	24
(LD_0 = 20 mg/kg)		10	33

10^6 P388 cells implanted i.p., i.p. or i.v. treatment (15 mice per group); $N_3P_3Az_6$ was dissolved in 0.9% NaCl solution; $N_4P_4Az_8$ and $N_4P_4Pyrro_8$ were suspended in 4‰ Klucel JF (Hercules Co.) water solution; median survival time of control: 9.9 days.

From Table 2, the following points may be emphasized.

(i) (III) appears to be the most active member of the series: indeed a single dose of 2.5 mg/kg leads to an ILS value which approaches the 25% level whereas an injection of 20 mg/kg ($LD_0/2$) enhances the ILS value to 101%. It may be noted that an i.p. dose of 40 mg/kg (LD_0) leads to an ILS equal to 166% (*not including 2 cured mice*) but the acute mortality (6 mice over 15 on days 5–6 after administration) under such conditions could not be considered as acceptable.

The therapeutic index of (III), defined as the ratio of the LD_0 value divided by the dose which gives an ILS of 40%, is about 6.

(ii) The use of the Q4D (1, 5, 9) schedule noticeably increases ILS figures with respect to the monoinjection D + 1 protocol: the ILS value is indeed multiplied by a factor 2 when passing from a (1.10) injection to a Q4D (3.10) one. A factor of 2 is also observed for (IV) when passing from a (1.40) injection to the Q4D (3.40) one.
(iii) Figure 4 shows the linear dose-activity relationships for (III) by i.p. and i.v. routes. The i.v. route affords ILS values of about 70% in monoinjection without side-toxicity. It may be noticed that an ILS value of 91% was obtained for an i.v. dose equal to 40 mg/kg but the accompanying side-toxicity (6 mice over 15 on days 5–6) was not acceptable.
(iv) The dose-activity relationship for (IV) in a monoinjection protocol by i.p. route (Fig. 4) in Klucel appears to be linear between 10 and 20 mg/kg, with a levelling-off trend occurring for higher doses *without any significant side-toxicity*. The therapeutic index of (IV) (as defined above) is *ca. 4.*

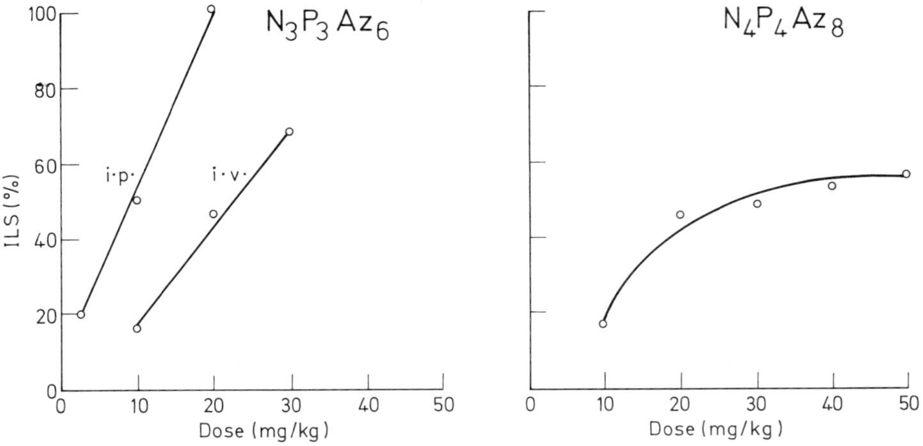

Fig. 4. Activity-dose relationships for $N_3P_3Az_6$ and $N_4P_4Az_8$ on P388 leukemia

From the foregoing results, $N_3P_3Az_6$ seemed the most promising antitumor agent of the series tested against the P388 leukemia and has the additional advantage of a high solubility in water and of being active both by the i.p. and the i.v. route.

4.5.2 Effects on the L1210 Leukemia

The tests on L1210 leukemia (as well as on B16 melanoma) were confined to the members of the series which exhibited a significant activity on the P388 tumor.

The activities for (III) and (IV) under various conditions using the i.p. route are shown in Table 3. (VI) was found to be non-significantly active, even within a Q3D (1, 4, 7) schedule.

ILS Figs. are definitely smaller for the L1210 than for the P388: (III) is the only compound which exhibits a significant (i.e., % ILS > 25) activity in a monoinjection protocol. A Q3D (1, 4, 7) schedule — chosen in order to take into account the fact that the median survival time of control animals is only about

8.5 days for L 1210 *vs* 9.9 days for P388 — has to be used to get significant ILS values for (IV).

However, as with the P388 tumor, ILS figures are approximately twice as large, for a given dose, using the Q3D polyinjection schedule compared with the D + 1 monoinjection protocol.

Table 3. Antitumor activity of some cyclophosphazenes against 1210 leukemia

Compound	Schedule	Dose (mg/kg/day)	%ILS
$N_3P_3Az_6$ (LD_0 = 40 mg/kg)	Once, day 1	10	28
		20	45
	Q3D; days 1, 4, 7	10	44
$N_4P_4Az_8$ (LD_0 = 75 mg/kg)	Once, day 1	40	8
		50	22
		60	17
	Q4D; days 1, 5	40	20
	Q3D; days 1, 4, 7	40	44
$N_4P_4Pyrro_8$ (LD_0 = 20 mg/kg)	Once, day 1	5	9

10^5 L 1210 cells implanted i.p., i.p. treatment (15 mice per group); $N_3P_3Az_6$ was dissolved in 0.9% NaCl solution; $N_4P_4Az_8$ and $N_4P_4Pyrro_8$ were suspended in 4‰ Klucel JF (Hercules Co.) water solution; median survival time of control: 8.5 days.

Table 4. Antitumor activity of some cyclophosphazenes against B16 sub-cutaneous melanoma

Compound	Schedule	Dose (mg/kg/day)	%ILS
$N_3P_3Az_6$ (LD_0 = 40 mg/kg)	Once, day 1	30	17
		40	53
	Q3D; days 1, 4, 7, 10, 13	10	22
	Q4D; days 1, 5, 9	20	28
	Q3D; days 1, 4, 7, 10, 13	20	39
$N_4P_4Az_8$ (LD_0 = 75 mg/kg)	Once, day 1	50	10
	Q4D; days 1, 5, 9	40	30
$N_4P_4Pyrro_8$ (LD_0 = 20 mg/kg)	Q4D; days 1, 5, 9	10	10

B 16 cells implanted s.c., i.p. treatment (15 mice per group): $N_3P_3Az_6$ was dissolved in 0.9% NaCl solution; $N_4P_4Az_8$ and $N_4P_4Pyrro_8$ were suspended in 4‰ Klucel JF (Hercules Co.) water solution; median survival time of control: 22.0 days.

4.5.3 Effects on the B16 Sub-Cutaneous Melanoma

B16 sub-cutaneous melanoma is a slow-growing tumor when compared to the L1210 and P388 leukemias. This is a very tedious tumor on which very few drugs were found to be significantly active, (i.e., % ILS > 40%).

From Table 4, (III) is the only cyclophosphazene giving ILS approaching or exceeding the 40% level, either using a D + 1 monoinjection (1.40) protocol of treatment or within a Q3D (1, 4, 7, 10, 13) (5.20) schedule. Furthermore, the quantity of the first inoculation (i.e., 40 mg/kg on the day D + 1) seems to be determinant, the ILS value (53%) being larger than that (39%) obtained with the Q3D schedule.

For some of these treatment schedules (Table 5) both the number of tumor-bearing mice and the tumor diameters on the 12th and on the 14th day were recorded: when compared to the control, the number and the size of tumors for the treated mice were considerably smaller, indicating the drugs were genuinely effective against s.c. B16 melanoma. Moreover, these two parameters are in good accord with the ILS determinations.

Table 5. Comparative effect of some cyclophosphazenes on B16 tumor evolution

	12th day		14th day		
Compound and schedule	Number of tumor-bearing mice	Size of tumors (cm)[a])	Number of tumor-bearing mice	Size of tumors (cm)[a])	%ILS
Control	14/15	1.2 ± 0.2	15/15	1.4 ± 0.1	—
$N_3P_3Az_6$					
(1.40)	5/13	0.4 ± 0.2	7/13	0.6 ± 0.2	53
(5.10)	10/15	0.8 ± 0.2†	10/15	1.0 ± 0.2	22
(5.20)	6/14	0.5 ± 0.2	8/14	0.5 ± 0.1	39
$N_4P_4Az_8$					
(3.40)	9/15	0.6 ± 0.2	11/15	0.7 ± 0.2	30
$N_4P_4Pyrro_8$					
(3.10)	9/13	0.6 ± 0.1	12/13	1.0 ± 0.1	10

a Mean ±S.E. of the mean, calculated on the total number of mice per group (a non-tumored mouse was counted as zero but involved into the calculation). The difference in median size of treated and control series were statistically significant (Student's t-test) at $P < 0.05$ unless for † where $P < 0.10$ only.

In conclusion of these antitumor tests on murine L1210, P388 and B16 tumors, it might be pointed out that the most effective in each case was the hexaziridinocyclotriphosphazene $N_3P_3Az_6$ whereas $N_3P_3Cl_6$ and $N_4P_4Cl_8$ appear to be poorly active, at least under our experimental conditions. Such a result requires some comments in the light of the ideas developed above.

(i) $N_3P_3Az_6$ is a molecule in which the planar N_3P_3 ring carries *pairs* of aziridino groups which may act as bifunctional alkylating agents of DNA by way of the classical reaction

$$-N\triangleleft \xrightarrow{H^+} -NH-CH_2-CH_2^+$$

With regard to the relative antitumor activities of $N_3P_3Az_6$ and $N_3P_3Cl_6$, aziridino groups seem consequently to be more active in that respect than chlorine atoms when substituted on a cyclophosphazene ring. It is noteworthy that the graft of such cyclic amino ligands on a phosphorus-nitrogen ring enhances drastically the intrinsic basicity of the endocyclic nitrogen atoms, from $pK'_{a_1} \sim -6.0$ for $N_3P_3Cl_6$ to pK'_{a_1} larger than 8 for $N_3P_3Az_6$ [5]. Thus, the second assumption we made above nicely fits the trend of activity we observe in vivo when passing from the chlorine derivatives to their amino isologs.

(ii) $N_3P_3Az_6$, containing aziridino groups, could be considered as belonging to the same class of antitumor agents as Thiotepa ($SPAz_3$), TEM (trisaziridinomelamine, $N_3C_3Az_3$), aziridinyl benzoquinones [6] or (1-aziridinyl) 2,4-dinitrobenzene [7], i.e., *alkylating antitumor agents*. Actually, we shall see below how $N_3P_3Az_6$ differs drastically in several respects from these classical alkylating agents.

Anyhow, $N_3P_3Az_6$ appeared to be interesting enough to urge us to engage multidisciplinary and international collaboration, such as is required for the development of any drug, for further clinical purposes. We shall detail in the next paragraph the strategy of development which we undertook in the case of $N_3P_3Az_6$.

5 Strategy of Development of $N_3P_3Az_6$ as Anticancer Agent

5.1 Further Antitumoral Tests at the Animal Level

As soon as our preliminary results on P388, L1210 and B16 tumors were announced in the literature [4], the E.O.R.T.C. (European Organisation for Research on Treatment of Cancer) decided to include $N_3P_3Az_6$ in its 1978–1980 plan for testing many new animal (mice and/or rats) tumors. At that time (May 1978), we had to provide a code name for this new anticancer drug: MYKO 63 was chosen because of the striking similarity between the X-ray crystal structure of this compound and the

Table 6. Activity in vivo of MYKO 63 on several animal tumors

Tumor	Schedule	Dose (mg/kg)	T/C %
P388 leukemia	day 1–9	10	260
L1210 leukemia	day 1, 5 and 9	40	240
B16 s.c. melanoma	day 1	40	153
Yoshida sarcoma	day 1	10	cure
P815 mastocytoma	day 1–9	5	cure
ependymoblastoma	day 1, 5 and 9	20	143
line 26 colon carcinoma	day 1, 5 and 9	20	180
line 16 mammary carcinoma	day 1, 5 and 9	20	170
M5076 ovarian carcinoma	day 1, 3, 5, 7 and 9	16	153
osteosarcoma	day 1, 3, 5, 7 and 9	20	cure

From Laboratoire Structure et vie, Toulouse University and from EORTC (Manchester and Milan Laboratories) .

MYKONOS wind-mills [8)] (see below); the first figure refers to the number of atoms within the phosphazenic ring (6 = N_3P_3, 8 = N_4P_4) and the second figure is connected with the number of atoms within the cyclic amino ligand (3 = Az, 5 = Pyrro).

Incidentally, the code names of $N_4P_4Az_8$ and $N_4P_4Pyrro_8$ are MYKO 83 and MYKO 85.

MYKO 63 was simultaneously tested on several animal tumors mainly by 3 laboratories belonging to the pharmacology screening group of EORTC: (i) Mario Negri Institute (Dr. F. Spreafico), Milano, Italy; (ii) Christie Hospital and Holt Radium Institute, Paterson Laboratories (Dr. B. W. Fox), Manchester, U.K. and (iii) Experimental Screening group of Jules Bordet Institute (Dr. G. Atassi), Brussels, Belgium. We are greatly indebted to these groups for their valuable and fruitful help with this project.

During this 1978–1980 campaign, MYKO 63 was found to be effective *on all the new 18 animal tumors it was tested on* (Table 6), mainly on Yoshida sarcoma (scrotum tumor) (cured within a monoinjection protocol at 10 mg/kg), P815 mastocytoma (mammary tumor with metastases) (cured within a chronical Q1D treatment at 5 mg/kg daily), M5076 ovarian carcinoma and osteosarcoma (bones tumor) (cured within a Q2D schedule at 20 mg/kg). It may be emphasized that MYKO 63 appeared on these tests to be as effective as Rosenberg's cisplatinum DDP on ovarian carcinoma. This is one of the most promising results through the series of data presented in Table 6 owing to the fact that DDP is the only drug used nowadays by clinicians on human ovarian carcinomas.

Moreover, in view of such an "universal" activity of MYKO 63, EORTC decided in May 1980 to extend to 1982 the period during which this drug will be tested on animal tumors. To date new results increase still further the spectrum of effectiveness of MYKO 63: successful experiments were performed — or are in progress — on tumors such as lymphomas and lymphosarcomas, either radio-resistant or radio-sensitive, on neuroblastomas (acute brain tumors, mainly in children), fibroblastomas (muscles tumors) and HeLa (from Helen Lattimer, the first patient who died from this tumor) cervix carcinoma (uterus tumor).

Such a wide activity — without any example demonstrating the opposite up to now — is so scarce in cancer chemotherapy that it stimulated several scientists to take a part in further concerted investigations which had to be carried out.

Thus, 16 different groups of research were involved from 1978 to 1981 in the MYKO 63's project. Many problems, of course, had to be solved and we wish to explain now what main results were obtained during the last three years both in vivo and in vitro.

5.2 Mass Spectrometry and Neutron Activation: Two Techniques for Testing the Purity of MYKO 63 and Relatives [9)]

As we are chemists, it is our job and privilege to provide pure samples of MYKO 63 for EORTC and other preclinical studies.

Thus, the chemical synthesis was repeated on a large scale and we checked carefully that each new sample had, of course, a reproducible effectiveness on P388, L1210 and B16.

This was always the case except once where a new sample, which we shall call B, gave an anomalous answer on P388 when compared to the classical response obtained with any sample A. As elemental analysis, melting point, infrared and ^{31}P nuclear magnetic resonance data were identical for the samples A and B, we would have expected similar antitumor activities for both; actually, we noticed that sample B had a slightly higher activity on P388 but exhibited significantly higher toxicity. In other words, a monoinjection of 30 mg/kg of A gave rise to an ILS equal to 154% (no cured mice), 1 mouse out of 15 treated having died on the day 6 of the experiment (side-toxicity), whereas B, under the same conditions, allowed an ILS equal to 166% to be reached — excluding 2 cured mice on the 45th day — but with a side toxicity (day 6) of 40%.

We thought that this enhancement of toxicity was due to an impurity. Its low level required the use of suitable techniques for detection and quantification. Due to their low volatility and to the lack of derivatization possibilities for forming volatile products, samples A and B could not be analyzed by gas chromatography (GC). Initially reverse-phase high performance liquid chromatography (HPLC) was applied. However, due to the lack of ultraviolet absorption above 200 nm, a differential refractometer was used as detector, which considerably limited the sensitivity. In order to obtain improved sensitivity we decided to examine the possibility of using the direct inlet of a mass spectrometer as a method for detecting and quantifying the toxic impurity.

The instrumentation we used for HPLC and mass spectrometry analysis is reported elsewhere [9]. Let me only give some details about detection and quantification of the impurity.

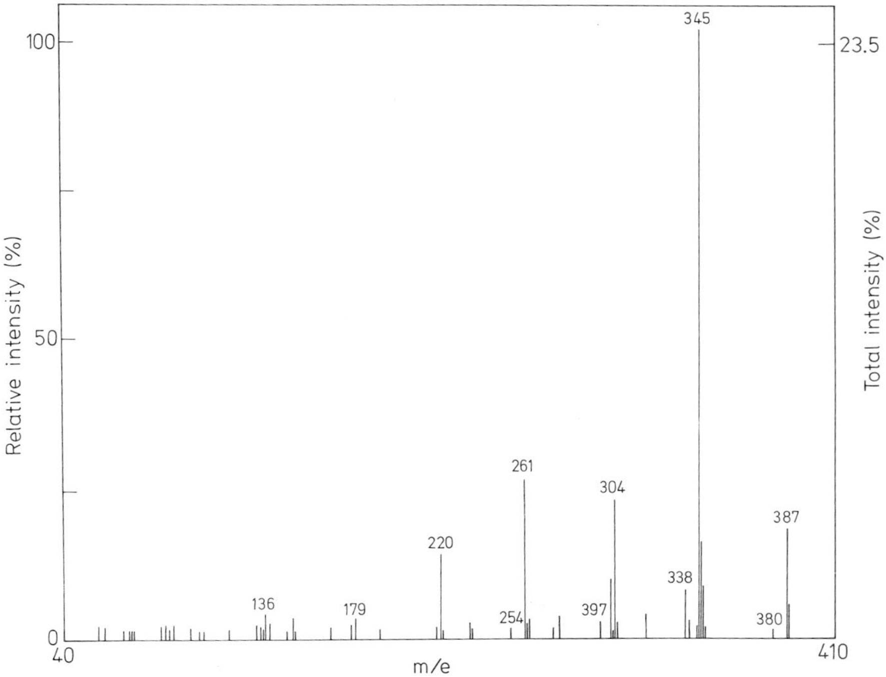

Fig. 5. Mass spectrum of the pure MYKO 63 (sample A; mol. wt = 387)

5.2.1 Detection of the Impurity in a Toxic Sample of MYKO 63

The 70 eV electron impact mass spectrum of pure MYKO 63 (sample A) is presented in Fig. 5. The molecular ion is observed at m/z 387. 2 fragmentation routes are detected. In the first, an aziridino radical is expelled giving the base peak at m/z 345. Alternatively, the loss of acetylene produced the lower intensity peak at m/z 361 and subsequent loss of an aziridino group gave the peak at m/z 319. From these ions, further consecutive losses of the aziridino substituent are associated with H-transfers. Thus, the expulsion of an aziridine molecule (43 mass units) from m/z 345 gave the ion at m/z 302 or, alternatively, the loss of 41 mass units (C_2H_3N = Az minus H) produced the ion at m/z 304. The elimination of a 3[d] substituent gave the ions at m/z 261 and 263 and the ion at m/z 220 finally resulted from the loss of a fourth substituent. Further elimination gave less characteristic peaks. A similar series is observed from the m/z 319 ion giving first m/z 276 and 278, then m/z 235 and finally m/z 194.

From the simplicity of this spectrum, it would be expected that the superimposition of peaks due to possible contaminants in the spectrum of the "toxic" sample B would be clearly observed. To facilitate their detection, we constructed the differential spectrum between the toxic sample B and the pure A one, in the normalized forms. This spectrum is presented in Fig. 6.

In the high mass range, the intensity ratio between two peaks at m/z 382 and 380 indicates the presence of a chlorine atom. The largest doublet at m/z 338 (base peak) and 340 arose from the loss of an aziridino radical. The two peaks at m/z 254 and

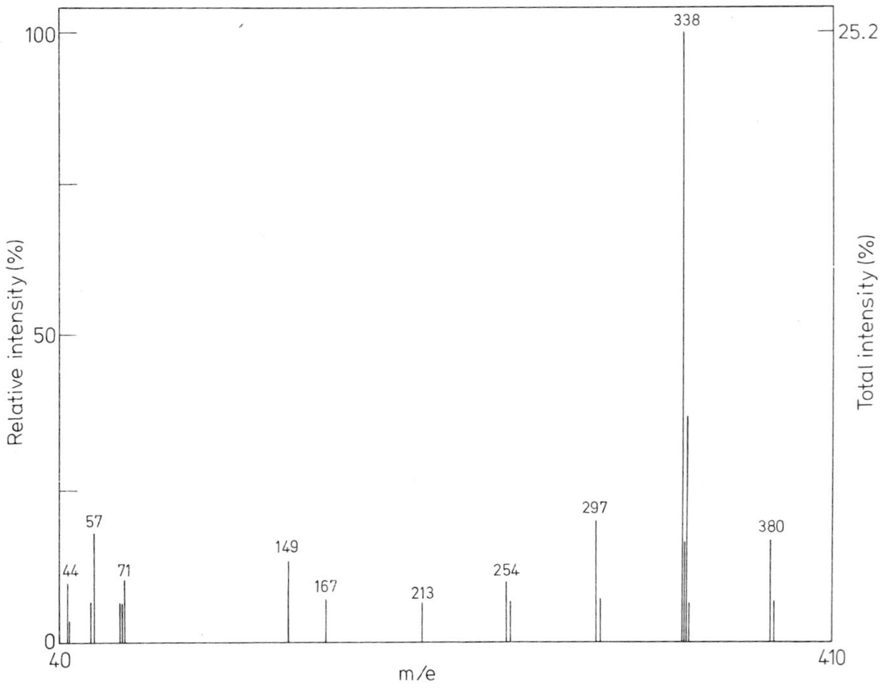

Fig. 6. Differential spectrum between MYKO B ("toxic") and MYKO A ("pure") forms

256 indicate the loss of a second substituent associated with H-transfer. Then m/z 213 arose from loss of a third aziridino substituent (in this case, the isotopic pattern of chlorine has probably been modified by back-ground subtraction). Thus, this differential spectrum corresponds definitely to the monochloropentaziridinocyclo-triphosphazene, $N_3P_3Az_5Cl$, M = 380 which is the searched for impurity. This pentaziridino derivative comes from an incomplete substitution of the $N_3P_3Cl_6$ chlorines by aziridino groups and it is clear now from antitumor tests performed with A and B that $N_3P_3Az_5Cl$ must be both more active and much more toxic than MYKO 63.

$N_3P_3Az_5Cl$ was extracted in a pure state by liquid chromatography from a complex mixture of partially substituted aziridino derivatives of $N_3P_3Cl_6$. Its LD_0 was found to be 12 mg/kg (about 4 times more toxic than MYKO 63) and its activity on P388 was only about 15% higher than MYKO 63's one. Thus, $N_3P_3Az_5Cl$ was not considered favourably as an improvement of MYKO 63.

5.2.2 Quantification of $N_3P_3Az_5Cl$ in MYKO 63 Samples

Selected ion monitoring can be used for the determination of the relative amount of each component of a mixture, introduced into the mass spectrometer by the direct inlet probe [10]. However, such a determination requires reference mixtures of known composition for calibration. In the present experiment, since the monochloro pentaziridino derivative had not yet been isolated in the pure form, it was necessary to determine its concentration, by an auxiliary method, in a sample which could then be utilized as a reference mixture for further experiments. In order to do this we titrated chlorine in the toxic sample of MYKO 63 (B) by the classical method. The results indicated that the amount of $N_3P_3Az_5Cl$ was between 0.5–1.5%. The large statistical error is due to the low chlorine content in the sample examined. Thus, we used the remarkable possibilities provided by neutron activation analysis when the impurity to be quantified is a chlorinated moiety. It is well-known indeed that the $^{35}Cl + 2n \rightarrow {}^{37}Cl$ peak is amongst the most easily detectable by neutron activation. We are warmly indebted to D. E. Ryan, Trace Analysis Research Center, Dalhousie University, Halifax, Nova Scotia, Canada, who performed the dosage of $N_3P_3Az_5Cl$ in the B sample. He found that the content of B in $N_3P_3Az_5Cl$ is 1.0% and we were thus allowed to come back to a quantification of this impurity in any toxic sample of MYKO 63 by mass spectrometry.

For the measurements by SIM, we selected the doublet at m/z 338–340 in the chlorine derivative. A constant ratio between these two peaks would indicate the absence of any interfering ions in the different samples. The molecular ion at m/z 387 was selected for the $N_3P_3Az_6$ compound. The samples for the calibration curve were obtained by adding known amounts of the toxic MYKO 63 (B) to the purest sample of this drug. During the slow vaporization of the samples into the ion source, the intensity of these three ions was recorded continuously. The corresponding curves were found to be exactly homothetic (which indicated a similar vapour pressure for both the impurity and the pure sample). The area ratio of these curves was plotted against the composition of the mixture (Figs. 7 and 8) and the observed linear relationship indicated that neither a non-reversible adsorption process nor interfering ions are produced. By calibration with the known concen-

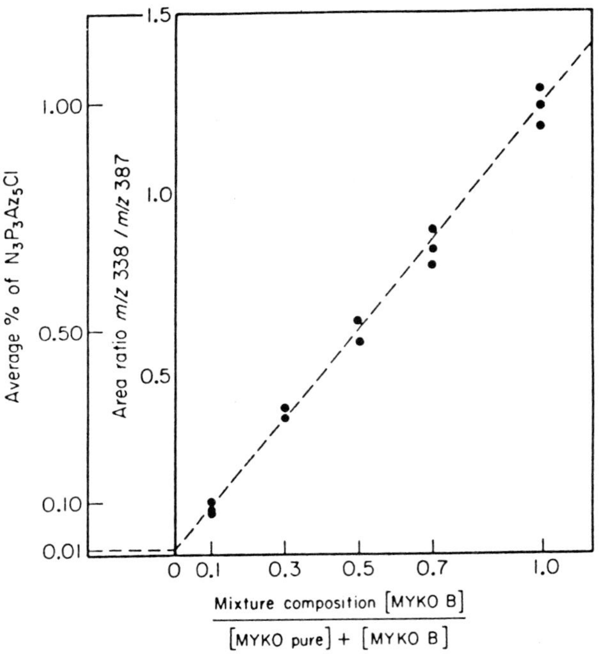

Fig. 7. Quantification of N_3P_3-Az_5Cl: relationship between the abundance ratio of the ion m/z 338 ($[M - Az]^+$ of the impurity) and m/z 387 (molecular ion of MYKO 63) and the composition of the mixture

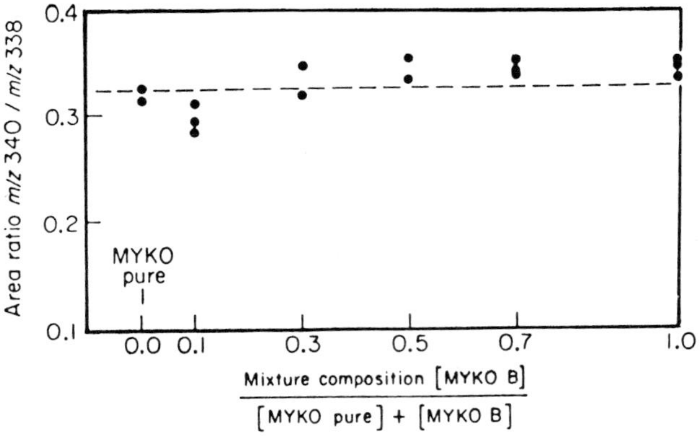

Fig. 8. Variation of the abundance ratio of the m/z 340 and 338 ions (isotopic pattern of the $[M - Az]^+$ ions from $N_3P_3Az_5Cl$) relative to the composition of the mixture

tration of the chlorine derivative in the MYKO 63 (B) mixture, a level of about 0.01% of this impurity was determined in the purest preparation of this drug.

5.2.3 Mass Spectra of MYKO Analogues [9)]

We recorded mass spectra of MYKO relatives in order to check if the simplicity of the fragmentation pathway found in the MYKO 63 spectrum was also observed with other kinds of substituent on the phosphorus atoms, or when the central ring system was modified.

Figure 9 shows the spectrum of $N_3P_3Pyrro_6$ (MYKO 65), an analogue of MYKO 63 in which the aziridino groups are replaced by pyrrolidino ones. This spectrum is almost simpler than the preceding one. The successive elimination of the substituents

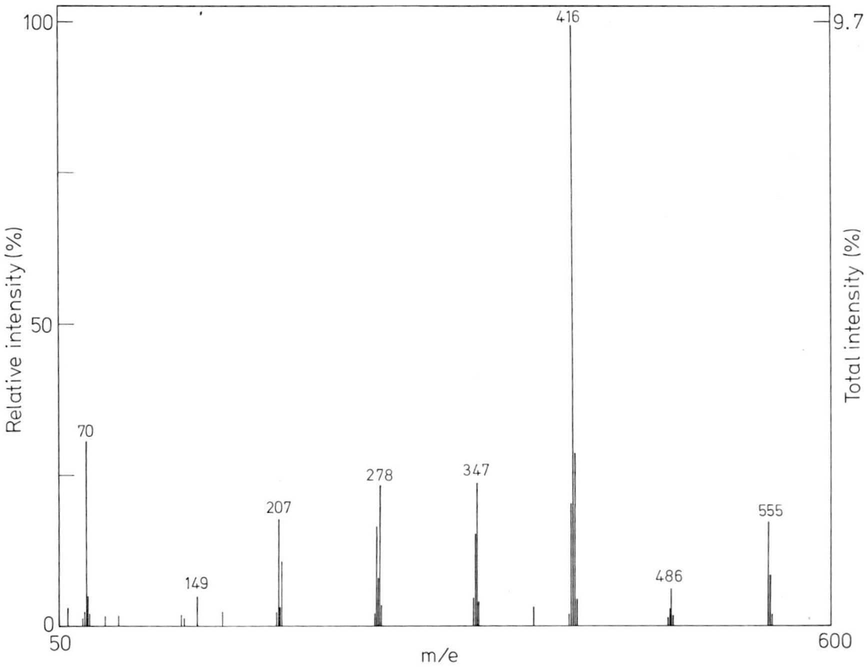

Fig. 9. Mass spectrum of $N_3P_3Pyrro_6$ (mol. wt = 555)

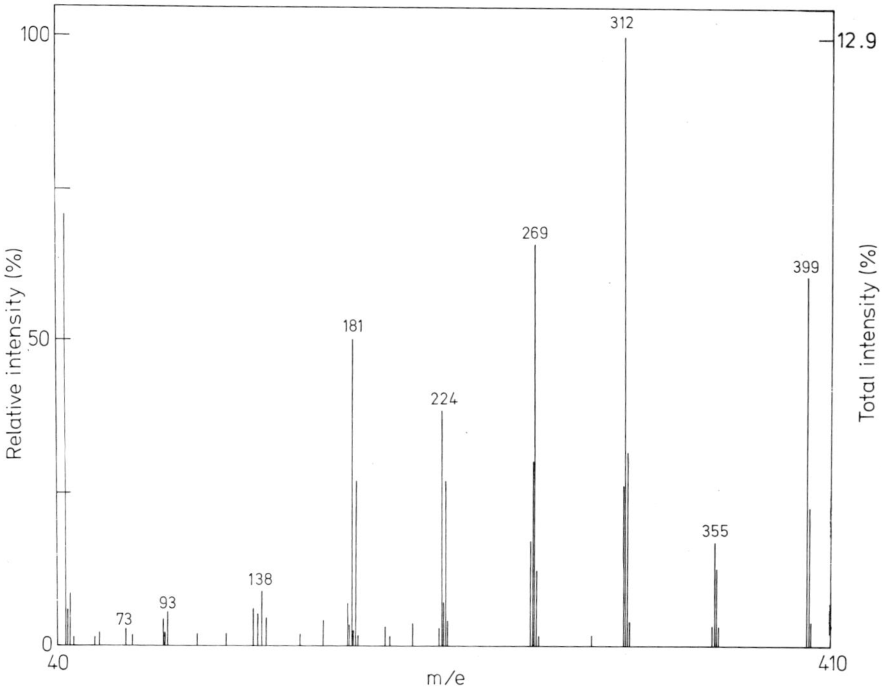

Fig. 10. Mass spectrum of $N_3P_3(NMe_2)_6$ (mol. wt = 399)

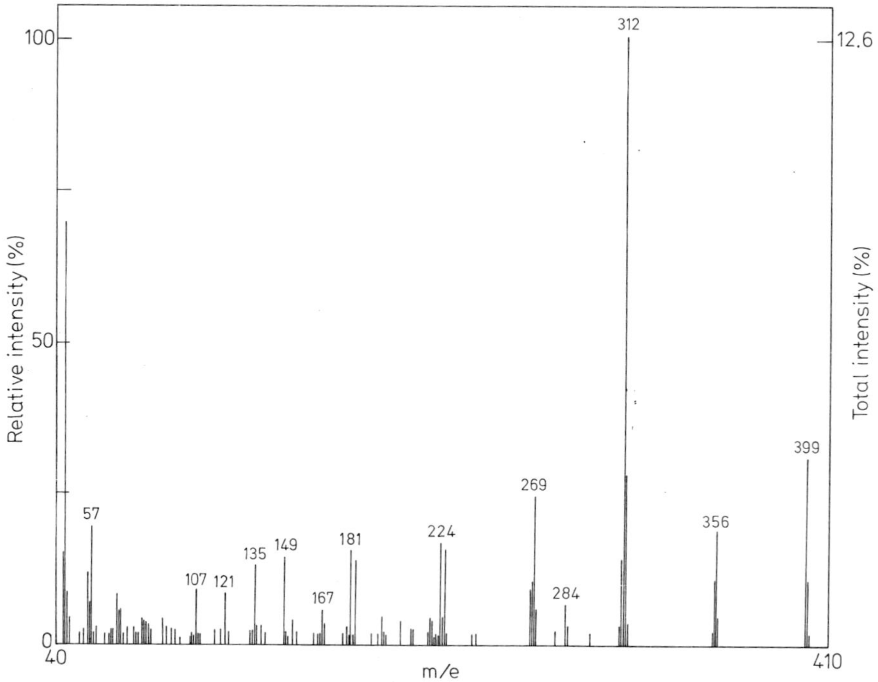

Fig. 11. Mass spectrum of $N_3P_3(NHEt)_6$ (mol. wt = 399)

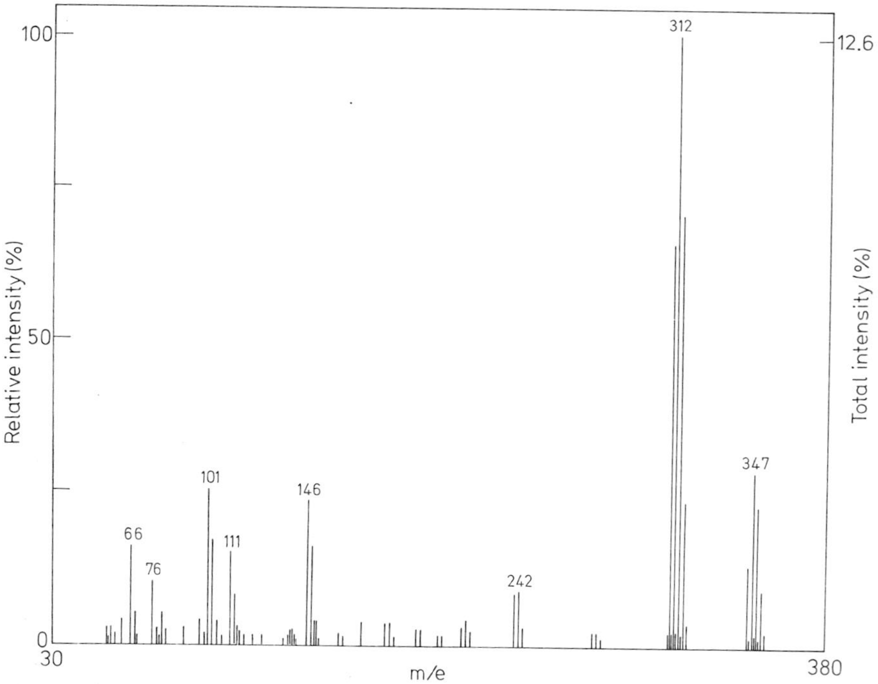

Fig. 12. Mass spectrum of $N_3P_3Cl_6$ (mol. wt = 347)

arose by two main pathways: a simple cleavage of the (P—N) bond and a cleavage associated with a hydrogen transfer. Each consecutive loss of a pyrrolidino group gave a set of peaks, corresponding to $[M - xPyrro + yH]^+$ where x lies between 1 and 5 and y between 0 and 4. The base peak corresponded to the loss of two pyrrolidino moieties (one as a pyrrolidino radical and the other as a pyrrolidine molecule).

The same behaviour was observed for the $N_3P_3(NMe_2)_6$ compound (Fig. 10) corresponding to the replacement of the aziridino groups by dimethylamino ones. The spectra of the $N_3P_3(NHEt)_6$ derivative, containing six ethylamino groups linked to the three phosphorus atoms of the cyclophosphazene 6-membered ring (Fig. 11) resembles the preceding one closely. The successive losses of ethylamino groups, whether or not associated with H-transfer, produced the main peaks. However, the spectrum is complicated by the elimination of the alkyl substituents from each fragment ion. As expected, the $N_3P_3Cl_6$ compound, which is the starting point for all these syntheses, gave a very simple spectrum with the successive loss of the chlorine atoms [11] (Fig. 12).

The analogues of these compounds which contain an 8-membered cyclophosphazene ring gave the following spectra: the MYKO 83, containing eight aziridino groups, behaved like MYKO 63 (Fig. 13); the first loss of the aziridino groups proceeded without an H-transfer, giving an ion at m/z 474. The other ions are associated with H-transfers. At least seven successive eliminations could be observed.

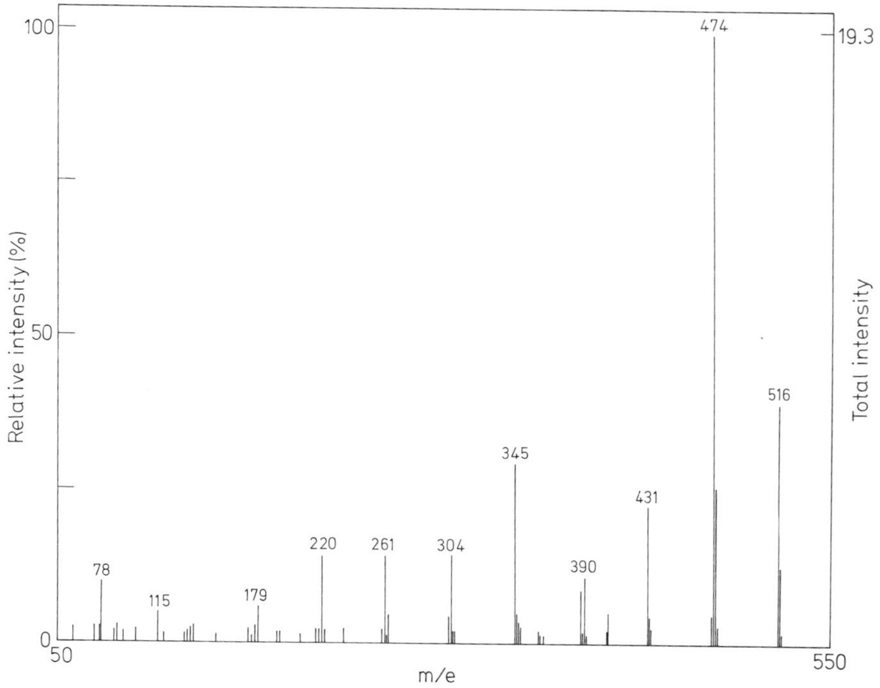

Fig. 13. Mass spectrum of $N_4P_4Az_8$ (mol. wt = 516)

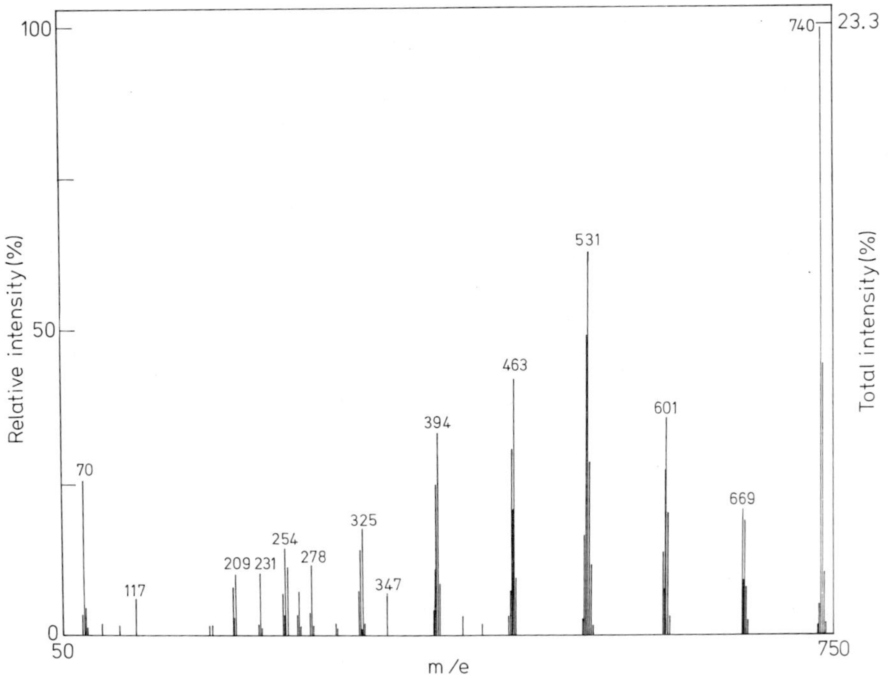

Fig. 14. Mass spectrum of $N_4P_4Pyrro_8$ (mol. wt = 740)

In the $N_4P_4Pyrro_8$ compound (Fig. 14), several sets of peaks are observed at $[M - n\text{-Pyrro} + xH]^+$, n varying from 1 to 7 and x from 0 to n, together with $[M - n\text{-Pyrro} - yH]^+$ with y = 1 or 2, these values depending on the n ones.

In all these compounds, both the 6-membered and the 8-membered phosphazene rings appeared to be remarkably stable toward fragmentation. This is probably due to the "Dewar's islands" structure, stabilized by transannular (P ... P) interactions in these rings, as already demonstrated [12].

In conclusion, due to the simplicity of its mass spectrum, the search for impurities by mass spectrometry using a direct inlet method was carried out successfully on a MYKO 63 sample. Similar studies may be equally easy for other analogues because of the similarity of their decomposition pathways. Thus, mass spectrometry appears to be an adequate tool for controlling the purity of MYKO 63 and other derivatives for clinical uses as antitumor agents in an unambiguous way.

5.3 X-Ray Crystal Structures of MYKO 63 and Relatives

In order to attempt an explanation of the intimate origin of universal antitumor activity of MYKO 63, we investigated X-Ray crystal structures of our drugs with the aim both (i) of elaborating some structure-activity clear relationships and (ii) of having some ideas about the geometrical and conformational structures of our drugs, i.e., about the "morphology of the keys", with respect to the well-known distribution of dialkylating sites on DNA, ("morphology of the lock").

X-Ray crystal structures of MYKO 63 and relatives were investigated in close collaboration with J. Galy, R. Enjalbert (G.I.T.E.R., Toulouse, France) and T. S. Cameron (Dalhousie University, Halifax, Nova Scotia, Canada). They appeared actually more intricate than expected, depending on so many intangible factors, that we must detail them step by step in the following.

5.3.1 X-Ray Molecular Structure of Genuine MYKO 63

When MYKO 63 is crystallized from saturated solutions in m-xylene, carbon disulfide or water, the single crystals so obtained belong to various space groups but the molecular geometry of the drug by itself appears strictly identical in the three cases (Fig. 15). Incidentally, these crystals do not contain in their unit cell any molecule of solvent [13].

As expected, the N_3P_3 ring is found to be perfectly planar. On the other hand, exocyclic nitrogen atoms have a highly pyramidal character, in contrast with the

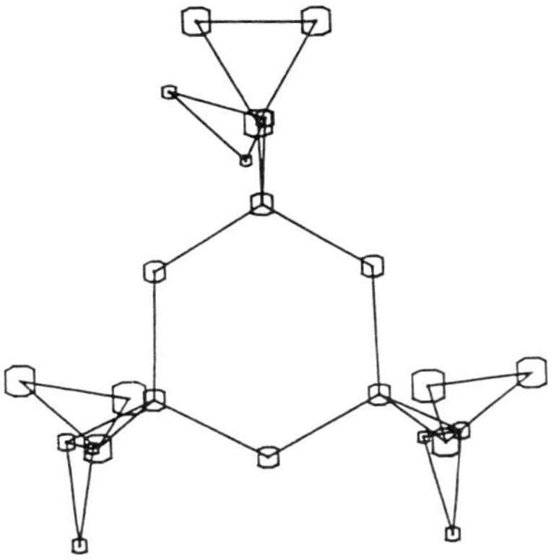

Fig. 15. X-Ray molecular structure of genuine MYKO 63

sp^2 situation which is generally observed for aminophosphine derivatives [14]. The distance of these N atoms from the corresponding PCC planes remains the same (0.69 Å) whatever the relative orientation of aziridino "wings" with respect to the ring plane is. In other words, the spatial distributions of the 6 wings may be analyzed in terms of *preferred conformations* defined in terms of rotations of rigid aziridino elements around the corresponding phosphorus-nitrogen bonds.

5.3.2 X-Ray Molecular Structure of MYKO 63 from its Benzene and CCl_4 Clathrates

When MYKO 63 is crystallized from saturated solutions either of benzene or carbon tetrachloride, the unit cells of the corresponding single crystals retain several molecules of solvent: the actual content of these two peculiar structures may be described as $(2N_3P_3Az_6 \cdot C_6H_6)$ [15,16] and $(N_3P_3Az_6 \cdot 3\,CCl_4)$ [17] respectively and the genuine geometrical patterns (Figs. 16 and 17) of MYKO 63 in these two cases look surprisingly quite different from the one visualized in Fig. 15, at least with regard to the spatial distribution of the conformations of aziridinyl groups.

The comparison of the three different MYKO 63 molecular structures so obtained calls for the following remarks.

(i) The symmetry point group of the drug molecule varies sharply from one struc-

ture to the other and appears to be highly sensitive to the existence in the crystal of very weak intermolecular forces: van der Waals interactions in benzene clathrate [15], (N ... Cl) halogen bonds in CCl_4 anticlathrate [17]. MYKO 63 appears in this manner as a *versatile* molecule owing to a potential high flexibility of its aziridinyl wings.
(ii) It is noteworthy that the genuine structure of MYKO 63 as represented in Fig. 15 does not have ternary symmetry and probably no symmetry element at all, challenging to some extent the rules of Group Theory. The expected 3-fold axis actually appears in the structures of Figs. 16 and 17, i.e., when the drug molecule is stressed by a D_{6h} (benzene) or T_d (CCl_4) solvent field.

Thus, X-Ray investigations of MYKO 63 molecular structures make a remarkable flexibility of its aziridino wings conspicuous. Having in mind that these wings may generate the suitable carbocations for dialkylating DNA, we may assume that MYKO 63, owing to its versatile character, will have the capability of easily

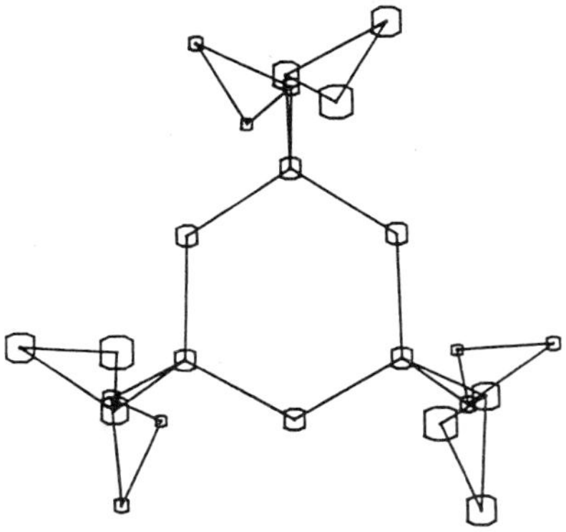

Fig. 16. X-Ray molecular structure of MYKO 63 from its CCl_4 "anticlathrate"

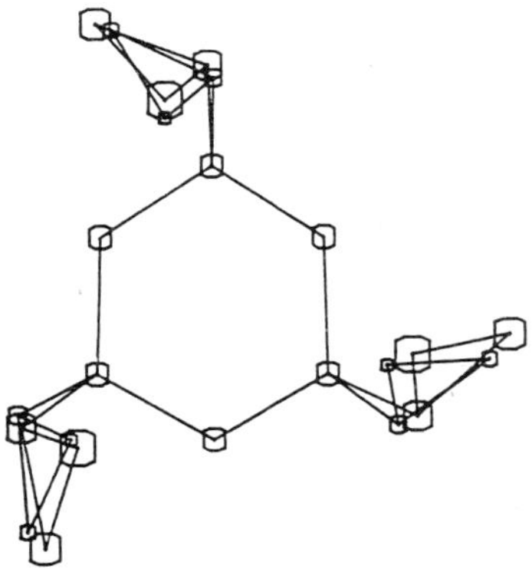

Fig. 17. X-Ray molecular structure of MYKO 63 from its C_6H_6 clathrate

adapting its molecular structure to several dialkylating sites of DNA and, consequently, will be able to have effects on many kinds of vicious DNA. This could be the key which could explain the wide (universal) antitumor activity of the drug.

Similarly, X-Ray crystallography allowed us to approach an understanding of the poor anticancer activity of pyrrolidinocyclophosphazenes through an X-Ray investigation of $N_4P_4Pyrro_8$ molecular structure.

5.3.3 X-Ray Molecular Structure of $N_4P_4Pyrro_8$ [18,19]

$N_4P_4Pyrro_8$ crystallizes in the tetragonal system, space group $P\bar{4}2_1c$. The unit cell contains two discrete molecules. The saddle-shaped eight-membered N_4P_4 ring has $\bar{4}$ symmetry. The P and N atoms for the puckered ring are alternatively and respectively 0.29 and 0.59 Å above and below the mean plane of the ring [(001) plane] (Fig. 18).

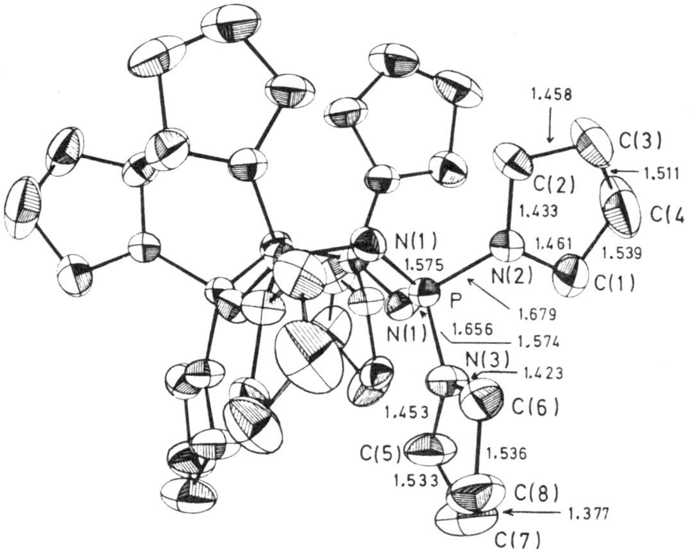

Fig. 18. View of a $N_4P_4Pyrro_8$ molecule perpendicular to the c axis

The exocyclic P—N(2) and P—N(3) bonds, 1.679(6) and 1.656(7) Å, are not equal within experimental error. Moreover, the sum of the three bond angles around N(2) and N(3) are close to 360°, conferring on N(2) and N(3) a definitely sp^2 character, in contrast with the situation in MYKO 63.

Furthermore, a careful analysis of bond lengths and valency angles within the 8 pyrrolidino groups clearly shows that the four upper entities (with respect to the saddle of the ring) are opened like the petals of a flower, whereas the other four are coiled around the four-fold axis like faded sepals [19]. Such a coiling makes the four "below" pyrrolidinyl groups quite different, from a geometrical point of view, from the four "above". The most remarkable difference between the below and above groups stays at the level of C(3)C(4) and C(7)C(8) bonds, 1.51(1) and 1.38(2) Å respectively. The latter very short bond length was clearly attributed to a "flip-flop" motion of C(7) and C(8) through the C(5)—N(3)—C(6) plane [11]. In other words, the four above pyrrolidinyl groups keep a rigid planar configuration whereas the four below ones are destabilized from this point of view, probably because of the packing induced by the saddle-shape of the N_4P_4 ring.

Therefore, this intramolecular packing within the pair of pyrrolidinyl groups linked to a given P atom induces a real asymmetry at the level of an eventual dialkylation of DNA through carbocations provided by pyrro groups. Both the evidence of such an asymmetry for dialkylation and the well-known low rate of production of carbocations from pyrro groups may explain the poor antitumor activity of N_4P_4-Pyrro$_8$ with respect to MYKO 63. However, the slight destabilization of 4 over 8 pyrro groups in N_4P_4Pyrro$_8$ favours production of carbocations with respect to what happens in N_3P_3Pyrro$_6$ the pyrro groups of which are "normal" owing to the strict planarity of the N_3P_3 ring. Thus, we may understand why N_4P_4Pyrro$_8$ is poorly — but significantly — active when N_3P_3Pyrro$_6$ was found ineffective [4].

Anyhow, X-Ray crystallography provided some information which may be of help for deriving a structure-activity relationship within the MYKO 63 series. However, such an approach implies two conditions to be valid:
(i) the drug must be active by itself without any dramatic modification of its molecular structure (metabolisation) and (ii) assuming (i) to be correct, the major point to be elucidated now is to discriminate the target(s) MYKO 63 will have to reach in vivo to be active.

The search for target(s) in vivo is practically insuperable. Thus, the only possible approach is to investigate in vitro eventual interactions of the drug with biological targets such as DNA, RNA, proteins and others. The first step of such in vitro studies consists in observing interactions with DNA, owing to the fact that this nucleic acid is generally considered as the most deeply involved in the intimate processes of life and death. This explains why we focused firstly our attention and efforts on the study of MYKO 63-DNA eventual interactions by two suitable techniques for this purpose, namely the Scatchard technique [20] and Raman spectroscopy [21].

5.4 In vitro Investigations of MYKO 63-DNA Interactions

5.4.1 Scatchard Technique

The scatchard technique consists of studying the difficulties encountered by a fluorescent dye, ethidium bromide (Fig. 19), with respect to its intercalation between the base pairs of DNA when the secondary structure of the macromolecule is modified by binding (or intercalation) of a drug.

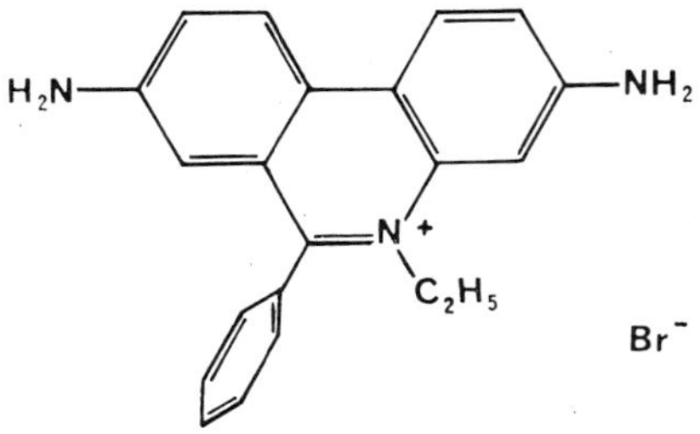

Fig. 19. Ethidium Bromide (EtdBr)

5.4.1.1 Nucleic Acid

Pure salmon sperm DNA (41% dG + dC), from Worthington Biochemical Co., Freehold, New Jersey, USA, was used. Protein content in it was determined by the method of Lowry et al. [22] and found to be less than 1%. RNA content, determined by Defrance and Delesdain's method [23], was also less than 1%. DNA absorption per mole of nucleotide at 258 nm was 6360 M^{-1} cm^{-1} (from Worthington). DNA concentrations were measured by absorbance.

5.4.1.2 Salts

The different salts, KNO_3 and $NaClO_4$, have to be purchased in an extremely pure state (Fluka, Buchs, Switzerland). Pure ethidium bromide (3,8-diamino-6-phenyl-5-ethylphenantridinium-bromide) was bought from Sigma Chemical Co., St. Louis, Missouri, USA; its molar absorption coefficient was 5450 M^{-1} cm^{-1} at 480 nm in water [24].

5.4.1.3 Cyclophosphazene-DNA Complexes

A stock solution of DNA was prepared by dropping a 10 nM solution of $NaClO_4$ on the DNA fibers to give an approximate concentration of 1 mg ml^{-1}. This solution was gently agitated at 4 °C for 48 h, centrifuged at 8000 r.p.m. for 10 min and its concentration determined spectrophotometrically.

Cyclophosphazene compounds were dissolved just before use in 10 nM $NaClO_4$ and added to DNA. The final DNA concentration was 0.10 mg ml^{-1} and the reactions were run at room temperature in the dark for several days (2–5 days, depending on the compound). The different cyclophosphazene-DNA complexes corresponded to values of r_i (number of cyclophosphazene molecules introduced per nucleotide) varying from 0 to 100 depending on the solubility of the drugs.

5.4.1.4 Apparatus

Ultraviolet and visible spectrophotometric and spectrofluorometric measurements [24, 25] were performed with a Zeiss PMQ II spectrophotometer equipped with a ZFM 4 fluorescence attachment. Excitation light (λ_{ex} = 546 nm) was provided by a mercury lamp and selected with a M 546 filter. Emitted fluorescence was measured at 590 nm.

Fluorescence measurements I_1/I_0 (I_1 = fluorescence intensity of the cyclophosphazene-DNA-EtdBr complex-fluorescence intensity of pure EtdBr; I_0 = fluorescence intensity of the DNA-EtdBr complex-fluorescence intensity of pure EtdBr) were performed at 25 °C in 0.4 M KNO_3 in order to avoid the second fixation site of EtdBr to DNA [25]. The DNA and EtdBr concentrations were respectively 10 µg ml^{-1} and 40 µg ml^{-1}.

The association constants to a site (K) and the ratio (r) of bound EtdBr per nucleic acid phosphate were determined for the different cyclophosphazene-DNA complexes using Scatchard's method [20]. The number of binding sites per base (n) for the non-complexed DNA has been found equal to 0.20 and the value of K to 1.2×10^5 M^{-1}, in agreement with previously reported data [26] and [27].

5.4.1.5 Spectrofluorometric Study of EtdBr Binding: the Case of the $N_3P_3Az_6$-DNA Complex [28]

The following is a detailed discussion of the experimental procedure applied to the above complex.

Figure 20 shows the saturation curve (x) which may be plotted when increasing the concentration of EtdBr to a constant concentration of DNA. The levelling-off of this curve corresponds to the I_0 value of the fluorescence intensity (arbitrary unit). As soon as increasing quantities (r_i) of $N_3P_3Az_6$ are introduced into the medium, a decrease of the maximal fluorescence intensity I_1 is observed as a function of r_i. As a consequence, this decrease of fluorescence (due to the increased

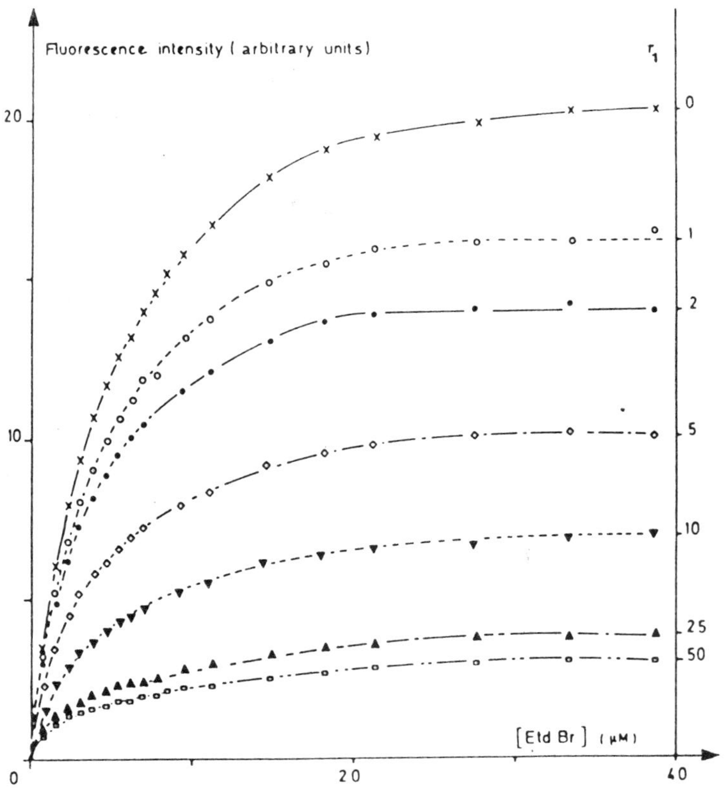

Fig. 20. Saturation curves of fluorescence intensity for a series of $N_3P_3Az_6$-DNA complexes (r_i varying from 0 to 50) obtained with the addition of EtdBr. Fluorescence of the $N_3P_3Az_6$-DNA-EtdBr concentrations added

difficulty of EtdBr intercalation when the secondary structure of DNA is modified upon interaction with $N_3P_3Az_6$) may be visualized more readily by plotting the I_1/I_0 ratio as a function of r_i (Fig. 21). Moreover, Fig. 21 indicates that the curvature of the curve depends on the incubation time and that 4 or 5 days are required to get the limit curve. It may be noted that a large excess of $N_3P_3Az_6$ may be added without giving rise to any conspicuous quenching.

Now, we may plot on the same graph (Fig. 22) the limit curves thus obtained for the different cyclophosphazene-DNA matrices. This graph reveals that the cyclophosphazenes studied herein belong to four classes: 1) those ($N_3P_3Morph_6$ and $N_4P_4Morph_8$) which neither induce a fluorescence decrease nor exhibit any antitumor activity; 2) those ($N_3P_3Az_6$, $N_4P_4Az_8$ and $N_3P_3(MeAz)_6$) which give rise to a significant fluorescence decrease and are highly active; 3) those ($N_3P_3Cl_6$ and $N_4P_4Cl_8$) in which lack of antitumor activity is paralleled by a huge fluorescence decrease; 4) finally $N_4P_4Pyrro_8$ which is a moderate anticancer agent and which does not cause any fluorescence decrease.

Thus, it is not possible to show a direct relationship between fluorescence decrease and antitumor activity of the cyclophosphazenes studied, and the question arises as to

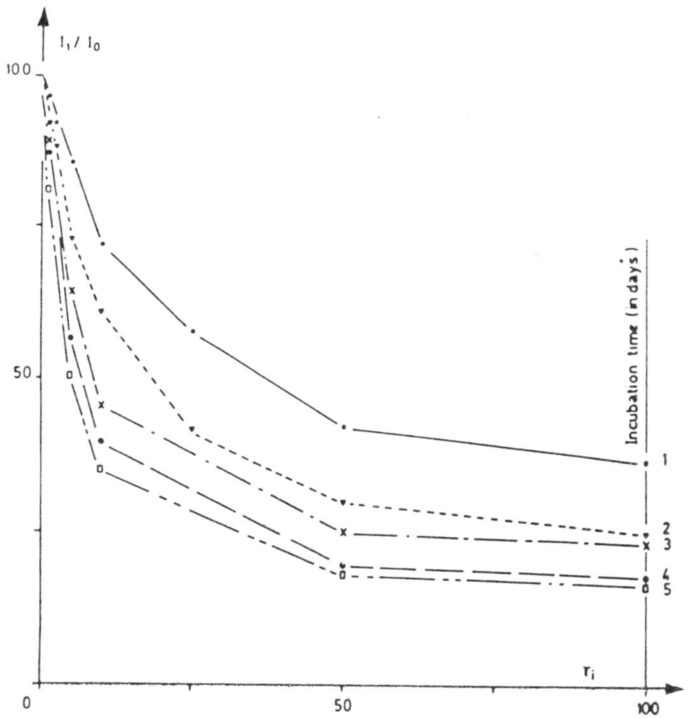

Fig. 21. Relationship between I_1/I_0 and r_i as a function of the incubation time

whether the drugs investigated are competing with EtdBr or dialkylating the DNA upon complexation.

Scatchard plots (Fig. 23) show that the more $N_3P_3Az_6$ is added, the less EtdBr is bound. Moreover, the binding constant of EtdBr to the $N_3P_3Az_6$-DNA complex is similar to that of the DNA-EtdBr complex, i.e., $K = 10^5\ M^{-1}$ at 25 °C in the presence of 0.4 M KNO_3.

These results indicate that there is no competition between $N_3P_3Az_6$ and EtdBr and suggest that $N_3P_3Az_6$ does not interact with DNA through an intercalation process but through covalent binding. This is supported by the fact that there is no modification of the fluorescence decrease after 10 dialyses against 200 volumes of 10^{-2} M $NaClO_4$. Furthermore, it is possible to separate on a gel column (Sepharose 6B) the $N_3P_3Az_6$ which has reacted with DNA from the free compound and in that way to follow the kinetics of the complexation (data available on demand).

The same technique was applied to the other cyclophosphazenes studied, allowing the same conclusions as above to be drawn: in each case where a decrease of the I_1/I_0 ratio was observed, there was a decrease in the number of sites for EtdBr (n), the K values remaining practically constant (0.6 to $1.0 \times 10^5\ M^{-1}$).

5.4.1.6 UV Spectrophotometric Study of EtdBr Binding to Cyclophosphazene-DNA Complexes [28]

Figure 24 shows the bathochromic and hyperchromic shifts of the λ_{max} of DNA on addition of increasing r_i quantities of $N_3P_3Az_6$ to the medium. These effects

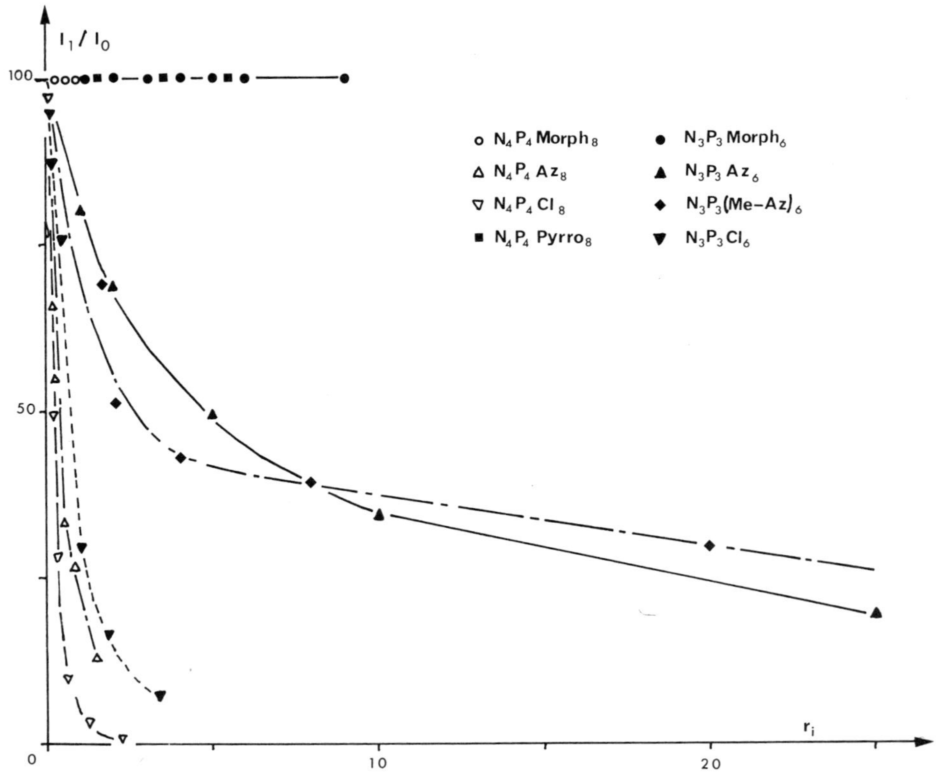

Fig. 22. Limit curves (incubation time = 1 week) of fluorescence decrease for some cyclophosphazenes

were observed in the same way for $N_4P_4Az_8$, $N_3P_3Cl_6$ and $N_4P_4Cl_8$; in contrast, $N_3P_3Pyrro_6$, $N_3P_3Morph_6$, $N_4P_4Pyrro_8$ and $N_4P_4Morph_8$ did not cause any shift of the λ_{max} in question.

The red shift of the absorption spectrum is an unambiguous proof that the bases are being attacked since the absorption spectrum of DNA arises from its aromatic bases.

These results may be visualized more simply when plotting the ratio [(OD 250/OD 270) (cyclophosphazene-DNA)]/[(OD 250/OD 270) DNA] as a function of r_i (Fig. 25) whereby the trend of the behaviour of the various cyclophosphazenes with respect to EtdBr binding, as revealed by UV spectrophotometry, looks virtually identical to the one provided by the spectrofluorometric study (Fig. 22), especially concerning the existence of the four classes previously mentioned. The batho-hyper-shift is observed only for the terms of the series which induce a fluorescence decrease.

5.4.1.7 General Conclusions

A set of cyclophosphazene-DNA matrices was investigated in order to obtain more information about the possible structural modifications of DNA due to the different possible modes of binding of the drugs and to thus gain insight into the origin of their antitumor properties.

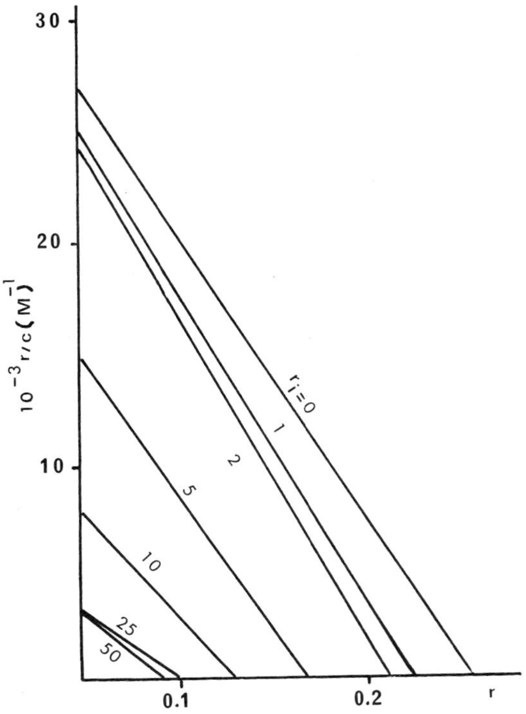

Fig. 23. Scatchard plots for EtdBr binding to different $N_3P_3Az_6$-DNA complexes. These experimental curves correspond to r_i varying from 0 to 50

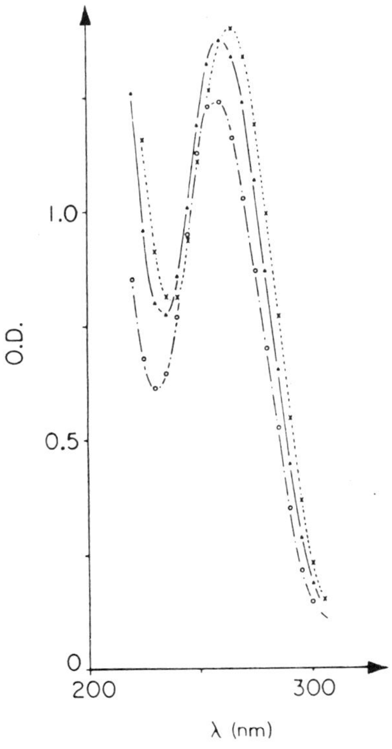

Fig. 24. Batho-hyperchromic shifts of λ_{max} of DNA under addition of increasing r_i quantities of $N_3P_3Az_6$. (o) r_i = 0, (Δ) r_i = 10, (x) r_i = 100

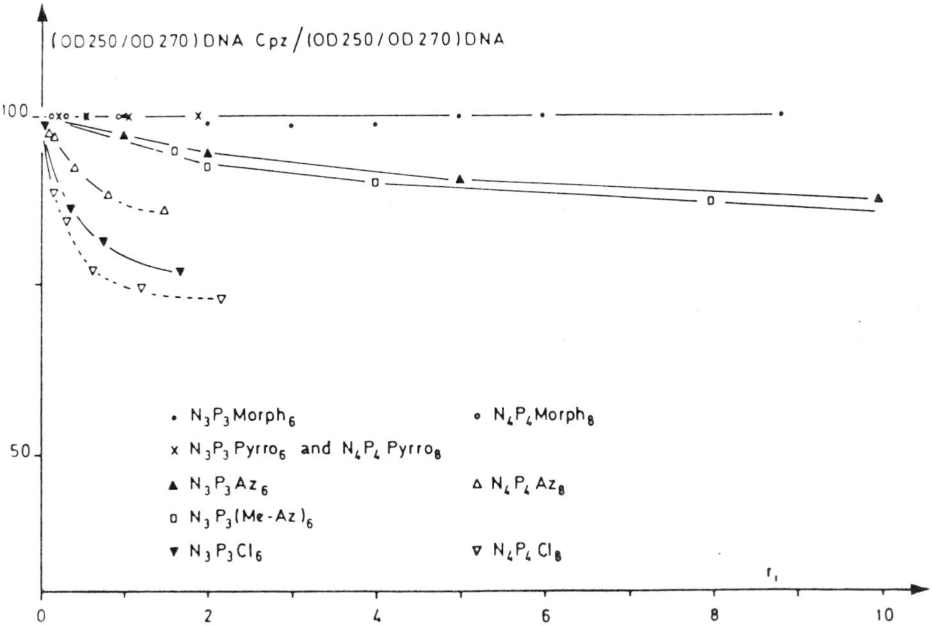

Fig. 25. Limit curves (incubation time = 1 week) of UV red shift for some cyclophosphazenes

Using UV spectrometry with EtdBr as a fluorescent probe of the DNA structure, the following results were obtained:

1) $N_3P_3Az_6$ and $N_4P_4Az_8$, which exhibit significant activity against either L1210 and P388 leukemias or B16 melanoma, induce a noticeable fluorescence decrease and bathochromic and hyperchromic shifts;

2) the expected pattern is observed for $N_3P_3Pyrro_6$, $N_3P_3Morph_6$ and $N_4P_4Morph_8$ which are biologically inactive and which do not cause any fluorescence decrease or any shift of the absorption spectrum;

3) in contrast, $N_3P_3Cl_6$ and $N_4P_4Cl_8$, which are also biologically inactive, provoke a huge fluorescence decrease and very important batho- and hyperchromic shifts of the absorption spectrum;

4) lastly, the significant antitumor activity of $N_4P_4Pyrro_8$ is not accompanied by any fluorescence decrease or UV shift.

These results indicate that the mechanism of the antitumor activity of cyclophosphazenes may be much more complicated than could have been expected a priori: if indeed the case of $N_4P_4Pyrro_8$ had not been included in our investigations, one could have explained the "$N_3P_3Cl_6$—$N_4P_4Cl_8$" exception to the rule that antitumor activity = fluorescence decrease and UV red shift, by assuming that the alkylating sites on DNA for these two chemicals were very different from the ones for the cyclophosphazenes having antitumor activity, this difference being responsible for observed biological activity or inactivity. However, the fact that $N_4P_4Pyrro_8$, which is an antitumor agent, does not exhibit any fluorescence decrease, UV shift and mutagenicity, renders impossible the design of any direct relationship between fluorescence decrease (or UV shift) and antitumor activity within the cyclophosphazenic family.

However, this negative conclusion must be qualified by taking into account the fact that the results presented here were obtained using r_i quantities of cyclophosphazenes added to the medium and not the real r quantities bound to the DNA.

Nevertheless there exist strong covalent interactions in vitro between MYKO 63 and its relatives and DNA. However, the Scatchard technique does not give the actual sites the drug will graft on. In other words, we had to look for a more local technique which would allow to reach the location in DNA of these sites. Raman spectroscopy was chosen for this purpose because it had previously proved successful for studying in this way interactions of many biological systems with nucleosides or DNA [21].

5.4.2 Raman Spectroscopy [29]

5.4.2.1 Nucleic Acid and Salts

Stock solutions of calf thymus DNA (Sigma Chemical Co.) and of salmon sperm DNA (Worthington Biomedical Co.) were prepared by dropping a 10 mM solution of $NaClO_4$ (Fluka) on the DNA fibers to give an approximate concentration of 1.5 mg ml^{-1} for the salmon sperm (set I) and 4.5 mg ml^{-1} for the calf thymus (set II) DNA. These solutions were gently agitated at 4 °C for 48 h., centrifuged at 8000 r.p.m. for 10 min and their concentrations determined spectrophotometrically.

5.4.2.2 $N_3P_3Az_6$-DNA Complexes

Hexaziridinocyclotriphosphazene-DNA complexes were made by mixing solutions of the drug and DNA in a molar ratio of approximately one base of DNA per drug molecule. The reactions were run at room temperature in the dark for 1–6 days. The final DNA and MYKO 63 concentrations in the complex were 1 mg ml^{-1} and 3.2×10^{-3} M (set I), and 3 mg ml^{-1} and 9.6×10^{-3} M (set II).

In the experiments, it was ascertained that laser irradiation had no effect on the UV absorbance spectra of the samples.

5.4.2.3 Raman Instrumentation

Raman spectra were obtained on a Coberg PHO Raman spectrometer equipped with a Coherent Radiation Model 52B Ar^+ laser using 800–1200 mW of power from the 488.8-nm line; a small twoprism monochromator was used to remove background plasma lines.

The Raman scattering was detected with a cooled EMI 9558 QB photomultiplier. Signal pulses from the photomultiplier were amplified and then counted digitally and stored in the computer memory. The computer (Alcyane, MBC, France) automatically coordinates the scanning of the spectrometer (steps of 2 cm^{-1}). The maximum counting time for each step was 2 s. In general, data accumulation (time averaging) was used to obtain the spectra and to improve the signal-to-noise ratio. This computer enables several operations to be carried out on the stored spectra, particularly the difference between two spectra (see Ref. [30] for all the applications available).

The sample was placed in a quartz cell in a thermostated holder (25 ± 0.5 °C). The laser line was focused by a lens and mirror system and the scattered light collected at right-angles to the incident radiation. The normal slits were 6 cm^{-1}. The Raman frequencies reported here are accurate to ± 1 cm^{-1} for intense and/or sharp lines and to ± 2 cm^{-1} for weak and/or broad lines. The Raman band intensities were measured by their peak heights.

5.4.2.4 Analysis Methods

Analysis of binding experiments required a careful comparison of (i) the MYKO 63 bands, either in the presence or absence of DNA bands and (ii) the DNA Raman bands, either in the presence or absence of MYKO 63 bands. This comparison was achieved by computer-subtracting variable amounts of one spectrum from another. Previously, the various spectra were normalized to the same relative Raman intensity, with the 934 cm^{-1} band (ClO_4^- symmetric stretch) as an internal standard. The intensity of the ClO_4^- scattering measures the combined effect of such experimental factors as counting time, optical alignment and laser power.

Comparison of the complete spectra of the solvent (10 mM $NaClO_4$ in water) and of the solutions (MYKO 63 or MYKO 63-DNA in the solvent) showed that the region around 1850 cm^{-1} can be considered to have nearly "zero Raman intensity" at least for the lower wave numbers. Thus, for each spectrum, a horizontal base-line is drawn from this point, then the solvent spectrum is subtracted taking into account a coefficient determined from the compound concentration of the solution in order to keep the horizontal base-line unchanged for the new spectrum.

Figure 26 compares the spectrum of free MYKO 63 with that of MYKO bound to calf thymus DNA. Both spectra have been normalized to the same Raman scattering intensity and drug concentration. The spectrum of free MYKO 63 was obtained by subtracting the solvent spectrum from the MYKO solution spectrum. The spectrum of the bound drug was obtained by a two-step process: (i) the solvent spectrum was subtracted both from the spectrum of the MYKO-DNA complex and from the DNA solution spectrum; (ii) subtraction of these two new spectra was

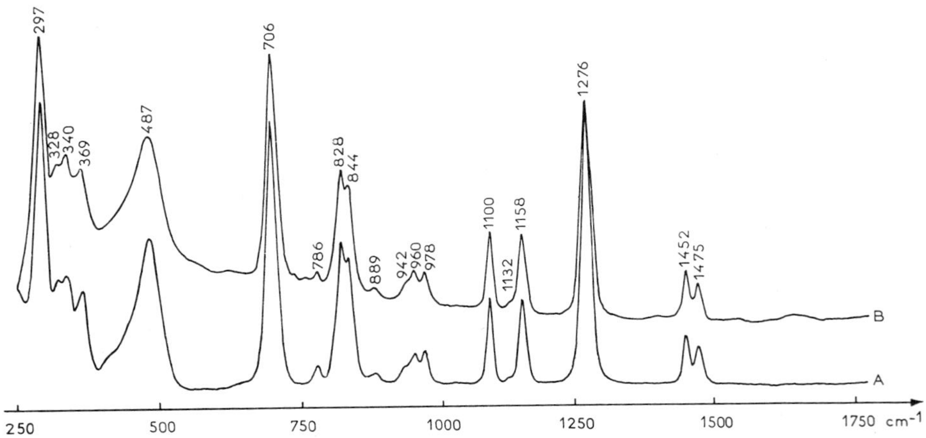

Fig. 26. Raman spectra of unbound (A) and bound (B) MYKO 63 after subtraction of calf thymus DNA and $NaClO_4$ backgrounds and normalization to the same Raman scattering intensity and drug concentration

then performed until most of the DNA bands were removed. Step (ii) is applicable only if the spectral differences between unbound and partially bound DNA are negligible. This is strictly correct only for DNA bands unaffected by drug binding. As can be seen from Fig. 26, the background and the base-line of the spectra around 1850 cm^{-1} are not affected by the subtraction process.

For comparison of bound and unbound DNA (see Figs. 27 and 28) the procedure was performed in the same way. The spectrum of partially bound DNA was obtained by subtracting the free MYKO difference spectrum (Fig. 26A) from the spectrum of MYKO-DNA complex (Fig. 27, set I and Fig. 28, set II).

Raman scattering were obtained for the first time for $N_3P_3Az_6$ [31] and for $N_3P_3Az_6$-DNA complexes (sets I and II).

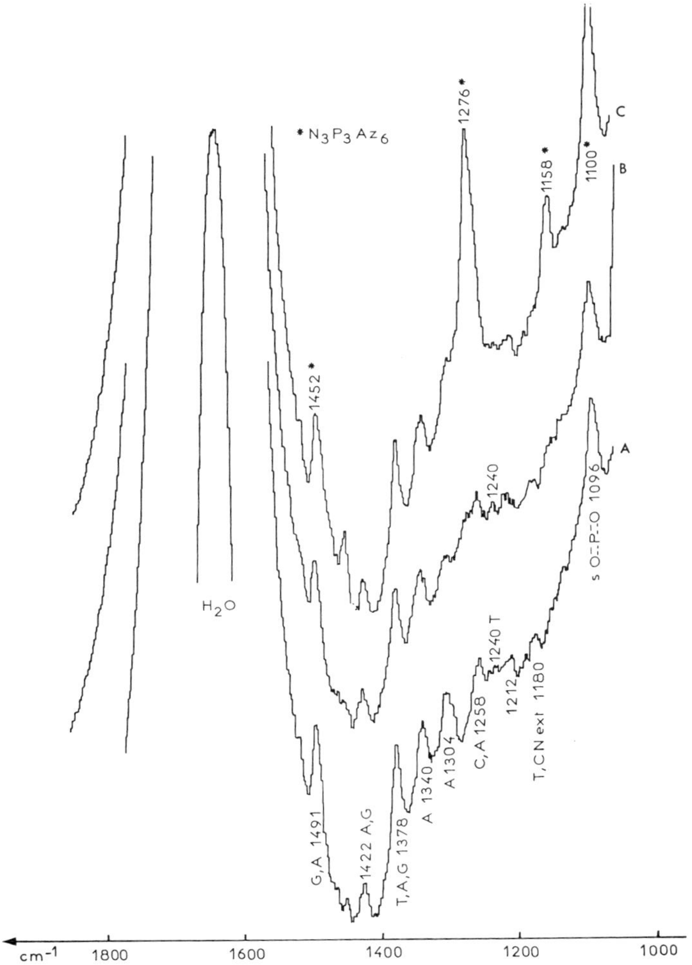

Fig. 27. Raman spectra of set (I): (A) salmon sperm DNA (1 mg ml^{-1}) in 10 mM $NaClO_4$ solution; (B) partially bound DNA after subtraction of free MYKO 63 and solvent (10 mM $NaClO_4$ in water); (C) MYKO-63-DNA complex (one base of DNA per drug molecule)

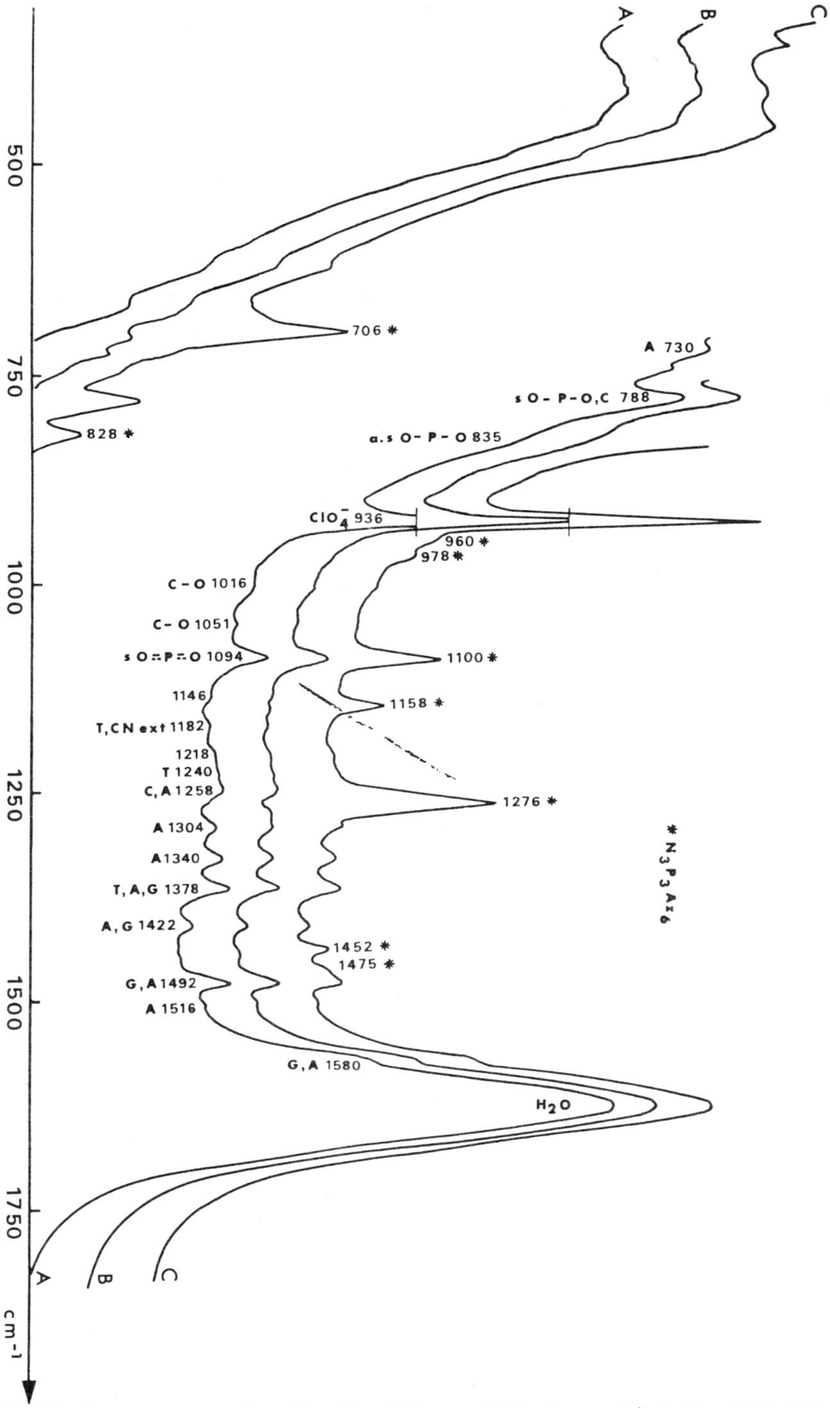

Fig. 28. Raman spectra of set (II): (A) calf thymus DNA (3 mg ml^{-1}) in 10 mM $NaClO_4$ solution; (B) partially bound DNA after subtraction of free MYKO 63 and solvent; (C) MYKO-63-DNA complex (one base of DNA per drug molecule)

5.4.2.5. Raman Spectrum of Hexaziridinocyclotriphosphazene ($N_3P_3Az_6$)

Figure 29 illustrates the spectrum of $N_3P_3Az_6$ in 10 mM $NaClO_4$ solution (concentration 3.87×10^{-2} M). In order to draw some conclusions about binding changes by Raman spectroscopy it was necessary to assign the observed Raman bands to the corresponding normal modes of vibration. Taking into account the previous frequency assignments appearing in the literature for $N_3P_3Cl_6$ [32] and $N_3P_3F_6$ [33] as well as for aziridine itself [34], we have classified the group frequencies of $N_3P_3Az_6$ as follows. (i) Framework frequencies: N_3P_3 ring (local symmetry D_{3h}); $\Gamma(fr) = 4A + 4E$. (ii) Ligand frequencies: NC_2H_4 (local symmetry C_s); $\Gamma(lig)(-NC_2H_4) = 34A + 34E$; $\Gamma(lig)(-NC_2) = 10A + 10E$; $\Gamma(lig)(CH_2) = 24A + 24E$. (iii) Ligand framework couplings: $\Gamma(cpl) = 8A + 8E$.

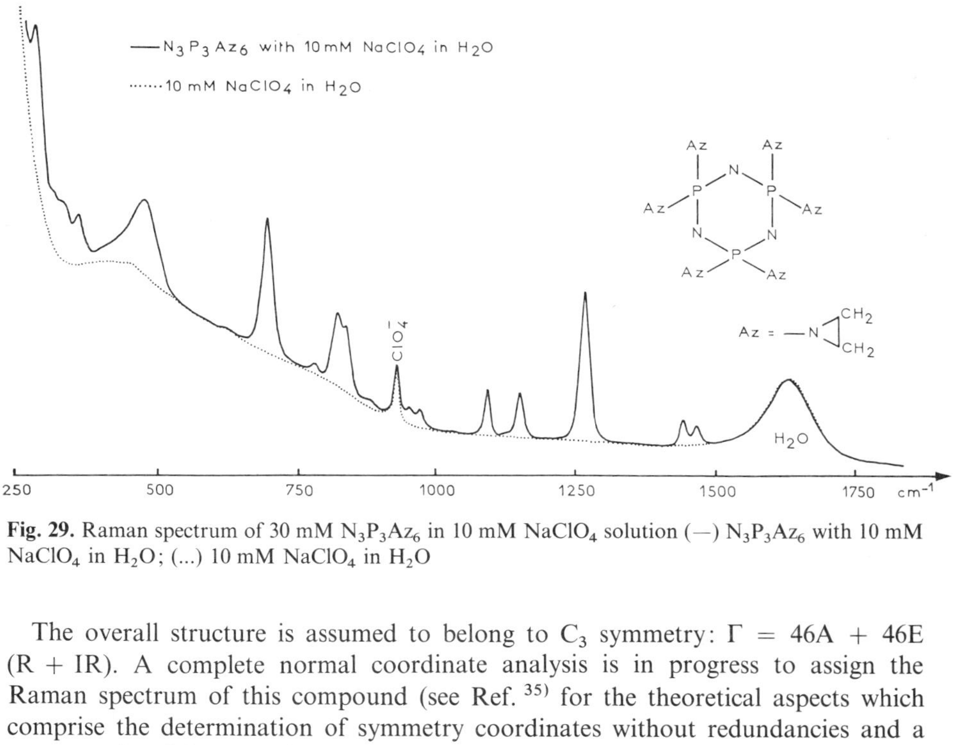

Fig. 29. Raman spectrum of 30 mM $N_3P_3Az_6$ in 10 mM $NaClO_4$ solution (—) $N_3P_3Az_6$ with 10 mM $NaClO_4$ in H_2O; (...) 10 mM $NaClO_4$ in H_2O

The overall structure is assumed to belong to C_3 symmetry: $\Gamma = 46A + 46E$ (R + IR). A complete normal coordinate analysis is in progress to assign the Raman spectrum of this compound (see Ref. [35] for the theoretical aspects which comprise the determination of symmetry coordinates without redundancies and a treatment involving complex representations). The most important group frequencies which are of use in the present work are listed in Table 7. When assigned, each Raman band serves as an indicator of what is happening to specific parts of the $N_3P_3Az_6$ molecule when it is interacting with DNA.

During the study of $N_3P_3Az_6$-DNA complexes by the Scatchard technique [28], it was shown that an incubation time of at least 5 days is necessary after mixing the MYKO 63 with DNA on Day D before complete interaction is observed. These results are clearly confirmed by the present work as no fundamental discrepancy was observed between the spectra of the free and bound MYKO 63 when the latter was deduced from the recorded spectrum of the MYKO 63-DNA complex on Day (D + 1).

Table 7. Main-group frequencies of $N_3P_3Az_6$ and their assignments

ν_i (cm^{-1})	I_{ν_i} [a]	ΔI [b]	Assignment	ν_i (cm^{-1})	I_{ν_i}	ΔI	Assignment
487	37	18	N_3P_3—Az coupling	1100	21.5	7	Az $\varrho_\tau(CH_2)$, $\varrho_\omega(CH_2)$
706	65	11.5	N_3P_3 ring mode	1158	21	7	N_3P_3 ring mode
				1276	71	21	Az ring breathing
828	35	11.5	Az $\varrho_r(CH_2)$	1452	12	~0	Az $\delta_{as}(CH_2)$
844	31	11.3	Az sym. ring. def.	1475	9	~0	Az $\delta_s(CH_2)$

a Relative intensity (%) of free MYKO 63, referred to $\nu_s ClO_4^-$ (936 cm^{-1}, I = 100).
b Decrease in intensity (%); (I_{ν_i} (free) — I'_{ν_i} (bound))/I_{ν_i} (free).

Figure 26 shows the spectra of free and partially bound $N_3P_3Az_6$ obtained by the process described above. The spectrum of the bound MYKO was obtained from the spectrum of the complex recorded on Day (D + 6). The main results provided by the comparison between the free and bound MYKO spectra are as follows.
(i) The intensity and position of some lines are practically unaltered ($\Delta I \leqq 7\%$, see Table 7) on binding: this is the case for the 1452- and 1475-cm^{-1} lines assigned to the $\delta(CH_2)$ deformation mode (ligand), and for the 1100- and 1158-cm^{-1} lines attributed to the twisting and wagging of CH_2 (ligand) and to one of the degenerated stretch modes of the N_3P_3 ring skeleton, respectively.
(ii) A relatively weak decrease in intensity is observed for the 706-cm^{-1} line (12%) (N_3P_3 ring mode) as well as for the 828–844-cm^{-1} group (11.5%) (rocking CH_2 and NC_2 ring modes of Az).
(iii) The main alterations in intensity are seen for the 487-cm^{-1} line assigned to the framework-ligand coupling (N_3P_3-Az) and for the 1276-cm^{-1} band corresponding to the ring breathing of aziridine; these decrease by 18% and 21% respectively. Thus, the largest decrease in intensity in the 1276-cm^{-1} band is due to the mechanism of interaction of aziridino ligands in the complex as described previously [36], following the scheme:

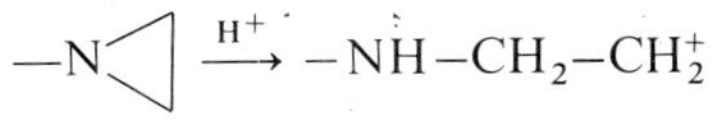

Furthermore, in the case of the bound MYKO, a small broadening of that band is observed; subtraction of the bound and unbound spectra around the 1276-cm^{-1} wavelength yields a very weak band shifted to the high-frequency side. This can be seen in Fig. 28B where the 1304-cm^{-1} band of DNA shows a shoulder on the low-frequency side. This effect is interpreted as a type of aziridino-ligand-DNA interaction involving neither a binding to DNA nor an opening of the NC_2H_4 ring.

5.4.2.6 Raman Studies on the Reaction of MYKO 63 with Salmon Sperm and Calf Thymus DNA

Figures 27A and 28B illustrate the spectra of salmon sperm DNA (1 mg ml^{-1}) and calf thymus DNA (3 mg ml^{-1}) at pH 6.9, 25 °C, in the presence of 10 mM $NaClO_4$. Contributions to the various bands by the individual bases are indicated (G = guanine, C = cytosine, A = adenine, T = thymine). The assignment of Raman bands is based on similar previous studies by Lord and Thomas [37], Small and Peticolas [38, 39], and Erfurth and Peticolas [40]. Figures 27C and 28C give the spectra of the MYKO 63-DNA complexes (sets I and II, respectively) in 10 mM $NaClO_4$ solution. Figures 27B and 28B represent the partially bound DNA spectra obtained by subtracting the spectrum of the free MYKO 63 from Figs. 27C, 28C respectively (see Analysis Methods).

The comparison of the unbound and bound DNA (set I, Fig. 27A, B, set II, Fig. 28A, C) leads to the following conclusions.

(i) The very small increase in intensity of the 1240-cm^{-1} band, which is the most sensitive line to unwinding of the DNA double helix, indicates that DNA is probably slightly denaturated in solution with MYKO 63 (a large increase in intensity is observed when DNA melts [40]). The calf thymus DNA solution spectrum (Fig. 28A) exhibits a weak band at 835-cm^{-1} which is the line that appears to be indicative of the normal B conformation of DNA [41] (i.e. it disappears on melting). In the complex (see Fig. 28B) this line is slightly shifted to lower frequency. This effect, connected with what was observed for the 1240-cm^{-1} line, indicates a very weak denaturation of the DNA by disruption of its B conformation.

(ii) The band at 1492-cm^{-1} seems to be unaltered on binding. As this Raman band of guanine in DNA is a strong indication of binding to the N(7) position of guanine (see Refs. [42,43]), it may be concluded that there is no interaction between MYKO 63 and guanine.

(iii) The major relative effects in intensity are seen for the 1340-cm^{-1} band, which decreases by 13% and 22%, for the 1304-cm^{-1} band, the corresponding decreases being 31% and 30%, for set I and set II, respectively. These two bands are characteristics of adenine; more precisely, the C(5) — N(7) stretching mode is predominant in the 1340-cm^{-1} vibration whereas the 1304-cm^{-1} band strongly involves the C(8) — N(7) bond stretch [44,45]. As these two bands were shown to be predominantly related to the bond vibrations attached to the N(7) position, it may be concluded that interactions between MYKO 63 and adenine occur in the N(7) position.

It should be noted that the N(3) position of adenine is not a binding site for MYKO: the 1580-cm^{-1} band, which is very sensitive to binding to the adenine base in the N(3) position [46], shows no significant shift or decrease either in frequency or in intensity in any of the spectra.

(iv) Finally, the spectra of the complexes in solutions (see Figs. 27B and 28B) exhibit a 20% decrease of the 1180–1182-cm^{-1} band. This band is related both to the stretch modes of vibration of the C—NH_2 groups of the individual bases and to thymine itself. An overall survey of all the spectra presented here prompts us to postulate that the decrease of the 1180–1182-cm^{-1} band could be attributed to perturbation of the C—NH_2 vibration mode of adenine. In other words, the nitrogen atom of the NH_2 group of adenine would be the second alkylating site of DNA for MYKO 63. Such a dialkylation on the N(7) and NH_2 sites of an adenine residue has been reported for K_2PtCl_4 by Theophanides et al. [47].

The fact that adenine appears to be the target for MYKO 63 through a dialkylating process on the N(7) and NH_2 sites is consistent with the very low kinetics of complexation of MYKO 63 to DNA mentionned above: Lawley and Brookes [48], Maxam and Gilbert [49] and Goodwin et al. [46] have clearly demonstrated that methylation on adenine happens very slowly, five times slower, for example, than on guanine.

5.4.2.7 Conclusions

The present Raman investigations show direct interactions between $N_3P_3Az_6$ and the N(7) and NH_2 sites of adenine. Further similar investigations are now in progress with several other anticancer inorganic ring compounds with the aim of elucidating the origin of their antitumor properties.

5.5 Biological Behaviour of MYKO 63 in vivo: Preclinical Estimation of Eventual Prejudicial Side-Effects

One of the main problems which arises when discovering a new anticancer agent is to know whether the curative properties of such a drug are not accompanied by too many prejudicial side-effects. In other words, as emphasized by G. Mathe, "the drug must cure the tumor-bearing patient without killing him by side-toxicity".

Many tremendous hopes, raised up in the past by the discovery of very powerful antitumor chemicals, vanished actually to some extent when it was shown that these drugs were not specific enough of the cancerized part of the body but were also attacking some other healthy organs in such a disastrous manner that "the patient dies cured . . ." This is the case, for example, with several platinum drugs which induce nephrotoxicity (kidney damage), of nitrosoureas (bone marrow depletion) and of some alkaloïds (heart diseases).

Thus bearing this in mind, we were urged to investigate the biological behaviour of MYKO 63 in vivo in order to estimate its side-effects, if any.

5.5.1 Mutagenicity

All the anticancer drugs so far known are mutagenic. This means that they induce some mutations within the gene distributions on DNA, such mutations being responsible for delayed relapses in patients who were cured of their initial tumor. Thus, mutagenicity appears to be a very prejudicial side-effect for any therapeutical drug, at least when it reaches too a high level.

Mutagenic character may be easily measured using Ames tests [50]. Let us describe briefly the method and the results we got with MYKO 63 and relatives.

Four Salmonella typhimurium strains, namely TA 98, TA 100, TA 1535 and TA 1538 [50], were used. The TA 100 strain for example bears the *his G 46* missense mutation, is excision-repair deficient (uvrB) and has enhanced permeability (rfa). Efficiency of reversion to histidine prototrophy was measured in each case by subtracting the background (ca. 140 spontaneous revertants) from plate counts.

In addition, these experiments may be performed in presence of rat liver microsomes (containing P450 and other cytochromes) which may modify the mutagenic character (measured by the number of revertants) if they induce a certain metabolisation of the drug.

By Ames' criteria, a chemical must be considered mutagenic if, at a dose of 100γ per plate, it more than doubles the number of spontaneous revertants, either with or without metabolizing microsomes.

For MYKO 63, this number was found in both cases to be equal to 320 at 100γ per plate. This indicates that MYKO 63 is only slightly mutagenic, when compared at least with drugs like cis-$Pt(NH_3)_2Cl_2$ [51]. Furthermore, it is noteworthy that MYKO 63 mutagenicity, staying unaffected in presence of microsomes, supports the idea that MYKO 63 does not seem to need any metabolisation to be effective as antitumor agent.

In conclusion, MYKO 63 exhibits a low mutagenic character when measured by the mean of Ames' tests. This is undoubtedly a point in favour of MYKO 63.

5.5.2 Teratogenicity

Any drug may induce some plus or minus desastrous organical lesions either at the adult level or in the foetus.

Preliminary investigations of the teratogenic character of MYKO 63 were performed on amphibians (pleurodeles and/or axolotls) by R. Deparis, A. Jaylet and A. M. Duprat at the Laboratoire de Biologie Générale, Paul Sabatier University, Toulouse. The main results of their studies are as follows.

(i) No morphological injuries were ever observed, even under repeated treatment schedules with large doses, on growing amphibians larvas: tail, branchiae, front and back legs in fact developed normally, in contrast with what is observed with other anticancer drugs such as adriamycine. Thus, teratogenicity of MYKO 63 does not occur "when the baby was born".

(ii) On the other hand, there is significant teratogenicity of the foetus, indicating that MYKO 63 (as all the other anticancer drugs), must be excluded for the treatment of pregnant, tumor-bearing women.

(iii) Moreover, MYKO 63 causes a large *reversible* (one month after end of treatment) inhibition of spermatogenesis, both in amphibians and mammalians. The reversible character of this effect lessens the prejudice in question, so much the more that clinicians may take the precaution of storing (for post-cure uses) the sperm of patients before treatment.

5.5.3 Other Pharmacological and Toxicological Data

Clinical uses of a new drug obviously need some deep investigations of its *pharmacological and toxicological* behaviour in vivo both in mammalians (mice, rats, dogs and monkeys) and in phase 1 humans. We shall detail below what the major factors are that have to be examined from this point of view.

5.5.4 Cumulative Toxicity

The entire work that has been done on MYKO 63 and its relatives as anticancer agents may be summarized as follows:

(i) MYKO 63 is highly soluble in water and in saline solutions. This makes it very convenient for any kind of inoculation.

(ii) Its chemical and biological purity may be checked in a very easy and accurate way by mass spectrometry and neutron activation.

(iii) MYKO 63 proved effective on any type of animal tumor it was tested on in vivo.

(iv) MYKO 63 interacts with DNA in vitro, principally at the N7 and NH_2 sites of adenine.

(v) This drug is only slightly mutagenic and not at all teratogenic in adult amphibians.

(vi) It causes a strong — but reversible — inhibition of spermatogenesis both in amphibians and mammalians.
(vii) The cost of the drug is extremely low (about US $ 300 per kg) in contrast with some other anticancer drugs such as alkaloïds of vinca rosea L. (vincristine, vinblastine, vindoline) the cost of which reaches about 2 millions US $ per kg or Rosenberg's DDP (about 40,000 US $ per kg). But however encouraging these results are, MYKO 63 does exhibit an unpleasant property: it induces cumulative toxicity in vivo when used within repetitive polyinjection schedules: for example, a Q1D chronical schedule with 10 mg/kg daily leads to complete lethality of treated mice after the 5th injection, i.e., on day 6 after the graft of the tumor.

Thus, *MYKO 63 appears to be a "very lazy" drug with respect to the kinetics of its attack on the tumor and, consequently, of its excretion rate.*

Chemically speaking, this observation may be translated in terms of a *very high chemical stability* of MYKO 63, at the level of either alkylating aziridinyl wings or of the inorganic ring carrier. So, we used our previous know-how of electronic structures within inorganic ring systems to attempt some designs of new drugs derived from MYKO 63 the structures of which would be destabilized slightly — but not too much! — in order to be still more aggressive towards the tumor and to be excreted faster, losing thereby the property of inducing any cumulative toxicity.

To do so, we pursued two logically complementary pathways: (i) the MYKOMET approach and (ii) the SOAz approach. These are described successively in the following sections.

6 Mykomet 63 [$N_3P_3(MeAz)_6$] as a Tentative Improvement on MYKO 63: A Disappointing Cul-de-sac [36]

Even if, from all the results obtained on MYKO 63, we could not unambiguously demonstrate that the drug is effective under an *in vivo* dialkylation on DNA, we may nevertheless make the assumption — based mainly on Scatchard [28] and Raman investigations [29] — that such a mechanism is the right one; in doing so, we shall design new drugs the antitumor activity of which will be better (or worse) than that of MYKO 63, thus supporting (or invalidating) the assumption we intended to start from.

Thus, let us come back to the way in which MYKO 63 would interact with DNA through pairs of carbocations provided by pairs of aziridinyl groups following the classical reaction path:

 $\xrightarrow{H^+}$ $-NH-CH_2-CH_2^+$ (A)

There is a possible way of designing new aziridinocyclophosphazenes which are more active yet than MYKO 63 by preparing derivatives in which we would perform a progressive methylation of the six Az groups at positions 2 and 2′; such a methylation does increase the rate of reaction (A) [52] and, as a consequence, we could expect to increase the antitumor activity.

We shall report here on the synthesis and the antitumoral tests against P388 and L1210 leukemias of the simplest term of the series, namely $N_3P_3(MeAz)_6$ (MeAz = 2-methyl aziridinyl) (Fig. 30). The EORTC code name of this drug was MYKOMET 63.

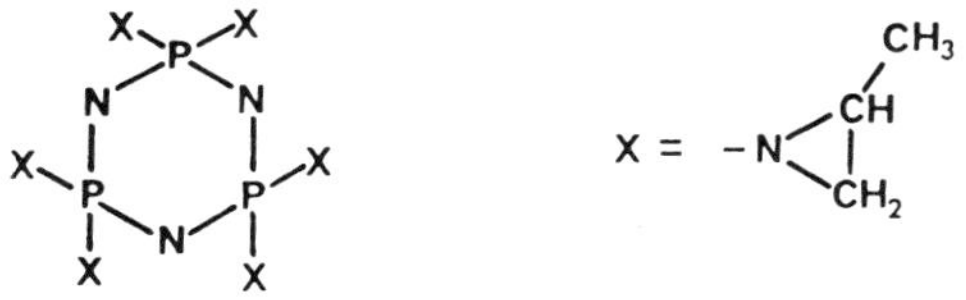

Fig. 30. Formula of MYKOMET 63

6.1 Synthesis of MYKOMET 63

Ratz et al. [2] reported on the synthesis of this cyclophosphazene in which the chlorine atoms of $N_3P_3Cl_6$ are substituted by 2- methyl aziridine (propyleneimine) groups in the presence of H_3N or Et_3N in benzene. However, the original paper provides only very few characteristics of the sample in question, i.e., elemental analysis and refractive index, $n_D^{25} = 1.5071$.

Using their procedure, we prepared in a very high yield (90%) a sample the refractive index of which was found to be $n_D^{20} = 1.5076$, close to Ratz's value. Incidentally, it may be noticed that the asymmetry induced — from a group theory point of view — in $N_3P_3Az_6$ when grafting one methyl group on each aziridinyl ligand modifies drastically the state of the chemical: $N_3P_3Az_6$ is in fact a high

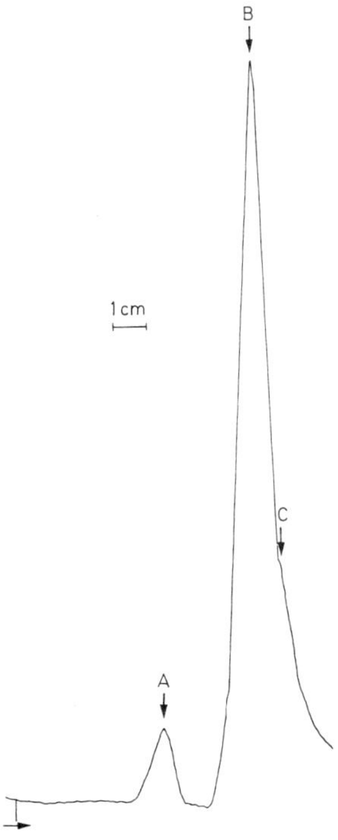

Fig. 31. HPLC chromatogram of the mixture obtained when using Ratz's procedure for the synthesis of $N_3P_3(MeAz)_6$ (Waters Liquid Chromatograph Instrument, μ Bondapack C_{18} reverse phase, pure methanol, RI detection, speed rate = 1 cm/mn) A = $N_3P_3(MeAz)_6$, B = $N_3P_3(MeAz)_5Cl$, C = $N_3P_3(MeAz)_4Cl_2$

melting point (147 °C) solid whereas $N_3P_3(MeAz)_6$ is a liquid at room temperature.

Thus, we could assume that the sample we had prepared was the right one, owing to the agreement of its refractive index with Ratz's data. Actually, further investigations did not support this conclusion: ^{31}P nmr measurements and H.P.L.C. (using methanol as a vector) did prove that the so-called $N_3P_3(MeAz)_6$ sample was a mixture of at least three components (Fig. 31), the main part consisting of $N_3P_3(MeAz)_5Cl$. These results were confirmed by mass spectrometry, a technique which was recently demonstrated to be a powerful tool for testing the purity of cyclophosphazenes [9].

Therefore, the refractive index mentioned above appeared to be characteristic not of a pure hexasubstituted cyclophosphazene but of a mixture in which the pentasubstituted trimer is predominant.

In the first step, we were able to separate this penta derivative by preparative H.P.L.C. and we subsequently treated it with an excess of propyleneimine in order to reach the required hexasubstituted compound. Under such conditions, we succeeded in preparing a $N_3P_3(MeAz)_6$ real sample (free of chlorine, as demonstrated by neutron activation analysis) identified by mass spectrometry (Fig. 32) and by ^{31}P nmr (Fig. 33) ($\delta = -36$ ppm with 85% H_3PO_4 as standard, to be compared with $\delta(N_3P_3Az_6) = -37$ ppm). The refractive index of this sample, $n_D^{25} = 1.4825$, appeared to be significantly far from Ratz's value.

To conclude this part of the work, we may say that Ratz et al. seem to have described a mixture which had nothing to do with the expected $N_3P_3(MeAz)_6$. This mixture has to be re-treated by an excess of propyleneimine in order to produce the right compound, the characteristics of which are reported here for the first time.

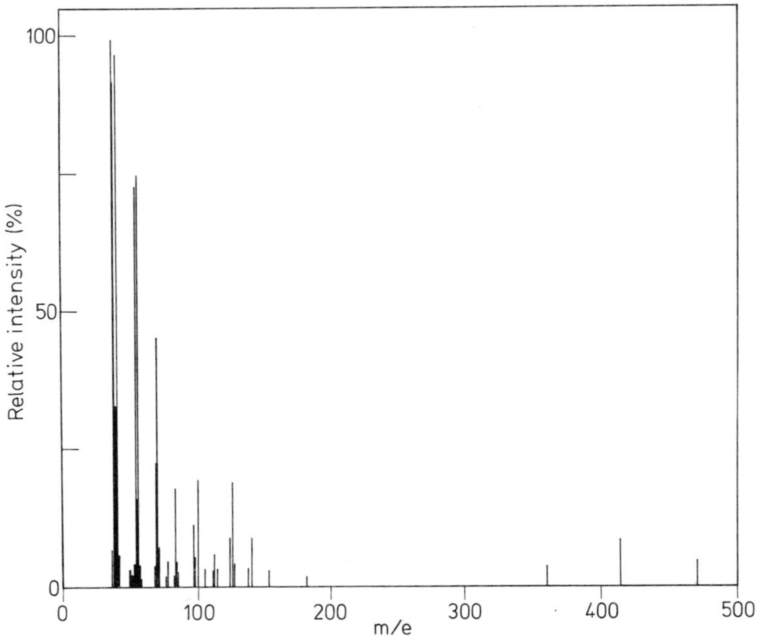

Fig. 32. Mass spectrum of $N_3P_3(MeAz)_6$ (R1010 Ribermag quadrupole mass spectrometer)

We wish now to mention briefly the difficulties encountered when preparing the higher homologues within the series.

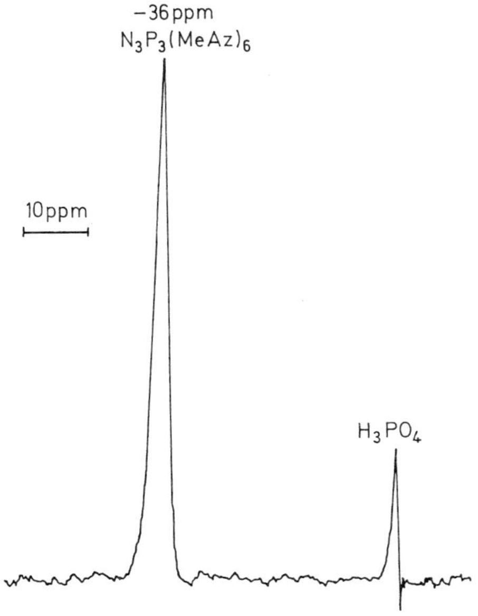

Fig. 33. ^{31}P nmr spectrum of $N_3P_3(MeAz)_6$ (Perkin-Elmer R10 Instrument; negative sign for δ indicates that the shift is towards the low field)

6.2 Synthesis of $N_3P_3(DiMeAz)_6$ (DiMeAz = gem-2,2-dimethyl-aziridinyl) and its Relatives

The previous situation was observed more than ever when the degree of methylation on the aziridinyl ligand increased and, in contrast again with Ratz's conclusions [2], we were never able to get more than a complex mixture of various substituted trimers when LH was the gem-2,2-dimethylaziridine or its higher isologues. However, the separation and the identification of the components in each case are now in progress in our laboratory.

6.3 Toxicity and Antitumor Properties of $N_3P_3(MeAz)_6$ [36]

Coming back to the possible origin of the antitumor activity of the aziridinocyclotriphosphazenes we described above, we would expect $N_3P_3(MeAz)_6$ (EORTC code name: MYKOMET 63) to be a more efficient drug than $N_3P_3Az_6$ (EORTC code name: MYKO 63) under the express condition that this drug would not provide too many reactive carbocations which would interact with side-targets before reaching the real target.

Toxicity measurements were performed in a classical way [4] on DBA/2 mice and LD_0 was found to be 250 mg/kg. We may notice that MYKOMET 63 is about six times less toxic than MYKO 63.

Table 8. Relative antitumor activities of MYKOMET 63 and of MYKO 63 against P388 and L1210 tumors

LEUKEMIA	DRUG	DOSE		MEDIAN SURVIVAL TIME		% ILS
		mg/kg	in term of LD_0	control	treated	
L 1210	MYKOMET 63	100	0.4	8.6 ± 0.7	10.7 ± 0.5	24
		150	0.6	8.6 ± 0.7	12.5 ± 0.5	45
		200	0.8	8.6 ± 0.7	13.6 ± 0.6	58
		1 × 200 + 1 × 100 (Q3D: 1, 4)	0.8 + 0.4	9.0 ± 0.3	14.3 ± 1.5	59
	MYKO 63	20	0.5	8.6 ± 0.5	12.5 ± 0.8	45
		30	0.75	9.2 ± 0.5	13.1 ± 0.8	42
		3 × 10 (Q3D: 1, 4, 7)	3 × 0.25	8.6 ± 0.5	12.4 ± 0.5	44
P 388	MYKOMET 63	100	0.4	10.5 ± 0.4	18.3 ± 1.1	74
		200	0.8	10.5 ± 0.4	22.4 ± 3.3	113
		3 × 100 (Q4D: 1, 5, 9)	3 × 0.4	9.7 ± 0.5	21.6 ± 1.7	123
	MYKO 63	20	0.5	9.9 ± 0.3	19.9 ± 1.5	101
		30	0.75	9.8 ± 0.4	23.0 ± 2.0	135
		40	1.0	9.8 ± 0.4	23.2 ± 3.7	137
		3 × 10 (Q4D: 1, 5, 9)	3 × 0.25	9.9 ± 0.3	19.8 ± 0.8	100

For L 1210: 10^5 cells implanted i.p.; for P 388: 10^6 cells implanted i.p.; i.p. treatment (10 to 15 mice per group) generally one day after the graft of cells; drugs were dissolved in 0.9% NaCl solution.

In Table 8 are shown the activities of MYKOMET 63 under various conditions against murine (DBA/2 mice) P388 and L1210 leukemias, compared with the ones reported previously [4] for MYKO 63.

The comparison of the two sets of %ILS values allows us to conclude that MYKOMET 63 appears to be actually as active as MYKO 63, but not more active. In other words, the graft of methylated aziridinyl ligands on a six-membered cyclophosphazene ring does not significantly improve the antitumoral activity with respect to the non-methylated isologues. However, because MYKOMET 63 possesses only a slight toxicity compared to MYKO 63, the new drug reported here may be of interest for eventual clinical use, no matter what difficulties are encountered in its synthesis.

Anyhow, the method of methylating the aziridinyl wings of MYKO 63 does not improve substantially antitumor activity as we might have expected. This route actually led to a rather disappointing cul-de-sac and therefore we abandoned the development of this kind of drug and directed our attention at the other route mentioned above, i.e., to design new anticancer inorganic ring systems in which the six-membered ring itself would be suitably destabilized.

We shall see in the next paragraph, that this approach brought about the improvements we were looking for.

7 SOAz and Relatives or "How to Play Double or Quits"

The slight destabilization required for the cyclophosphazene ring of MYKO 63 could be expected from the substitution of one or two sulphur atoms for phosphorus. Some previous quantum-mechanical studies [53–60] had indeed shown that such a process would induce a progressive decrease of ring stability since, according to the Dewar islands models [61] of cyclophosphazenes and cyclophosphathiazenes, the transannular (P ... P) and (P ... S) interactions play a dominant role in the ring stability, (P ... P) interactions being significantly stronger than (P ... S) ones [60] (Fig. 34).

Consequently, compounds of the series (I) and (II) (Fig. 35) were prepared {X = F (code names: SOF for (I) and DISOF for (II)), X = Phenyl (code name: SOPHi for (I)) and X = Aziridinyl (code name: SOAz for (I))}.

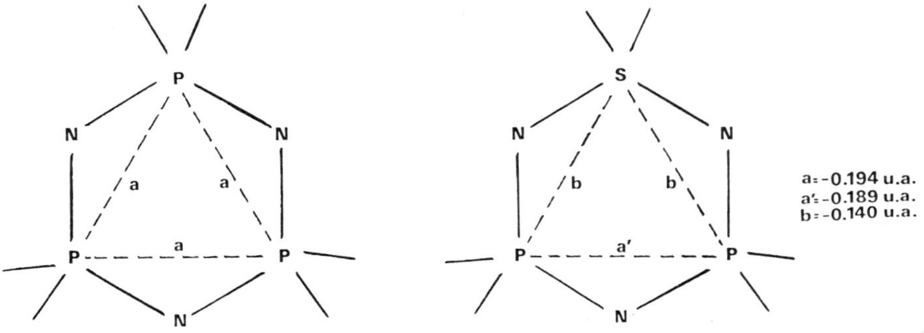

Fig. 34. Transannular (P ... P) and (P ... S) interactions in cyclophosphazenes and cyclophosphathiazenes

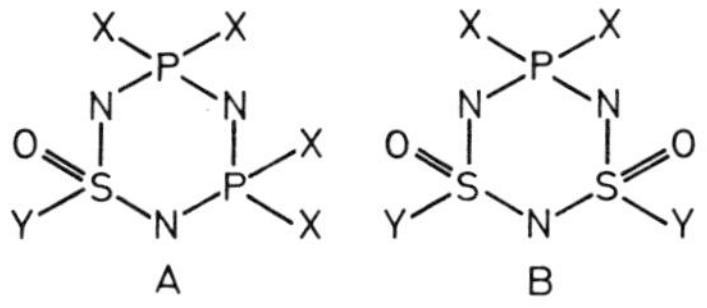

Fig. 35. Formulae of the cyclophosphathiazenes studied

7.1 Materials and Methods

7.1.1 Synthesis and Purity

The original synthesis of these chemicals has been reported elsewhere [62]. Their purity was checked by classical IR, ^{31}P and ^{1}H nmr [62] as well as by HPLC and mass spectrometry [63]. In the cyclophosphathiazenes mass spectra there are mass peaks caused by fracture of the NPS rings; this was particularly noticeable with DISOF and in complete contrast to the cyclophosphazenes analogous [9]. These observations are consistent with the proposed stability of the ring systems.

7.1.2 Solutions

SOF, SOPHi and SOAz are soluble in water and could consequently by used in 0.9% NaCl water solution for both toxicity measurements and antitumoral tests; DISOF, having very poor solubility in water, was inoculated as a suspension in 4‰ aqueous hydroxypropylcellulose (Klucel J. F.; Hercules Co.).

7.2 Toxicity and Antitumor Activity

7.2.1 Toxicity

The LD_0 values obtained for the four cyclophosphathiazenes studied are given in Table 9 in which two points are worthy of particular attention:

a) SOF and DISOF have the same LD_0 value, close to that of $N_3P_3Az_6$ (LD_0 = 40 mg/kg) [4]; thus, the replacement of PAz_2 moiety by a SOF entity does not alter the toxicity within the $(NPAz_2)_{3-n}(NSOF)_n$ series.

b) Within the two series, SOF and DISOF appear to be much more toxic than SOPHi and SOAz; this probably reflects the higher toxicity of F with respect to the

Table 9. Highest non-lethal doses LD_0 for the 4 cyclophosphathiazenes studied. (in mg/kg)

Compound	LD_0 (mg/kg)
SOF	50
SOPHI	>200
SOAz	200
DiSOF	< 50

phenyl and aziridinyl residues. Despite the large difference between *in vitro* and *in vivo* approaches, this assumption would be in agreement with the existence of many HF-containing fragments in mass spectrometry.

7.2.2 Antitumor Activity [64)]

a) *Effects on the P388 leukemia*

The activities of SOF, SOPHi, SOAz and DISOF under various conditions are shown in Table 10.

Table 10. Antitumor activity of some cyclophosphathiazenes against P388 leukemia

Compound	Schedule	Dose (mg/kg/day)	% ILS
SOF	once, day 1 (i.p.)	10	20
(LD_0 = 50 mg/kg)		25	44
		50	98
	Q4D; days (1, 5, 9)	10	51
		25	119
SOPHI	once, day 1 (i.p.)	150	57
(LD_0 = 200 mg/kg)		200	68
		250	107[a)]
SOAz	once, day 1 (i.p.)	50	35
(LD_0 = 200 mg/kg)		100	73
		150	134
		175	194
		200	195 + 40% mice cured[b)]
	Q4D; days (1, 5, 9)	50	73
		100	196
	Q4D; days (1, 5)	150	181 + 10% mice cured[b)]
DiSOF	once, day 1 (i.p.)	10	3
(LD_0 < 50 mg/kg)		25	3
		50	6

10^6 P388 cells implanted i.p., i.p. treatment (10 mice/group); median survival time of control: 10.0 days

a This ILS figure was actually vitiated by the mortality of 4 mice out of 10 which were not included in the calculation of %ILS.

b Cured mice (i.e. survival time greater than or equal to 60 days) were not included in the calculation of ILS figures.

It is worth noting that:

1) The three compounds of series (I), namely SOF, SOPHi and SOAz, are found to be active, but DISOF does not exhibit any antitumor activity at all. The delayed toxicity of DISOF may however conceal an eventual antitumoral activity. It appears therefore that at least *two geminal aziridino pairs* are necessary in these compounds for antitumoral activity.

2) SOAz appears to be the most active member of series (I): indeed a single dose of 200 mg/kg resulted in *40% cured mice on Day 60.*

3) Still more important perhaps is the fact that the activity-dose relationship for SOAz (Fig. 36) is *exponential*, the equation of this curve may be determined from the Tchebytchev polynomial:

$$\%\text{ILS} = 0.212 + 0.589\,d + 2.248\,10^{-2}\,d^2 - 1.068\,10^{-3}\,d^3 + 2.214\,10^{-5}\,d^4 - 2.390\,10^{-7}\,d^5 + 1.405\,10^{-9}\,d^6 + \ldots$$

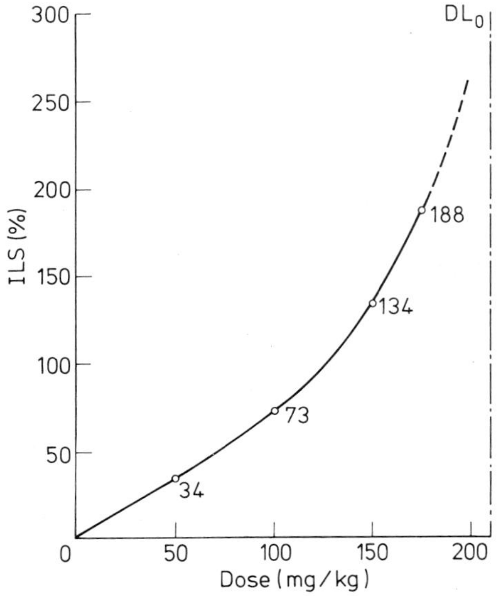

Fig. 36. Exponential activity-dose relationship for SOAz on P388 leukemia (DBA/2 female mice)

4) From Table 10, It can be seen that the Q4D schedule noticeably increases %ILS values with respect to the monoinjection D + 1 protocol: for SOAz there is a twofold increase in the value of %ILS between a (1.50) and a Q4D (3.50) injection and almost a threefold increase between a (1.100) and a Q4D (3.100). The Q4D (1, 5) (2.150) protocol led to an ILS of 181% (*not including one surviving mouse on Day 60*). It can also be seen that a polyinjection schedule allows more than the single LD_0 dose to be given without significant lethality: This suggests that SOAz acts quickly on the tumor cells and is excreted before the following injection; this is consistent with the idea of structure destabilisation referred to earlier.

5) The therapeutic index of SOF, SOPHi and SOAz, which is defined as a ratio of the LD_0 value divided by the dose which gives an ILS of 25% (extrapolated from the activity-dose relationships), is respectively about 2, 2 and 8. This may be compared with the values previously reported for $N_3P_3Az_6$ (TI = 6) and $N_4P_4Az_8$ (TI = 4)[4].

From the foregoing results, SOAz seems the most promising antitumor agent of the series. It is highly soluble in physiological serum, and allows the cure of the P388 leukemia within a monoinjection protocol.

b) *Effects on the L1210 leukemia*

The tests on L1210 (and B16 melanoma) were confined to the members of the series which exhibited a significant activity on the P388 tumor, namely SOF, SOPHi and SOAz.

The activities for the three drugs with the i.p. route under various conditions are shown in Table 11. ILS figures are definitely smaller for the L1210 than for the P388 tumor. However, the chemicals tested exhibit a significant (i.e., %ILS ≧ 25) activity in a monoinjection protocol, even where the dose administrated is equal to only $LD_0/2$. A Q3D (1, 4, 7) schedule appears to be useless for SOF but multiplies the ILS figures for SOAz by *ca.* 2.

Table 11. Antitumor activity of some cyclophosphathiazenes against L1210 leukemia

Compound	Schedule	Dose (mg/kg/day)	% ILS
SOF	once, day 1	25	27
(LD_0 = 50 mg/kg)		40	40
		50	48
	Q3D; (1, 4, 7)	10	4
		25	24
SOPHI	once, day 1	200	40
(LD_0 = 200 mg/kg)			
SOAz	once, day 1	100	28
(LD_0 = 200 mg/kg)		150	46
		200	69[a)]
	Q3D; (1, 4, 7)	50	22
		100	46

10^5 L1210 cells implanted i.p., i.p. treatment (10 mice per group); median survival time of control: 9.0 days.

a This ILS figure was actually vitiated by the mortality of 5 mice out of 10 during the treatment which were not included in the calculation of %ILS.

c) *Effects on the B16 melanoma*

Compared to the L1210 and P388 leukemias, B16 melanotic melanoma is a slow-growing tumor. From Table 12, it can be seen that for a monoinjection protocol, SOAz gives an %ILS value greater than the 40%, which is the value normally associated with B16. The result is encouraging in view of the generally weak sensitivity of this melanoma to most of the usual antitumor agents.

For some of the treatment schedules with SOAz (Table 13), both the number of tumor-bearing mice and the tumor diameters were recorded on the 14th day. Compared to the control, it can be seen that the number and the size of tumors for the treated mice are smaller, which indicates a real activity for the drug in question. Moreover, these two parameters were in good agreement with the ILS determinations.

Finally, the activity-dose relationship for SOAz on B16 melanoma seems to be linear and not of a higher polynomial degree as in the case of P388.

7.3 Discussion on the Antitumoral Activity of SOAz and Relatives as Compared with MYKO 63 Effectiveness

Among the four cyclophosphathiazenes studied, three, namely SOF, SOPHi and SOAz, exhibited a significant antitumor activity against P388 leukemia on i.p.

administration. The most active is SOAz which cures some mice bearing this tumor within a monoinjection (1.200) protocol. Furthermore, this compound is highly soluble in physiological serum (450 g/l). The ratio of the LD_0 value (200 mg/kg) to the minimum significantly active dose (~25 mg/kg) is approximately 8.

Moreover, SOAz shows a striking exponential activity-dose relationship by the i.p. route (Fig. 36). Such relationships are usually linear at best, but more often they start linear and level-off.

Against the L1210 leukemia, SOF, SOPHi and SOAz were all active, giving a significant %ILS value even within a monoinjection protocol. However, the ILS figures were much less than in the case of P388.

The same conclusions broadly apply to the B16 melanoma results. SOAz appears to be an efficient drug for this tumor. The additional check on the number of tumor-bearing mice and tumor diameters on the 14th day after the graft showed that tumor-growth is significantly delayed by the active doses referred to in Table 13.

Table 12. Antitumor activity of some cyclophosphathiazenes against B16 melanoma

Compound	Schedule	Dose (mg/kg/day)	% ILS
SOF	once, day 1	30	21
(LD_0 = 50 mg/kg)		40	0
		50	0
SOAz	once, day 1	50	32
(LD_0 = 200 mg/kg)		100	47
		150	64

B16 cells implanted s.c., i.p. treatment (10 mice per group); median survival time of control: 23.3 days.

Table 13. Effect of SOF and SOAz on B16 tumor evolution

Compound and schedule	14th day		% ILS
	Number of tumor-bearing mice	size of tumors (cm)	
Control	10/10	2.10 ± 0.1	—
SOF			
1×30	10/10	1.33 ± 0.1	21
1×40	10/10	1.47 ± 0.1	0
1×50	10/10	1.43 ± 0.1	0
Control	10/10	1.72 ± 0.1	—
SOAz			
1×50	10/10	1.48 ± 0.1*)	32
1×100	10/10	1.35 ± 0.1	47
1×150	5/10	0.41 ± 0.1	64

Mean ± S.E. of the mean, calculated on the total number of mice per group (a non-tumor-bearing mouse was counted as zero but incorporated into the calculation). The differences in median size of treated and control series were statistically significant (Student's t-test) at $P < 0.05$ unless for*) where $P < 0.10$ only.

Anyhow, SOAz has two main advantages over MYKO 63: (i) *in full agreement with our original challenge, it never induces any cumulative toxicity even within a chronical Q1D (1–9) polyinjection schedule at 150 mg/kg (i.e., 3/4 LD_0) daily*; (ii) Moreover, SOAz exhibits a totally unexpected, exponentially increasing activity-dose relationship for P388 grafted on *female* DBA/2 mice. Such a "more than linear" trend is to our knowledge quite unique in cancer chemotherapy: It may happen, for some synergic reasons, when two drugs or more are used in combined therapy but there is no a priori reason to observe such a phenomenon when one drug is used alone.

A possible explanation for this was that the antitumor activity of SOAz could be potentialized in vivo by some specific biological material and the sex of the treated animals could play a certain role in this respect. More precisely, since the exponential relationship was observed for *female mice*, we could assume that SOAz would be a hormonal-dependent drug (potentialized by oestrogens?) which would act in a different manner on *male mice*.

Thus, systematic antitumor tests against P388 were performed on both male and female mice with the help of F. Spreafico's EORTC team in Milano and of the Otsuka Pharmaceutical Company (see below) in Osaka, Japan. Conclusions are presented in Fig. 37 which clearly shows that SOAz is definitely a hormonal-

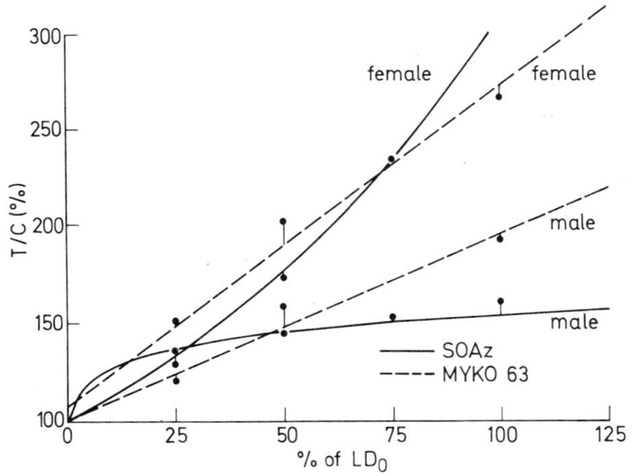

Fig. 37. Activity of SOAz and MYKO 63 *versus* P388 (*male and female mice*)

dependent drug, its efficiency being significantly higher for females than for males. Endocrinological investigations are now in progress for discriminating the female hormone(s) which are responsible for SOAz potentialization. If we succeed to reveal the hormones in question, it will be possible to inoculate them jointly with SOAz in male patients: They will benefit from such a combined therapy, even if they have to suffer from some slight increase in their breast size . . .

Anyhow, one may understand why EORTC showed an even greater interest in SOAz than in MYKO 63 itself. SOF, SOPHi and SOAz were included in December 1979 in EORTC planning for the 1980–1982 period. Furthermore, SOAz appears month after month globally to be as active — if not more active — than MYKO 63 in monoinjection protocols, but fundamentally more active within polyinjection schedules, owing to the absence of the cumulative toxicity we mentioned above.

The effectiveness of SOF, SOPHi and SOAz on several tumors was protected on July 4, 1979 by the French ANVAR patent n° 79-17336 [65] which was extended on July 4, 1980 to 28 foreign countries:

Austria, Belgium, Greece, Italy, Luxemburg, Netherlands, West Germany, Great-Britain, Sweden, Switzerland, Liechtenstein, Hong-Kong, Malaysia, Singapore	n° 80 401002.3		
		Ireland	1308/80
		Denmark	146 188
		Spain	493 454
		U.S.A.	06/162 759
		Canada	354 877
		Mexico	8908/80
		Israel	60 383
		Hungary	1657/80
		Japan	092159/80
		Taiwan	69 11769
		South Korea	80 2638
		USSR	2940708/04
		Portugal	71488
		Australia	59776/80
		New Zealand	194 229

The licensees for development and exploitation are Otsuka Chemical and Pharmaceutical Co (Osaka — Japan) and Roger Bellon S.A. (Paris — France) through a joint-venture agreement.

Parallel to these industrial applications, our duty was to perform the same fundamental research which we described above in connection with MYKO 63. Let us now present the structural and biological parameters we obtained in this way.

8 Strategy of Development of SOAz as an Anticancer Agent

8.1 Further Antitumoral Tests in vitro and in vivo

SOAz and relatives are now currently being investigated by the EORTC screening pharmacology group on all the animal tumors on which MYKO 63 had already been tested (see above). So far, SOAz appears as active as MYKO 63 in each case, in any monoinjection protocol, and we may expect even better activities within polyinjections QnD schedules with large doses daily.

Parallel to this, several investigations of antitumor activities *in vitro* of SOAz were performed both at the Department of Radiopathology of Groningen University (H. Lamberts) and at Toulouse University Hospital. These experiments provide the doses of SOAz which are able to give a complete inhibition of growth for tumor cells in culture. These doses are gathered in Table 14 which calls for the following remarks.

(i) SOAz appears as very effective on both radioresistant and radiosensitive lymphomas and lymphosarcomas, Lewis Lung carcinoma, neuroblastoma, Rhabdomyosarcoma and Zajdela rat tumor.

Table 14. Minimal doses of SOAz for complete growth inhibition in vitro of several tumor cells in culture

Tumor	Dose for 100% Inhibition of Culture Cells Growth (μg/ml)	Tumor	Dose for 100% Inhibition of Culture Cells Growth (μg/ml)
S-Lymfoma	2 to 4	Zajdela (RAT)	8
R-Lymfomas	2	Osteosarcoma	16
Lymfosarcoma	2	Wermehes (MB)	32
Lewis lung	2	Melanoma (ardon deal)	33
Rhabdomyosarcoma (RMS)	4	Hazenberg (ovarian carcinoma)	65
Erlich ascites	8	Swieringa (ovarian carcinoma)	65
L-Fibroblastoma	8	Hela (cervix carcinoma)	125

From Groningen University, Department of Radiopathology, and Toulouse University.

(ii) In contrast, SOAz is poorly active *in vitro* on ovarian carcinoma (Hazenberg and Swieringa strains) and on HeLa cervix carcinoma. Thus, there exists a discrepancy between *in vitro* and *in vivo* SOAz activities on ovarian carcinomas (poor activity in vitro *versus* promising activity in vivo). This indicates how careful we have to be when attempting to deduce in vivo data from in vitro approaches. From this point of view, anticancer inorganic ring systems do not behave as many other kinds of drugs (for example ellipticines [66] and nitrosoureas) for which in vitro results generally correlate with in vivo activities.

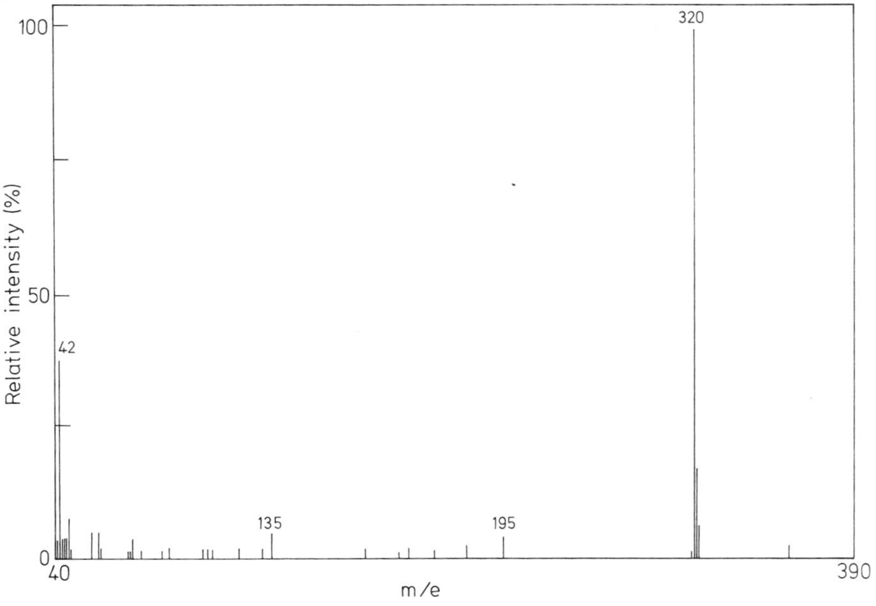

Fig. 38. Mass spectrum of SOAz

The EORTC 1980–1982 campaign will surely provide plenty of further exciting results which will allow us to discriminate the specific tumors which clinical trials will deal with afterwards.

8.2 Mass Spectrometry for Testing the Purity of SOAz

The mass spectrum of SOAz is shown Fig. 38 and its pattern is quite different from that of MYKO 63: in fact we no longer observe the "fall of the Az leaves" which characterizes any mass spectrum within the MYKO 63 series. The base peak is at m/z 320 and there are very few other secondary peaks till m/z 50. No chlorinated impurity could be detected either by mass spectrometry or by neutron activation.

Incidentally, the noticeable difference between mass patterns for MYKO 63 and SOAz is fully consistent with the relative stability of N_3P_3 and N_3P_2S rings as computed within CNDO/2 approach [60]. Anyhow, the extreme simplicity of the SOAz mass spectrum would certainly facilitate the detection of any impurity. The mass spectrum of SOF is also reported without any further comment in Fig. 39.

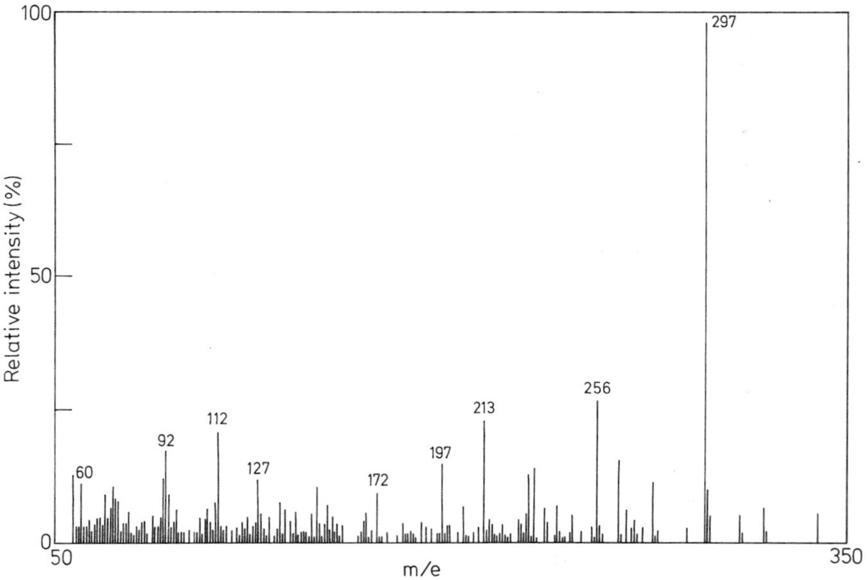

Fig. 39. Mass spectrum of SOF

8.3 X-Ray Crystal Structures of SOAz and Relatives

8.3.1 SOAz Molecular Structures [67]

8.3.1.1 Single Crystals

As mentioned previously, SOAz was prepared following the procedure which was recently described [62, 65]. When repeating such a preparation many times, we obtained, surprisingly, two kinds of samples, SOAz (I) and SOAz (II), whose analytical, physicochemical and biological data were strictly identical, except their melting points: 84 °C and 103 °C respectively. Incidentally, we may note that we have never obtained any sample of SOAz with melting points different from these two particular values.

Chemically speaking, SOAz (I) and SOAz (II) are strictly identical and pure; their actions on animal tumors are also identical. Thus, we would expect that the difference in their melting points to be due to some structural peculiarities, either with regard to their space group (if the two kinds of crystals do not contain any insertion of solvent) or possibly by some inclusion of solvent in the unit cell (clathrate structure) as in the case of MYKO 63 when crystallized from C_6H_6 or CCl_4 (see above).

8.3.1.2 X-Ray Analysis

Carefully selected crystals from both experiments, with regular prismatic shapes (all dimensions are within 0.4 to 0.6 mm) were glued on glass fibers and mounted on a Stoe "reciprocal lattice explorer" camera. Zr filtered molybdenum Kα radiation was used.

This photographic study revealed that two allotropic forms of the compound in question were obtained: SOAz (I) and SOAz (II).

SOAz (I): This form crystallizes in the othorhombic system; the space group is $P2_12_12_1$. There is one independent molecule in the cell.

SOAz (II): This second crystalline form adopts the monoclinic system, systematic reflection absences indicating $P2_1/c$ as the space group. The volume of the cell is twice that of SOAz (I), indicating that the two crystallographically independent molecules exist.

8.3.1.3 Data Collection and Computer Treatment

The crystals were transferred to a CAD4-Enraf Nonius PDP8/M computer-controlled diffractometer. In each case, 25 reflections were used in order to settle the crystals, to check their quality and to refine the cell dimensions.

The conditions for the data collections are summarized in ref. [67]. The intensities of selected reflections [$I > 2\sigma(I)$] were corrected for Lorentz and polarisation factors but not for absorption.

Atomic scattering factors corrected for anomalous dispersions were obtained from Cromer and Waber [68] and Cromer and Liberman [69].

8.3.1.4 Structures Solutions and Refinements

Both structures were determined using direct methods included in the Multan 78 version [70].

The refinement includes an anisotropic vibration for all the non-hydrogen atoms. Difference-Fourier syntheses phased by the phosphorus, nitrogen, sulfur, oxygen and carbon atoms revealed the hydrogen atoms in their expected positions.

In the last refinement cycles the hydrogen atoms were positioned geometrically following the indications concerning the bonding of such atoms in aziridinyl groups, i.e. (C—H) = 0.97 Å and H—C—H = 116° according to Dermer and Ham [71] and isotropic thermal parameters were assigned following the rule $B(H_i) = B_{eq}(C_i) + 1$ Å^2.

In the final stages of refinement, an overall extinction parameter was calculated: 0.257×10^{-7} for SOAz (I) and 0.129×10^{-7} for SOAz (II).

The final R values were 0.051 and 0.035, corresponding in both cases to final difference maps exhibiting peaks lower than 0.4 eÅ^{-3}.

Important bond lengths and bond angles of the molecule in its different conformations, i.e., SOAz (I), SOAz (II A) and SOAz (II B), are given in Table 15.

8.3.1.5 Description of the Various SOAz Structures

A perspective view of one of the molecules isolated, $(NPAz_2)_2(NSOAz)$, is shown in Fig. 40; the chosen model belongs to SOAz (I). In this Fig., obtained using the program Ortep of Johnson [72] the numbering of the atoms is indicated, and is identical for both structures SOAz (I) and SOAz (II) (molecules A and B).

Table 15. Comparison of molecular parameters with e.s.d.'s in parentheses, for SOAz(I), SOAz(II A), SOAz (II B) and MYKO 63 from its CCl_4 anticlathrate [17]

	SOAz(I)	SOAz(II)		$N_3P_3Az_6$
		A	B	
P(1)-N(1)	1.598(7) Å	1.607(4) Å	1.620(4) Å	1.592(3) Å
P(1)-N(2)	1.578(7)	1.580(4)	1.583(4)	1.588(2)
P(2)-N(2)	1.574(7)	1.569(4)	1.580(4)	1.588(2)
P(2)-N(3)	1.623(7)	1.601(4)	1.616(4)	1.592(3)
S-N(1)	1.557(7)	1,544(4)	1.553(4)	1.587(3)
S-N(3)	1.535(7)	1.548(4)	1.550(4)	1.587(3)
P(1)-N(5)	1.648(8)	1.666(5)	1.651(4)	1.676(4)
N(5)-C(3)	1.464(13)	1.470(7)	1.467(7)	1.468(6)
N(5)-C(4)	1.452(15)	1.463(7)	1.476(7)	1.467(6)
C(3)-C(4)	1.452(18)	1.460(9)	1.455(8)	1.465(7)
P(1)-N(6)	1.642(8)	1.654(5)	1.655(4)	1.677(3)
N(6)-C(5)	1.501(15)	1.471(8)	1.471(7)	1.469(5)
N(6)-C(6)	1.491(14)	1.479(9)	1.456(7)	1.461(6)
C(5)-C(6)	1.453(18)	1.460(11)	1.458(8)	1.466(7)
P(2)-N(7)	1.672(9)	1.656(4)	1.645(4)	1.677(3)
N(7)-C(7)	1.478(13)	1.468(7)	1.493(7)	1.469(5)
N(7)-C(8)	1.448(13)	1.486(7)	1.477(7)	1.461(6)
C(7)-C(8)	1.435(16)	1.479(9)	1.476(9)	1.466(7)
P(2)-N(8)	1.662(7)	1.648(5)	1.663(5)	1.676(4)
N(8)-C(9)	1.455(15)	1.474(8)	1.471(7)	1.468(6)
N(8)-C(10)	1.463(12)	1.469(8)	1.476(7)	1.467(6)
C(9)-C(10)	1.432(18)	1.481(10)	1.467(9)	1.465(7)
S-O	1.444(6)	1.433(3)	1.430(4)	
S-N(4)	1.666(7)	1.677(5)	1.657(5)	1.675(4)
N(4)-C(1)	1.468(12)	1.474(7)	1.456(7)	1.468(6)
N(4)-C(2)	1.473(11)	1.459(7)	1.455(7)	1.468(6)
C(1)-C(2)	1.461(15)	1.470(8)	1.454(8)	1.464(8)

Table 15. (continued)

	SOAz(I)	SOAz(II)		$N_3P_3Az_6$
		A	B	
N(1)-P(1)-N(2)	115.5(4)°	115.2(2)°	114.7(2)°	116.6(2)°
P(1)-N(2)-P(2)	123.8(6)	122.9(3)	122.8(3)	123.4(2)
N(2)-P(2)-N(3)	114.9(4)	115.9(2)	115.5(2)	116.6(2)
P(2)-N(3)-S	124.4(4)	125.1(3)	123.9(3)	123.2(2)
N(3)-S-N(1)	116.2(4)	114.7(2)	115.0(2)	116.8(2)
S-N(1)-P(1)	124.3(5)	125.3(3)	125.6(3)	123.2(2)
N(5)-P(1)-N(6)	99.6(4)	99.8(2)	100.3(2)	99.0(2)
P(1)-N(5)-C(3)	120.5(8)	119.1(4)	119.3(4)	117.7(3)
P(1)-N(5)-C(4)	119.8(8)	117.8(4)	120.0(4)	118.1(3)
C(3)-N(5)-C(4)	59.7(8)	59.7(4)	59.3(4)	59.9(3)
P(1)-N(6)-C(5)	119.6(7)	119.9(4)	120.6(4)	118.4(3)
P(1)-N(6)-C(6)	119.0(7)	120.4(5)	120.1(4)	118.0(3)
C(5)-N(6)-C(6)	58.1(7)	59.3(5)	59.8(4)	60.1(3)
N(7)-P(2)-N(8)	101.7(5)	106.9(3)	101.4(2)	99.0(2)
P(2)-N(7)-C(7)	119.4(7)	119.5(4)	119.4(4)	118.4(3)
P(2)-N(7)-C(8)	116.4(7)	116.5(4)	117.8(4)	118.0(3)
C(7)-N(7)-C(8)	58.7(7)	60.1(4)	59.6(4)	60.1(3)
P(2)-N(8)-C(9)	118.0(8)	120.9(4)	120.5(4)	117.7(3)
P(2)-N(8)-C(10)	118.3(6)	122.4(5)	116.7(4)	118.1(3)
C(9)-N(8)-C(10)	58.8(8)	60.4(4)	59.7(4)	59.9(3)
O-S-N(4)	109.1(4)	104.0(2)	104.1(2)	98.9(3)
S-N(4)-C(1)	116.2(6)	116.5(4)	118.1(4)	117.8(3)
S-N(4)-C(2)	116.5(7)	118.1(4)	119.8(4)	118.2(3)
C(1)-N(4)-C(2)	59.6(6)	60.2(4)	59.9(4)	59.8(3)

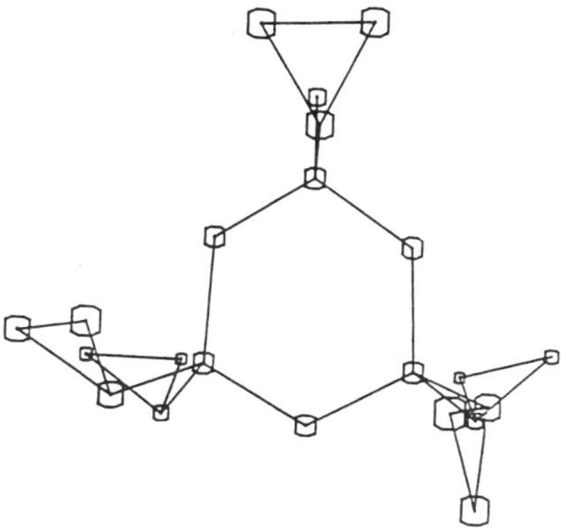

Fig. 40. A perspective view of the SOAz(I) molecule

These three molecules, with quasi-planar six-membered rings (Table 16), do not admit any symmetry. Table 15 allows an easy comparison of the details of these molecules, the main differences being at the level of the conformations, the aziridino groups "rotating" around the P—N and S—N directions.

A joint view, projection on the plane of the ring, of these three types of SOAz

Table 16. Equations of the best plane for the rings N_3P_2S and distances of the atoms of the molecule

SOAz(I) : $-0.50885x - 0.23208y - 0.82899z + 8.46557 = 0$
SOAz(II)A: $-0.73321x - 0.59239y - 0.33388z + 16.15988 = 0$
SOAz(II)B: $-0.85549x - 0.34134y - 0.38938z + 9.65378 = 0$

Distances of atoms (Å) to the best plane:

Atoms	SOAz(I)	SOAz(II)	
		A	B
S	−0.046	0.073	0.019
N(1)	0.030	0.039	0.060
P(1)	−0.014	0.064	−0.050
N(2)	0.019	0.029	−0.029
P(2)	−0.027	0.002	0.102
N(3)	0.041	−0.033	−0.010
O	0.880	−0.855	1.080
N(4)	−1.604	1.572	−1.347
N(5)	−1.297	1.388	−1.377
N(6)	1.208	−1.137	1.157
N(7)	−1.384	1.322	−0.956
N(8)	1.196	−1.332	1.559

molecules is given in Fig. 41; it readily illustrates the different conformations adopted in each case by the aziridino groups.

The packing of the molecules (without H atoms) in both allotropic forms SOAz (I) and (II) is visualized by means of the projection on to the planes (001) and (010) (Figs. 42 and 43); for SOAz (II), half of the cell content was drawn for the sake of clarity (the helicoïdal twofold axes have not been used).

Coming back to the molecular structures of SOAz (I), SOAz (II A) and SOAz (II B), it is noticeable that the conformations of their aziridinyl ligands differ drastically from one allotropic form to the other. In other words, the molecule of SOAz appears to be a versatile molecule with respect to its aziridinyl groups. There is, at the moment, no ready explanation for the adoption of either the one or the other conformation when the synthesis is repeated under exactly the same experimental conditions. However, the SOAz molecule seems to be extremely flexible with regard to the orientation of its "wings", even more flexible than that of $N_3P_3Az_6$ [13, 15, 16, 17].

8.3.2 SOF and SOPHi Molecular Structures

The determination of the molecular structures of SOF and SOPHi did not present any particular difficulties, SOF crystallizing in the $P2_1/c$ monoclinic space group ($Z = 4$: Fig. 44) [73] and SOPHi in the $P2_12_12_1$ orthorhombic system ($Z = 4$: Fig. 45) [73]. It may again be noted that these two structures differ from those of SOAz and MYKO 63 by the spatial arrangement of their aziridinyl wings.

In conclusion of our work on the X-Ray crystal structures of SOAz, SOF and SOPHi, these molecules appear to be as versatile as MYKO 63. The puzzle we had

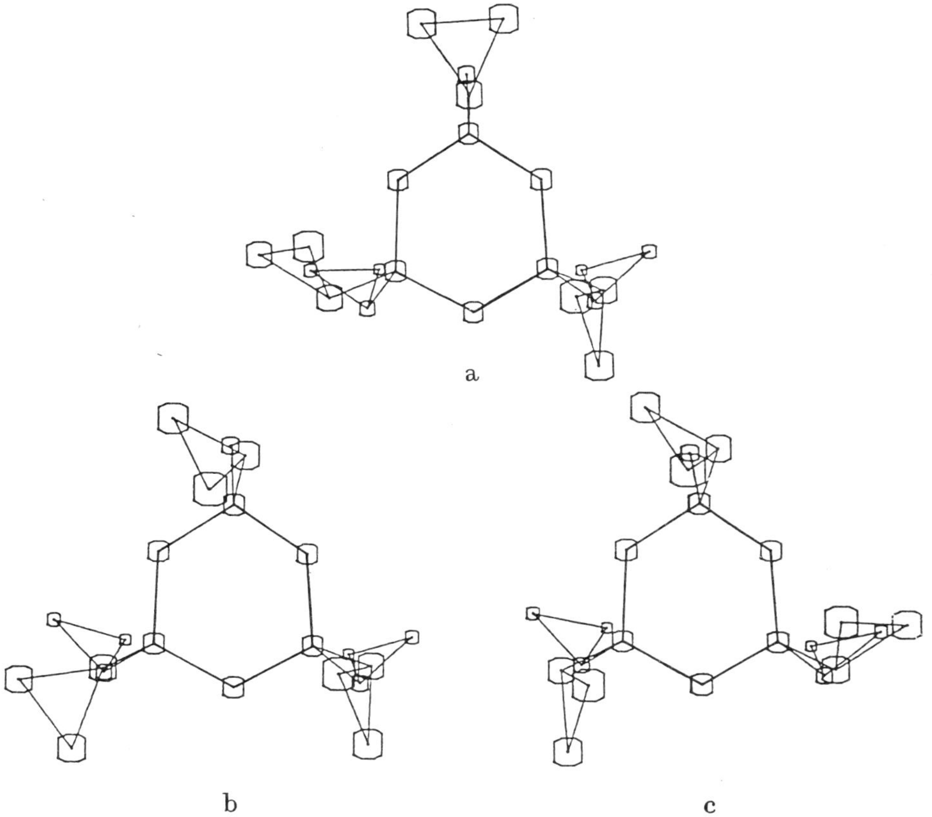

Fig. 41. Joint view of the three molecular structures of SOAz: **a** SOAz (I), **b** SOAz (II A), **c** SOAz (II B)

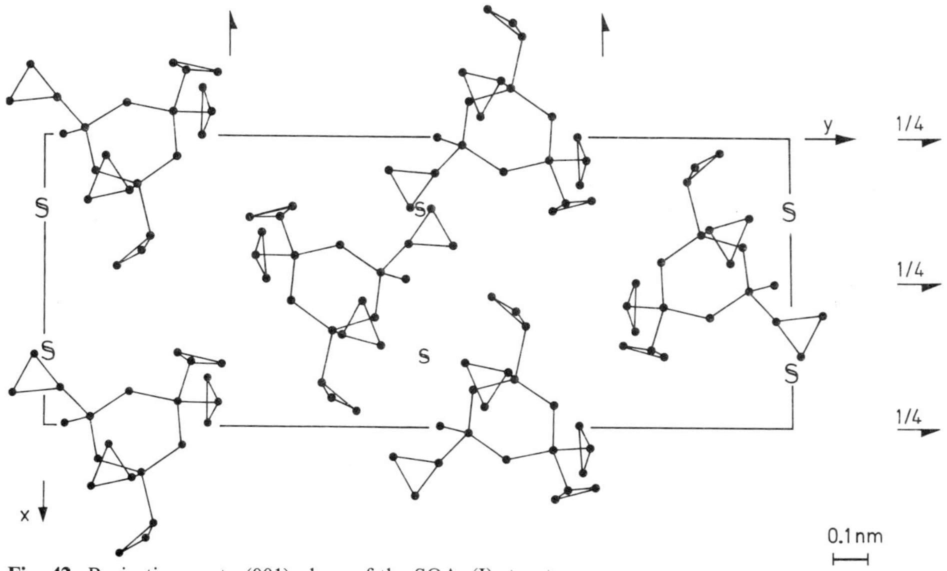

Fig. 42. Projection on to (001) plane of the SOAz(I) structure

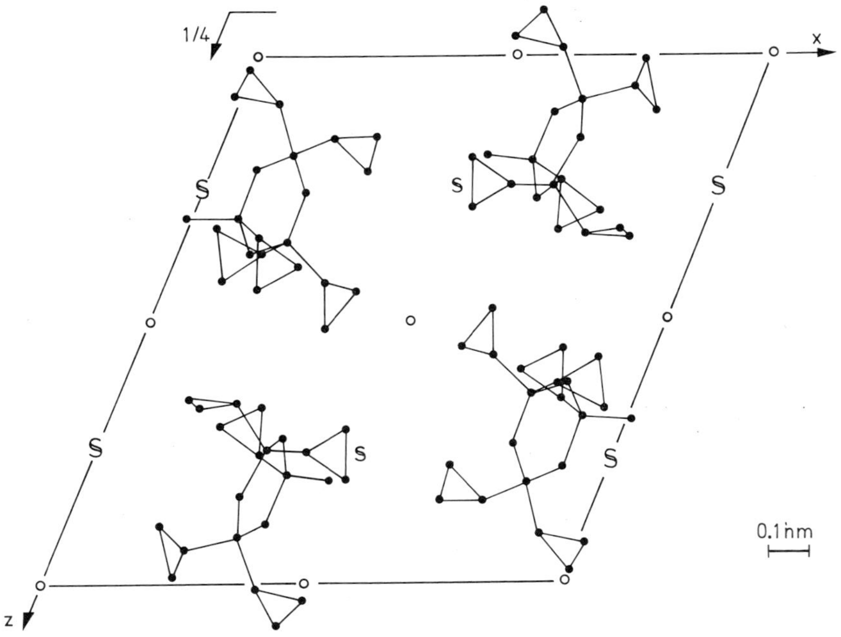

Fig. 43. Projection on to (010) plane of the SOAz(II) structure

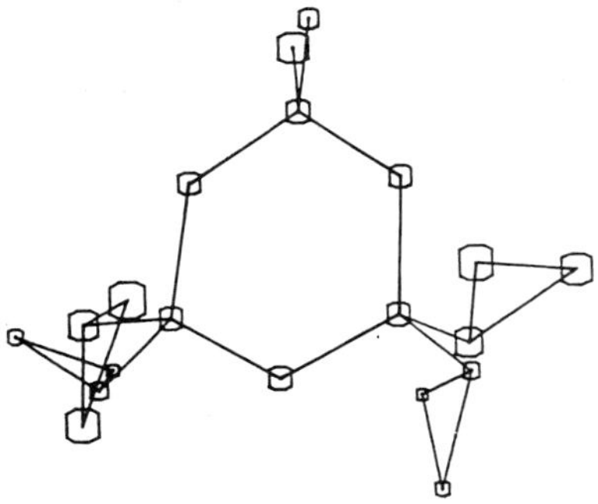

Fig. 44. A perspective view of the SOF molecule

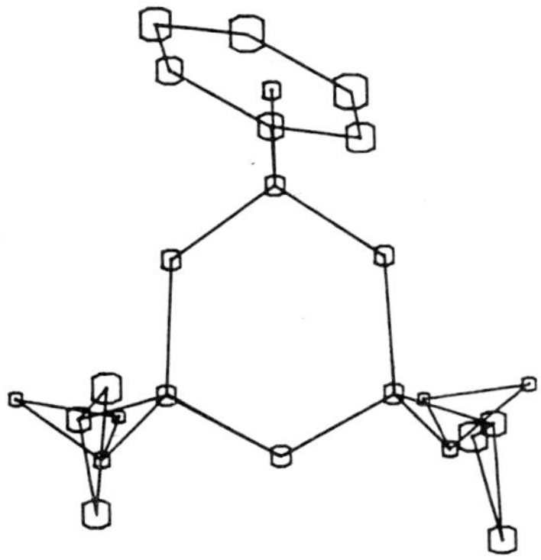

Fig. 45. A perspective view of the SOPHi molecule

consequently to solve consisted in analysing eventual relationships which must exist between the eight appearingly random sets of aziridinyl conformations which characterize the various structures of MYKO 63, SOAz and its relatives. In other words, we had to attempt a reduction of the 5-, 6- and 7-dimensional problems manifested by these drugs (at the level of their wing conformations) to a lower degree which would then enable us to perform in a classical way the conformational analysis of the whole system by quantum chemistry. We shall explain below (see paragraph 9) how we could achieve such a reduction by skilfully using a very simple micro-computing system.

8.4 In vitro Investigations of SOAz-DNA Interactions

8.4.1 Scatchard Results

The Scatchard technique was applied to SOAz-DNA interactions exactly in the same way as we did for MYKO 63-DNA ones (see paragraph 5.4.1.).

The main result we arrived at is that *SOAz does not induce any significant decrease of fluorescence or UV-shift in spectrophotometry.*

Would this result mean that SOAz does not interact at all in vitro with DNA? The answer is actually negative owing to the evidence for such interactions as demonstrated by Raman spectroscopy.

8.4.2 Raman Spectroscopy [74]

Raman investigations of SOAz-DNA interactions were performed in the same way as described above (see Sect. 5.4.2).

Figure 46 compares the spectrum of free SOAz with that of SOAz bound to calf thymus-DNA. Both spectra have been normalized to the same Raman scattering intensity and drug concentration.

The spectrum of free SOAz was obtained by subtracting the solvent spectrum from the SOAz solution spectrum. The spectrum of the bound drug was obtained

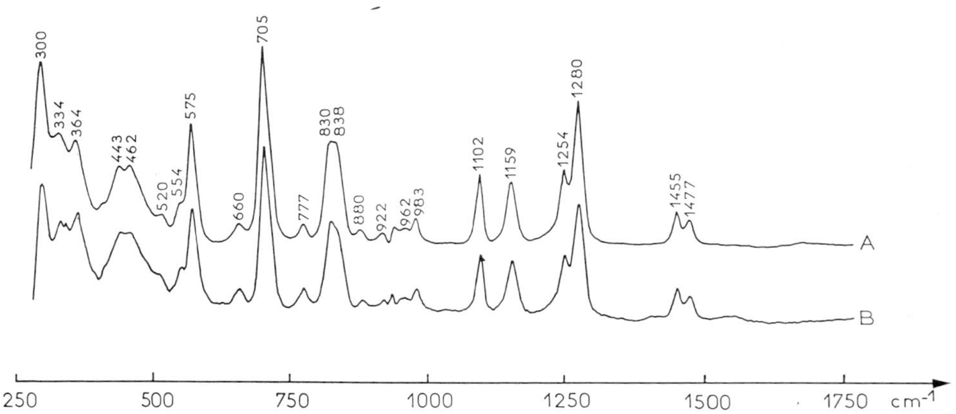

Fig. 46. Raman spectra of unbound (A) and bound (B) SOAz after subtraction of DNA and $NaClO_4$ backgrounds and normalization to the same Raman scattering intensity and drug concentration

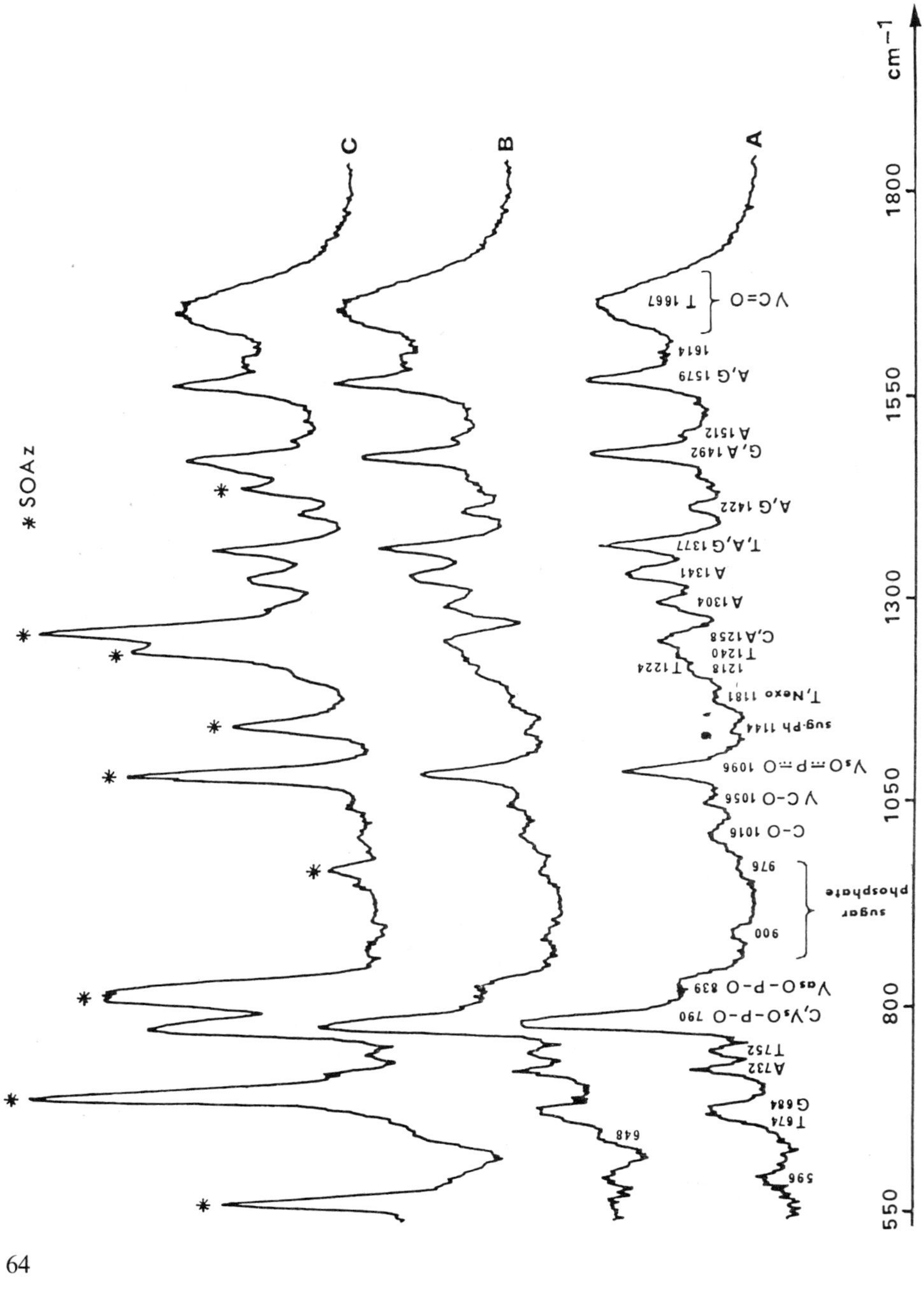

Fig. 47. Raman difference spectra of: (A) free DNA (2.8 mg ml^{-1}); (B) partially bound DNA; (C) SOAz-DNA complex (one drug molecule per base of DNA) (see text)

by a two-step process: (i) the solvent spectrum was subtracted both from the spectrum of the SOAz-DNA complex and from the DNA solution spectrum; (ii) subtraction of these two new spectra was then performed until most of the DNA bands were removed. Step (ii) is applicable only if the spectral differences between unbound and partially bound DNA are negligible. This is strictly correct only for DNA bands unaffected by drug binding. As can be seen from Fig. 46, the background and the base-line of the spectra around 1850-cm^{-1} are not affected by the subtraction process.

For comparison of bound and unbound DNA (see Fig. 47) the procedure was performed in the following way: the spectrum of free DNA (Fig. 47A) was obtained by subtracting the solvent (10 mM $NaClO_4$ in water) spectrum from the spectrum of the DNA solution. The spectrum of partiahy bound DNA (Fig. 47B) was obtained by a two-step process: (i) subtraction (see Fig. 47C) of the solvent from the DNA-SOAz complex solution and (ii) subtraction of the free SOAz difference spectrum (Fig. 46A) from the previous (i) difference spectrum.

Raman scattering were obtained for the first time for SOAz [75] and for SOAz-DNA complexes. By using the difference spectra of Figs. 46B and 47B, we may expect to discriminate both the modifications induced in SOAz by its linkage to DNA and vice et versa.

8.4.2.1 Modifications on SOAz

Taking into account the previous frequency assignments appearing in the literature for $N_3P_3Cl_6$ [32,76], $N_3P_3F_6$ [33], $N_3P_3Az_6$ [29] as well as for aziridine itself [34], we assigned the main group frequencies of SOAz, the most important of which are listed in Table 17. During a study of SOAz-DNA complexes by the Scatchard technique, it was shown that an incubation time of at least 5 days seemed necessary, after mixing the SOAz with DNA on day D, before complete interaction is observed. This result is roughly confirmed by the present work as no fundamental discrepancy was observed between the spectra of the bound SOAz when deduced from the recorded spectra of the SOAz-DNA complex on days (D + 3) and (D + 6).

The main results provided by the comparison between the free (see Fig. 46A) and bound (see Fig. 46B) SOAz difference spectra are as follows.

Table 17. Main-group frequencies of SOAz and their assignments

ν_i (cm^{-1})	ΔI[a])	Assignment	ν_i (cm^{-1})	ΔI[a])	Assignment
575	6.6	$\delta(OSN_{exo})$	1102	7.5	Az ϱ_τ, $\varrho_\omega(CH_2)$
705	12.8	N_3P_2S ring mode	1159	6.7	N_3P_2S ring mode
830 838	8.6	Az $\varrho_r(CH_2)$ sym. ring def.	1254	6.5	Az (on S) ring breathing
983	5.7	Az def. (CH_2)	1280	14.8	Az (on P) ring breathing
			1455 1477	0 0	Az $\delta_{as}(CH_2)$ $\delta_s(CH_2)$

a Decrease in intensity (%); $(I_{\nu_i}$ (free) — I'_{ν_i} (bound))/I_{ν_i} (free)

(i) The intensity and position of two lines are unaltered (see Table 17) on binding: there are the 1455- and 1477-cm^{-1} lines assigned to the $\delta(CH_2)$ deformation mode within the aziridino ligand.
(ii) A relatively weak decrease intensity ($\Delta I \sim 6$–8%) is observed for the 575-, 838-, 983-, 1102-, 1159- and 1254-cm^{-1} lines. The latter corresponds to the breathing mode of the aziridinyl ring linked to the sulphur.
(iii) The main alterations in intensity are seen for the 705- and 1280-cm^{-1} lines which correspond to a N_3P_2S ring mode and to the breathing mode of the aziridinyl ring linked to the phosphorus; these decrease by 12.8 and 14.8% respectively. The decrease in intensity in the 1280-cm^{-1} band may be attributed to the mechanism of interaction of aziridino ligands in the complex as described previously [29], the aziridinyl ring being opened for giving the corresponding — $NH—CH_2—CH_2^+$ carbocation.

Incidentally, we may note that a small broadening of this 1280-cm^{-1} line is observed in the case of the bound SOAz: when subtracting the unbound spectrum (Fig. 46A) from the bound one (Fig. 47C) around the 1280-cm^{-1} wavelength, we do get a very weak band shifted to the high frequency side. This can be seen in Fig. 47B where the 1304-cm^{-1} band of DNA shows a shoulder on the low-frequency side. Such effects can be attributed to a certain type of aziridino ligand-DNA interaction involving neither a linkage to DNA nor an opening of the NC_2H_4 ring.

Coming back to a comparative analysis of ΔI values, two important conclusions may be drawn.
(i) The aziridino ligands on P are much more affected upon complexation than the one linked to S. In other words, *it seems that this complexation occurs essentially through the four Az groups grafted on the two P of the molecule, the Az group brought by S remaining practically uninvolved.* Furthermore, the ΔI value observed here for the 1280-cm^{-1} line is lower than the one previously got for the corresponding 1276-cm^{-1} line ($\Delta I = 21\%$) in the case of the MYKO 63 [29]. However, it is noteworthy that these two ΔI values are approximately in the 2/3 ratio corresponding to the respective numbers of Az groups on P involved.
(ii) In contrast, the ΔI value for the 705-cm^{-1} line (12.8%) appears slightly — but significantly — greater than the corresponding one ($\Delta I = 11.5\%$ for the 706-cm^{-1} line) [29] for MYKO 63. This increase may be related to the respective stabilities of the N_3P_2S and N_3P_3 rings as expected from previous quantum calculations [60].

At least, the weak 520-cm^{-1} line (non-assigned) which exists in unbound SOAz is deeply affected upon complexation; up to now, no valuable explanation of this effect can be reasonably proposed.

8.4.2.2 Modifications on DNA

Figure 47A illustrates the spectrum of DNA at pH 6.9, 25 °C, in the presence of 10 mM $NaClO_4$. Contributions to the various bands by the individual bases (G = Guanine, C = Cytosine, A = Adenine, T = thymine) [37–39] are indicated. Fig. 47C gives the spectrum of the SOAz-DNA complex in the presence of 10 mM $NaClO_4$. Fig. 47B represents the partially bound DNA spectrum obtained as described above.

By comparison with our previous work on MYKO 63 [29], we surveyed in the first

step the 1000–1500 cm^{-1} area of Figs. 47A and 47B in order to detect what bases are affected or not upon complexation. Thus, we noticed the following points.
(i) The band at 1492-cm^{-1} stays unaltered on binding. As this Raman band of Guanine in DNA is a strong indication of binding to the N(7) position of it [21,42], it may be concluded that there is no interaction between SOAz and G.
(ii) The 1304-, 1341- and 1181-cm^{-1} bands which are characteristics of Adenine (more precisely of its C(5) — N(7), C(8) — N(7) and C—NH_2 stretching modes [29,44]) are much more weakly affected (9.8 and 8%) than in the case of MYKO 63 (30, 22 and 20%). Consequently, it may be concluded that interactions between SOAz and Adenine in the N(7) and NH_2 positions are not the determinant parts of the complexation mechanism (in contrast to MYKO 63) and that the main routes of SOAz-DNA interactions must be looked for outside of the set of bases of the nucleic acid.

For this purpose, we cleaned the "lower than 1000 cm^{-1}" and "higher than 1500 cm^{-1}" areas by subtracting the Raman spectrum of the solvent (see Analysis Methods) in order to make some possible modifications of the sugar-phosphate part of DNA upon complexation conspicuous. Furthermore, we could expect to refine the nature of the weak interactions of SOAz with DNA bases by observing the low frequencies characteristic of these bases.

Within the low frequencies area, the main effects are as follows.

The greatest ΔI values are exhibited by the 1016-cm^{-1} line (28%) and mainly by the 900-cm^{-1} line (34%) which were assigned to a mixing of stretching modes of the C—C_{sug} and of the C—O_{sug} for the former and to a stretching mode of the (C—C—O) entities of the ribose backbone [77,78]. *Thus, these two dramatic changes in intensity prove clearly that SOAz interacts drastically with DNA at the ribose backbone level*, probably by grafting of its —NH—CH_2—CH_2^+ carbocations on the oxygens of the (C—C—O) links. This presumption is supported by some complementary observations: (i) slight decrease in intensity and blue-shift of the 839-cm^{-1} line corresponding to a ν_{as}(OPO) when the 1096-cm^{-1} band, attributed to $\nu_s(PO_2^-)$, remains unaffected; (ii) the desoxyribose-dependent 1144-cm^{-1} line exhibits a significant increase in intensity and, still more important perhaps, (iii) the 1667- and 674-cm^{-1} lines, assigned to thymine, appear to be enhanced and weakened respectively upon complexation; (iv) at least, a band at 648-cm^{-1} appears in the spectrum of bound DNA.

What actually is the meaning of these four observations and how are they consistent with the assumption of an interaction of SOAz on the ribose backbone?

If SOAz dialkylates the oxygen atoms of the ribose backbone, the lone pairs of these atoms are now involved in a O—CH_2^+ linkage and, consequently, are less available to engage in p_π—d_π back-bonding towards the phosphorus; the multiplicity of the (P—O) bonds within the OPO block decreases and the frequency of the corresponding band (839-cm^{-1}) shifts to a lower value. Simultaneously, the chelation of SOAz on the oxygen atoms grips the OPO angle, inducing a decrease in intensity of the corresponding line.

As a consequence of these chelations, DNA goes through a kind of denaturation and it is well-known [40] that such an evolution must be accompanied by an enhancement of the 1667-cm^{-1} band and probably by a decrease of the 674-cm^{-1} line (destacking) of thymine. Thus, the (i) and (iii) complementary observations mentioned above are consistent with the chelation of SOAz on the ribose backbone as revealed by the large ΔI values on the 1016- and 900-cm^{-1} lines. Incidentally, such an explanation enables us to understand why the 1096-cm^{-1} line remains unaltered, the PO_2^- groups being quite far from the reaction sites. As to the nature of the 648-cm^{-1} line, no clear explanation of its origin could be given up to now.

The richness of the information collected from the low frequency area in the case of SOAz awakened our suspicion about the previous investigations we had performed on MYKO-DNA complexes in considering only the 1000–1500-cm^{-1} part of the spectra. We reconsidered these MYKO-DNA spectra after cleaning the low frequency zone in the same way as here and then we discovered that MYKO 63 also interacts with the ribose backbone of DNA, the magnitude of the interaction being approximately the same as with SOAz . . .

In other words, MYKO 63 as well as SOAz interacts with the ribose backbone of DNA but MYKO 63 also interacts to the same extent as a dialkylating agent (on N(7) and NH_2) of Adenine, whereas this kind of dialkylation appears to be of a second order of magnitude (with respect to chelation on the ribose backbone) in the case of SOAz.

Anyhow, a combination of the Scatchard technique and Raman spectroscopy shows (i) that SOAz actually interacts with DNA at the level of ribose backbones and (ii) that this kind of interaction does not drastically modify the DNA secondary structure, ethidium bromide encountering no more difficulty to intercalate between DNA plates SOAz being grafted or not on the nucleic acid. Thus, the behaviour of MYKO 63 and of SOAz appears quite different with respect to their mode of interaction with DNA despite their close chemical and molecular structure. This surprising observation may be of interest for understanding why SOAz does not induce any cumulative toxicity in vivo in contrast with MYKO 63.

8.5 Biological Behaviour of SOAz in Vivo: Preclinical Estimation of Mutagenicity and Teratogenicity

8.5.1 Mutagenicity

Ames tests reveal that SOAz does not induce *any mutagenicity at all*, even with toxic doses (100γ per plate and more) on any TA strain. This is a real — but fully unexpected — improvement on MYKO 63. We may in fact assume that patients cured by treatment with SOAz, will not have any further relapse. The origin of this absolute non-mutagenic behaviour has so far not been explained.

8.5.2 Teratogenicity

From this point of view, SOAz behaviour is strictly the same as MYKO 63's one. Inhibition of spermatogenesis is reversible here too but the time necessary for recovering healthy spermatozoa seems shorter than for MYKO 63, i.e., about 3 weeks versus one month.

8.6 Pharmacological and Toxicological Data

The licensee of our SOAz patent [65], namely Otsuka Chemical Co., was in charge of filling the pharmacological and toxicological file necessary for future marketing of a new anticancer drug.

The following experiments are now in progress in Osaka.

(i) Data regarding the acute toxicity of the medicament
- animal species: rat, mouse, dog, monkey;
- sex: male or female;
- administration route: intravenous, intraperitoneal, intracerebral, per os;
- autopsy findings on animals which died during the experimental period;
- measurement of the change of body weight and white blood-cell number.

(ii) Experimental data regarding the sub-acute toxicity of the medicament
- animal species: rat and mouse;
- sex: male and female;
- administration route: intraperitoneal;
- administration period: 12 weeks (4 weeks interruption);
- general toxic symptoms: body weight, feed intake amount, water intake amount;
- general blood inspection: red blood-cell number, reticulocyte number, white blood-cell number and blood platelet number;
- biochemical blood inspection: GOT, GPT, LDH, ALP, ChE, glucose, total cholesterol, total protein, albumin, A/G ratio, total bilirubine, BUN creatine, uric acid, Na^+, K^+, Cl^-, Ca^{++}, Mg^{++};
- uroscopy: uric sugar, protein, occult blood, pH, ketone body, urobilinogen, urinary output, osmotic pressure;
- pathological inspection: macroscopy of general organs, histodiagnosis, measurement of weight of main organs.

(iii) Data regarding embryo experimentation
- experiments regarding the effect on procreation;
- administration experiment on the organ formation of the embryo.

(iv) Experimental data regarding general effect
- on the central nervous system;
- on the respiratory and circulatory system;
- on the autonomic nerve system;
- on the digestive canal;
- on the immune function;
- on blood coagulation;
- on the rhenal function.

(v) Experimental data on absorption, distribution, metabolism and excretion.

Such an impressive check-list is given here with the aim of showing the long route that has to be followed from the discovery of a new anticancer drug to its extended clinical use. Redhibitory defects or side-effects may arise at any stage of this long journey and one may understand the anxiety facing discoverers of a new drug as they wait day after day for results which will show whether the drug is in harmony with Creation or not . . .

Although we are obliged not to disclose too much about the data already obtained by Otsuka, owing to the secrecy requested by the patent agreement, we may claim that *SOAz does not induce any nephrotoxicity or hepatotoxicity or cardiotoxicity.*

Moreover, phase 1 clinical trials started on May 21, 1981 at the Institut Jean-Godinot in Reims, France, with Dr. P. Coninx as supervisor. These trials will allow us to make a more precise determination of the LD_0 value for humans, i.e., the highest non-lethal dose which can be inoculated within mono and polyinjections protocols or schedules without any redhibitory side-effects for the patients.

Meanwhile, some phase 2 trials have already been performed on some naturally cancerous dogs, bearing tumors which are histologically close to the corresponding human ones, namely, osteolytic osteosarcomas and fibrosarcomas (bone cancers). Let us give some details about the protocol of treatment we used in the case of a dog weighing 70 kg and suffering from an acute osteolytic osteosarcoma (expected survival time: less than 2 weeks on March 26, 1981).

The LD_0 for dogs being equal to 10 mg/kg, the highest non-lethal dose within a monoinjection protocol for this dog was 700 mg.

We injected 500 mg by i.v. route (antebrachial vein) on day D and six times 250 mg by the same route within a Q2D schedule. The tumor was arrested on day D + 25, i.e., the bone was recalcified and the size of the oedema was stationary. The size of the oedema decreased afterwards slowly till day D + 45 on which the dog was clinically and radiologically considered as cured. To the best of our knowledge this was the first case in the world of a cure by chemotherapy of an osteosarcoma-bearing dog and the fact that SOAz could be applied for treating a cancerous dog clearly proves that SOAz does not induce any nephrotoxicity and hepatoxicity: it is well-known that dogs are not amenable to any anticancer drug owing to their extreme liver, and kidney, fragility.

Four other dogs were treated with SOAz and cured in the same way. This brings real hope for the clinical treatment of osteosarcoma-bearing patients, especially as there is no chemotherapy available at all for such tumors and that surgery (amputation) is the only technique which might be able to delay death. Incidentally, it may be noted that 16,000 men and women died from osteosarcomas in 1980 in E.E.C. countries . . .

Systematic phase 2 human treatment will start early September 1982 in Reims (Institut Jean-Godinot), in Paris (Hôpital Tenon, Hôpital St Louis, Hôpital Avicenne and others) under the responsibility of several well-known oncologists. This list of curative centres is not limitative and any clinician wishing to apply for treating patients with SOAz will be warmly welcomed.

9 Returning to the Molecular Versatility Exhibited by MYKO 63 and SOAz: A Microcomputer-Assisted Conformational Analysis

9.1 Introduction

In the preface of his book devoted to internal rotation in molecules [79], Orville-Thomas pointed out excellently that: "In the first half of this century, chemical methods proved a powerful tool in the elucidation of molecular configuration and conformation. Their importance then decreased as they gradually became superseded by physical methods of structural determination (. . .). The current position is that the emphasis has tilted even further in favour of the use of a greater range of physical methods and a new factor has emerged to the extent that theoretical methods are becoming of increasing importance and power in the elucidation of conformational problems".

We agree totally with these statements, especially as we were able to prove by ourselves during several years how powerful a concerted use of many physical methods could be (including quantum chemistry) for conformational analysis within the field of molecular inorganic chemistry [12]. However, some limits of such approaches do exist which are essentially due either to the intrinsic constraints of the experimental technique itself or to the time-and-money-consuming character of the appropriate quantum method to be used. For example, it is well-known that electron diffraction cannot reasonably detect more than three preferred conformations in equilibrium [80]. Microwave spectroscopy would be a priori able to solve this problem, but it is actually never used in such a way owing to the incredible number of isotopic derivatives which would need to be investigated in order to achieve the conformational analysis in question. When dealing with quantum methods, some limits also exist which are essentially related to the desperate balance to be found between good reliability of the technique for a given purpose and the cost of calculation: if it is in fact clear that the conformational analysis of simple molecules such as ethane or borazane is amenable to very accurate "double-zeta" (or more) CI Ab Initio calculations, this cannot be at all the case, for obvious reasons, when the conformational problems to be elucidated belong to the vitamin B12 series . . .

Thus, we could prove that the semi-empirical CNDO/2 quantum approach of Pople and Segal [81] was a convenient tool for solving conformational problems in the

Fig. 48. A picture of the microcomputing system

case of large inorganic molecules [12], i.e. up to 80 atoms involving 180 atomic orbitals at the maximum. However, the practical use of this powerful technique has limits, mainly when the number of knee-joints to be considered in the molecule becomes greater than three or four: the price one has to pay for the mathematical drawing of the n-dimensional $E = f(\alpha, \beta, \gamma \ldots)$ hypersurface increases exponentially with n and becomes rapidly impossible to be borne.

Consequently, quantum chemists are daily confronted with the following dilemma: either to waste the budget of five years of research to solve a given problem at any price with the help of the CYBER 205, the biggest computer in the world, or to join the Seven Wise Men . . . and to stop working.

The purpose of the present work is to claim that a third method exists in quantum chemistry for the dynamical conformational analysis of large molecules. We challenged that it would be possible to design such a new route by skilfully using a very simple micro-computing dynamic system in order to reduce the n-dimensional problem to a much smaller one, the size of which would be then compatible with the currently low budget of an academic laboratory.

9.2 Materials and Methods

9.2.1 Hardware: Apple II Plus and Hewlett-Packard Plotter

When the non-initiate enters the world of computers, he is very rapidly all at sea: myriad machines are offered to his choice and, believing the demonstrators, each of them has the best quality/price ratio. All the ability of the intelligent and patient purchaser goes into convincing the salesman of the possibility that his stuff has perhaps some rare but important vulnerable points.

The following lines describe the "winners" of our research in this field.

1) *The computer*: We chose the Apple II Plus microcomputer, which is microprocessed by a 6502 chip, one of the fastest known at the present time. It was bought in its maximal configuration, i.e. with a 48 Kbytes capacity of storage and the language card by which programs can be written using Integer Basic as well as Extended Basic or UCSD Pascal or even Fortran IV. Furthermore, machine language, assembler and macro-assembler are available. Apple II Plus is connected to an external console for the visualization of texts and graphics. This console can be any type of TV set (PAL, NSTC, SECAM, RVB or Black and White). In the text mode, the screen contains 24 lines of 40 characters. In the high resolution graphic mode, the picture is divided in 280×192 points, each point disposing of six different colors (which is particularly useful in the case of tangled curves network). Concerning the mathematical functions, most of them are present. Programs and data are saved on magnetic diskettes. The data files may be screen copies, a useful particular case for our application as it will be explained below. Apple II Plus allows up to eight peripheral connections (for example, a plotter, a printer, a modem, a digitizer, a couple of disk drives, a hard disk and many other features).

Ending this description, let us say that Apple II Plus, in size no larger than an electric typewriter, is more powerful than a big 15-year old concepted computer . . . and not more expensive!

2) *The plotter*: Within this field, the choice was much simpler for very few serious firms are developing this type of peripheral, and speaking money, all of them are in the same region. For various reasons, we chose the HP 7225 A Hewlett-Packard plotter. It can draw and write and, of course, all the figures of this article were entirely designed using the HP 7225 A plotter connected to the Apple II Plus microcomputer. The interface used for this connection is a RS 232 C Serial Interface.

3) *Other items*: the visualization console is a color monitor made by Thomson and the printer is a thermic paper Trendcom 100. The whole system is presented in Fig. 48.

9.2.2 Software: TRIDIM 80

The chain of programs necessary for the resolution of our problem is assembled on one diskette and called TRIDIM 80, the chosen language being Extended Basic (Applesoft). Its object is the tridimensional representation of molecules, with the possibility of drawing these molecules (on screen or on paper) from any angle and to vary, for a given configuration, up to 20 interatomic distances, bond angles or dihedral angles so that any deformation can be visualized. In the case of this work, a dihedral angle variation was enough to give us the key to the problem.

The heart program of TRIDIM 80 is that, knowing the cartesian coordinates of the atoms in space, it can draw at any chosen angle the projection of the corresponding molecule. It is analogous to the well-known ORTEP program (*O*ak *R*idge *T*hermal *E*llipsoid *P*lot), of which it is an extreme simplification; we named it ARTEP (*A*rrangements *R*igides *T*racés en *P*rojection = Projected Rigid Arragements Drawings). If the coordinates are obtained from an X-ray structure, or expressed in a polar system, they are transformed into cartesian coordinates by the program RX-XYZ or the program DAD-XYZ respectively ("RX" is for "*R*ayons-*X*" = *X*-*R*ay, "DAD" is for "*D*istances-*A*ngles-*D*ièdres" = Interatomic *D*istances-bond *A*ngles-*D*ihedral angles). Anyway, when the coordinates enter ARTEP, they are expressed in a cartesian system. If we want to compare two (or more) drawings obtained by ARTEP, a preliminary operation must be done which consists in transforming all the sets of cartesian coordinates to an unique well-defined reference position: this is the action of the program COPLAN, through which the cartesian coordinates go before entering ARTEP.

When ARTEP has generated a drawing of a molecule, this picture is stocked on a diskette named IMAGES. Then, if one has to compare two conformations, the FLIP-FLOP program is executed, which loads the two corresponding pictures from IMAGES and displays them alternatively on the screen as rapidly as required and in the manner of cartoon. Since COPLAN and ARTEP calculate these pictures at the same orientation and with the same centring, they are comparable, and all the modifications appearing between them are significant of real deformations and not due to the way the atomic positions were calculated.

On the other hand, if a progressive deformation of one molecule has to be studied, ARTEP connects to ROTATION which works as follows: the cartesian coordinates are converted into a DAD system; then, the value of the deformation divided by 10 is added to the preceding value of the parameters in question so that, a new D′A′D′ system is constituted; this new system is transformed into a cartesian one by

DAD-XYZ; and finally these cartesian coordinates arrive in ARTEP which generates the new picture. All these steps are repeated 10 times (for example, if one wants to study a 60° variation of a given dihedral angle, ROTATION adds 6° ten times to this angle and, via ARTEP, calculates the ten corresponding pictures). The 10 pictures are saved on a diskette named IMSTOCK and animated then by the MOUVEMENT (= Movement) program (Fig. 49).

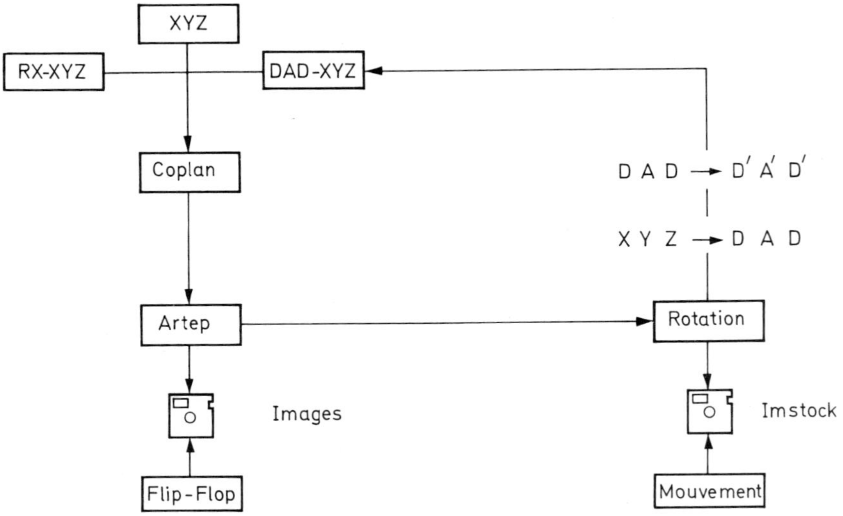

Fig. 49. TRIDIM 80 simplified chain of programs

Of course, each set of bonds and cartesian coordinates of a given molecule are gathered in data files, stored on diskettes under the name of the molecule in question and are easily reloaded when needed.

9.3 Reduction of n-Dimensional Problems

The puzzle we had to solve consists in deep analyses of the possible direct relationships which must exist between the 8 apparently random sets of aziridinyl conformations within Figs. 50 to 56. In other words, we had to attempt a reduction of the 5-, 6 and

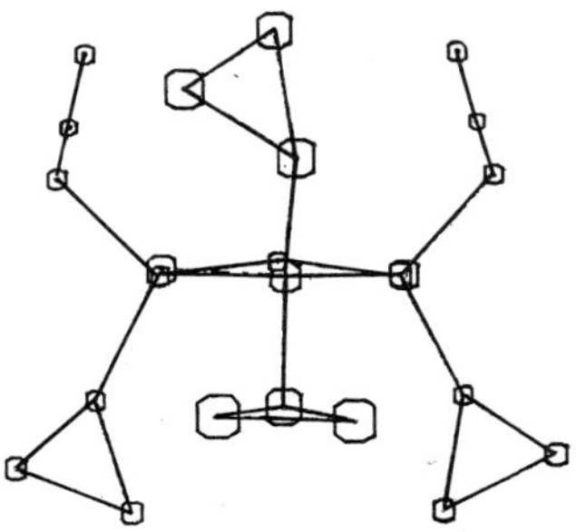

Fig. 50. X-Ray molecular structure of MYKO 63 from m-xylene solutions

7-dimensional problems manifested by SOF or SOPHi, SOAz and MYKO 63 (at the level of their wings conformations) to a lower degree which would then enable us to perform the conformational analysis of the whole system by quantum chemistry in a classical way. The code names we shall use below for the molecules studied are gathered in Table 18.

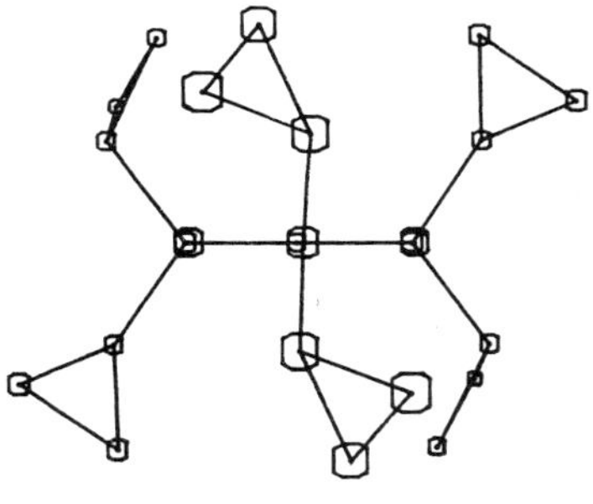

Fig. 51. X-Ray molecular structure of MYKO 63 from CCl_4 solutions

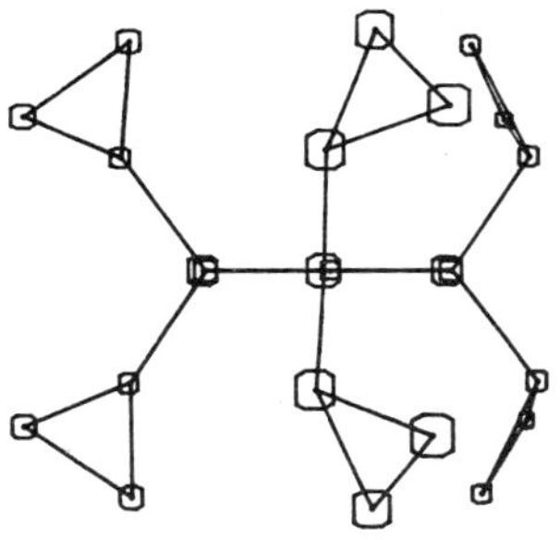

Fig. 52. X-Ray molecular structure of MYKO 63 from C_6H_6 solutions

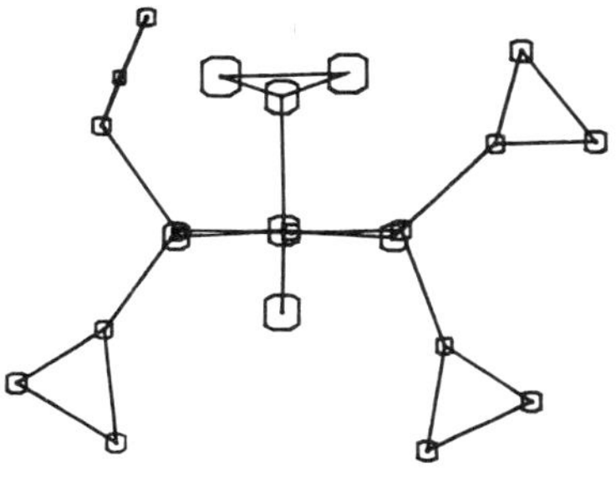

Fig. 53. X-Ray molecular structure of SOAz (I)

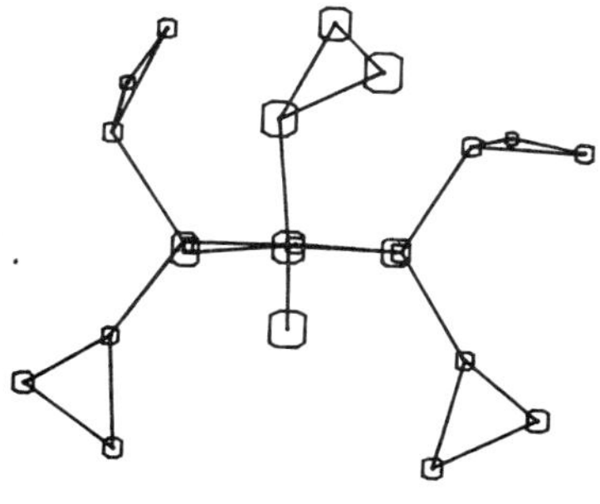

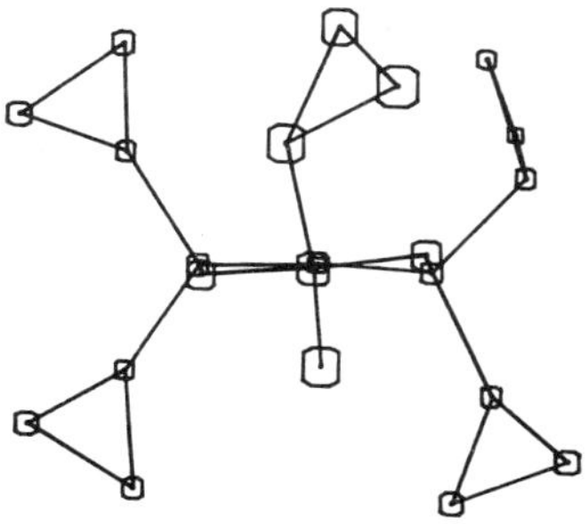

Fig. 54. X-Ray molecular structures of SOAz (IIA) and SOAz (IIB)

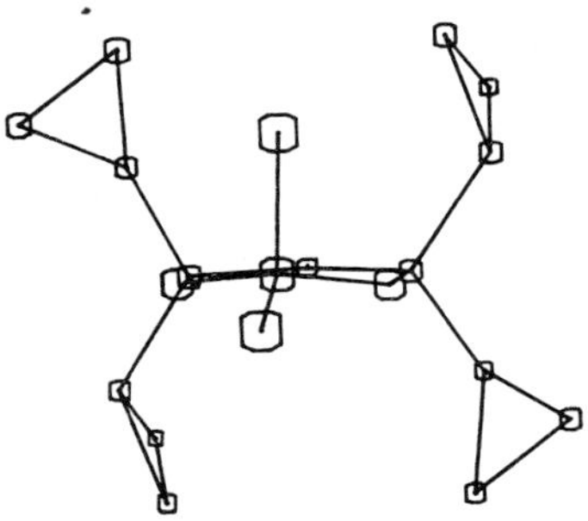

Fig. 55. X-Ray molecular structure of SOF

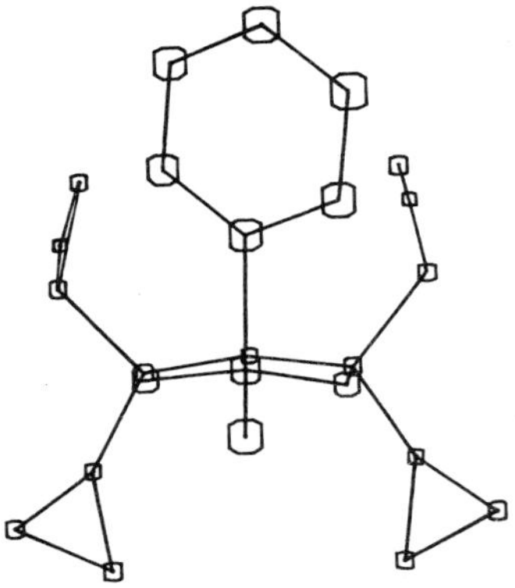

Fig. 56. X-Ray molecular structure of SOPHi

Table 18. Code names of molecules studied

Molecule	Code name
$N_3P_3Az_6$ (Az = Aziridinyl)	MYKO 63
$(NPAz_2)_2(NSOAz)$ MP = 83 °C	SOAz(1)
$(NPAz_2)_2(NSOAz)$ MP = 104 °C	SOAz(II A) and SOAz(II B)
$(NPAz_2)_2(NSOPh)$	SOPHi
$(NPAz_2)_2(NSOF)$	SOF
$(NPAz_2)_2(NPCl_2)$	MYCLAz 632
MYKO 63 crystallized from m-xylene	MYKO-XYL
MYKO 63 crystallized from benzene	MYKO-C_6H_6
MYKO 63 crystallized from carbone tetrachloride	MYKO-CCl_4

The presentations of Figs. 50 to 56 will be called "front-views" from now on. They were drawn assuming the observer is looking along one 2-fold axis of the hexagonal inorganic ring. Actually, these front-views are the best from a topographic point of view because they do not flatten out the threedimensional character of molecules studied, mainly with regard to the spatial distribution of the aziridinyl wings.

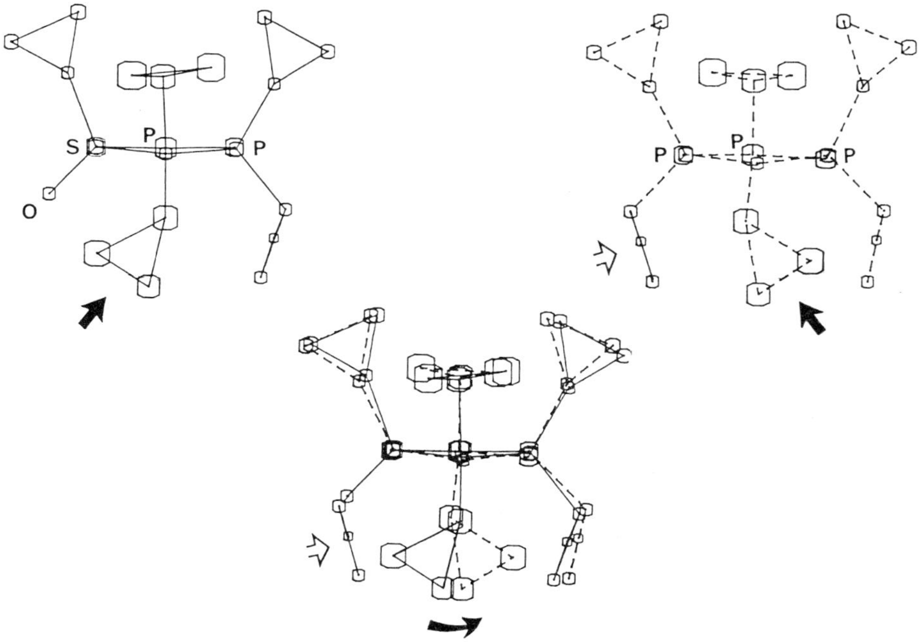

Fig. 57. SOAz (II A) *versus* MYKO-XYL

Consequently, there are three different front-views for every molecule and it is possible to choose at any time which one of them will be the most suitable for comparison of two given structures.

We performed in this way a systematic comparison two by two (FLIP-FLOP program) of all these front-views and the most important features we obtained are as follows:

(i) There is a clear conformational relationship between SOAz(II A) and MYKO-XYL: Fig. 57 does indeed show that, when superimposing the two corresponding front-views, a unique main variation of conformation happens on the lower aziridinyl wing linked to the central phosphorus, this wing being pushed in (curved black arrow) from the left to the right at the beginning of the wing labelled by a white arrow in the MYKO-XYL structure (the two straight black arrows indicate the wings involved in this movement).

(ii) A very similar situation occurs when comparing SOPHi and MYKO-XYL: Fig. 58 shows that the four aziridinyl wings linked to the two external P atoms stay unchanged when the central O—S—Ph moiety is superseded by a Az—P—Az entity.

(iii) In the same way, a comparison of SOAz(II B) and MYKO-C_6H_6 (Fig. 59) shows that the lower wing linked to the external right phosphorus is pushed in when the sixth aziridinyl group appears, the four other wings remaining approximately or strictly unaffected.

(iv) Same observation for the pair (SOAz(II B), SOAz(I) 'M')[3], the only signi-

[3] SOAz(I) 'M' is a mirror image of actual SOAz(I) through the O—S—N_{exo} σ_v plane of the N_3P_2S ring.

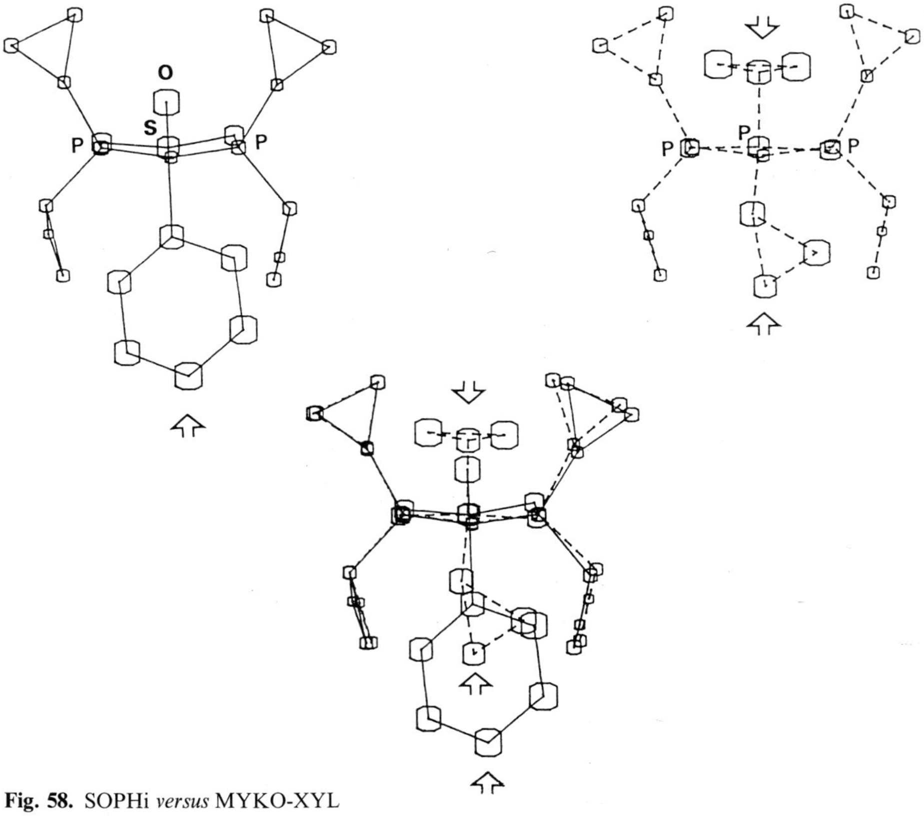

Fig. 58. SOPHi *versus* MYKO-XYL

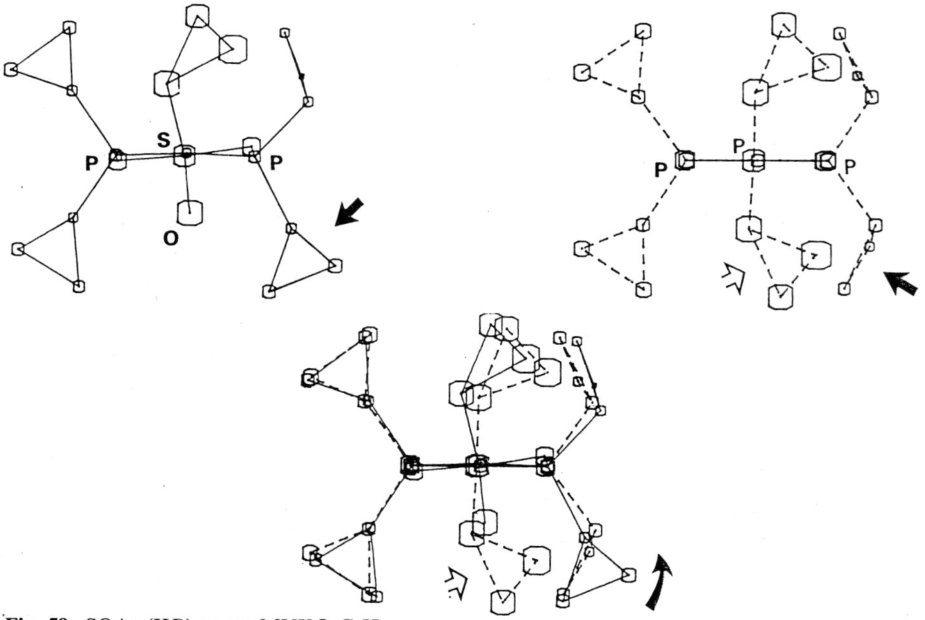

Fig. 59. SOAz (IIB) *versus* MYKO-C_6H_6

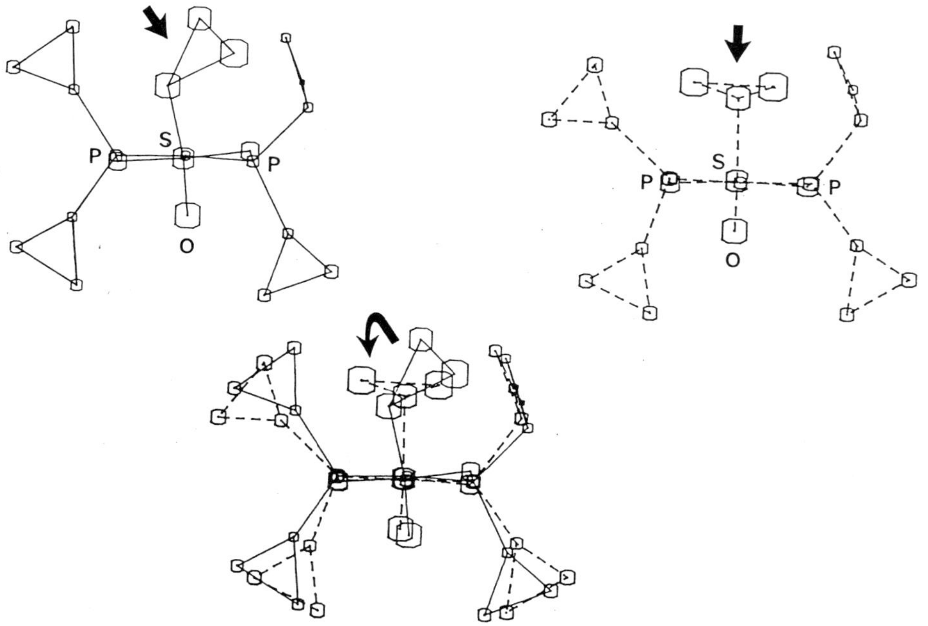

Fig. 60. SOAz (II B) *versus* SOAz (I) 'M'

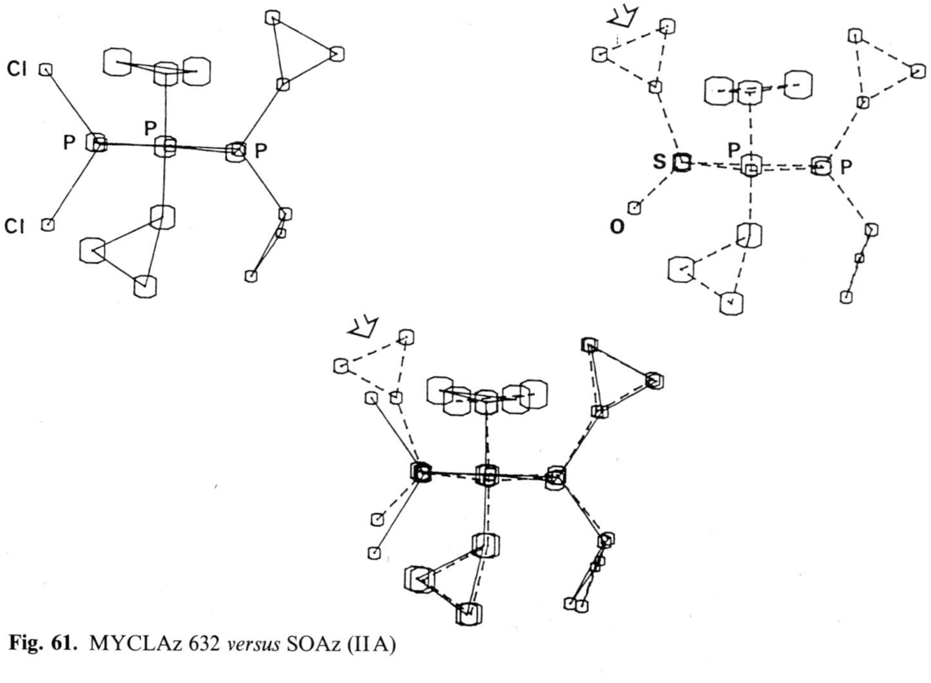

Fig. 61. MYCLAz 632 *versus* SOAz (II A)

ficant conformational modification occurring on the upper wing linked to the S atom (Fig. 60). This result will be of help for the quantum calculations detailed below.

Thus, we may conclude at this stage that there exists some direct connection *through the rotation of one wing only* between SOAz(II A) and MYKO-XYL, SOAz(II B) and MYKO-C_6H_6, SOAz(II B) and SOAz(I) 'M' respectively. *For these three pairs, the six or sevendimensional conformational problem could be then reduced to a monodimensional one.*

Furthermore, the passage from SOPHi to MYKO-XYL can be described in a simple way keeping four wings (over the six of MYKO-XYL) unchanged.

It is noteworthy that the structure of MYKO-XYL — which is the genuine crystalline form of MYKO 63 (see above) — can be directly deduced from the structure of SOAz(II A) which is the thermodynamically stablest form of SOAz (see below). Thus, we could predict that any $(NPAz_2)_2(NPY_2)$ (Y being any ligand except Az) molecule would have the same wing pattern as in SOAz(II A) with its four wings located on the two P atoms.

Figure 61 collects together the front-views of SOAz(II A) and of MYCLAz 632 $(NPAz_2)_2(NPCl_2)$ the X-ray structure of had been determined for this purpose [82]: This Fig. supports definitely our hypothesis about the unaffected situation of the four Az groups grafted on the P atoms when a O—S—Az moiety is superseded by a Cl—P—Cl entity.

Now, what happens through the (MYKO-XYL, MYKO-C_6H_6, MYKO-CCl_4) series? Figures 62 and 63 show that the relationships within the (MYKO-XYL, MYKO-CCl_4) and (MYKO-XYL, MYKO-C_6H_6) pairs are highly intricate, the conformational changes involving three aziridinyl wings in both cases which rotate

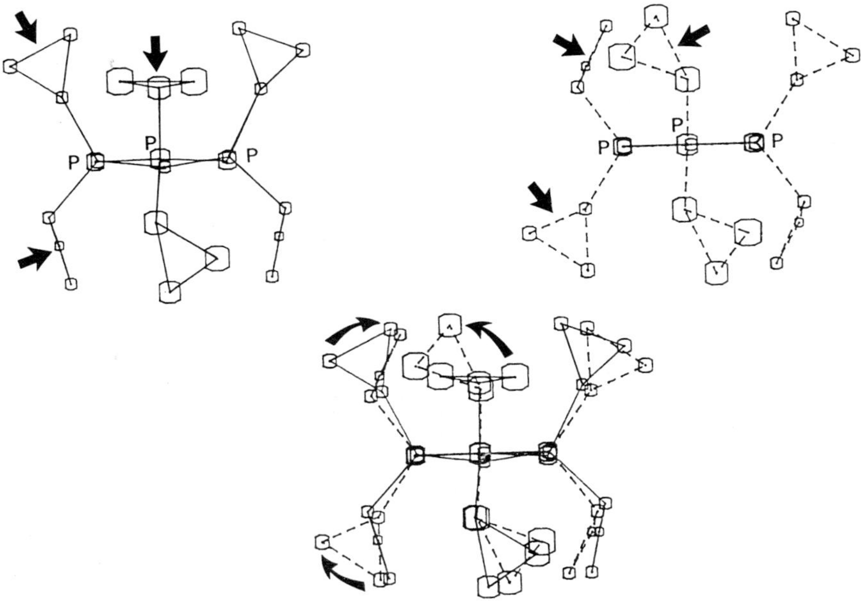

Fig. 62. MYKO-XYL *versus* MYKO CCl_4

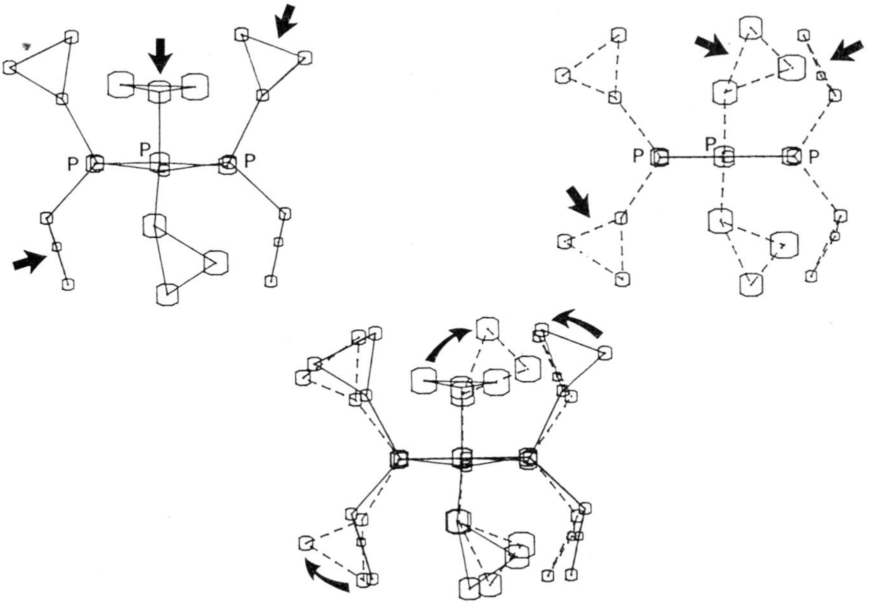

Fig. 63. MYKO-XYL *versus* MYKO-C_6H_6

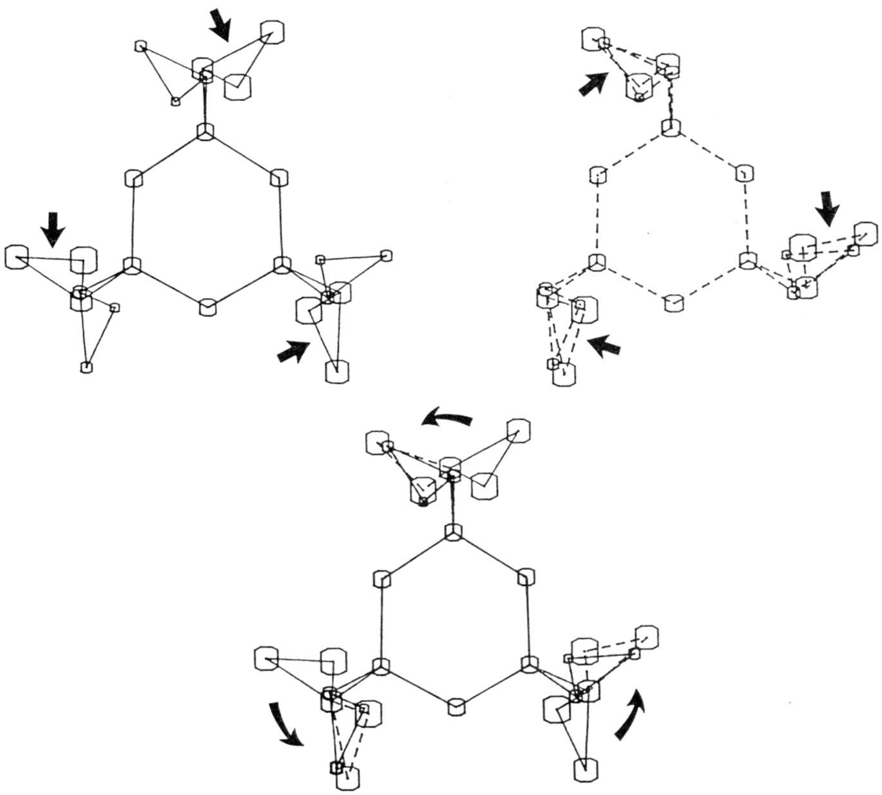

Fig. 64. MYKO-CCl_4 *versus* MYKO-C_6H_6

in a random manner. Three such rotations are also encountered when passing from MYKO-CCl_4 to MYKO-C_6H_6 (Fig. 64) but they occur through a very symmetrical propeller-like counterclockwise motion, the rest of the molecule (N_3P_3 ring and the three lower Az wings) remaining strictly unchanged.

The non-existence of a direct single rotation relationship within the MYKO series can be attributed to the influence on geometrical structures of CCl_4 and C_6H_6 molecules which are clathrated in the corresponding unit cells. In other words, MYKO-C_6H_6 and MYKO-CCl_4 are subjected to highly symmetrical D_{6h} and T_d solvent fields respectively which induce in both cases a ternary symmetry which does not exist in genuine MYKO-XYL.

Incidentally, it may be emphasized that (i) the MYKO-XYL structure constitutes a challenge to group theory in that it has no ternary symmetry and that (ii) the structure of MYKO-C_6H_6 exhibits a tricky counter-example to Gillespie's VSEPR model, the lone pairs of the two exocyclic N atoms on each P being strictly parallel . . .

9.4 Quantum Calculation on the Reduced Systems

The micro-computer-assisted approach we have developed shows that the 6-dimensional $E = f(\alpha, \beta, \gamma, \delta, \varepsilon)$ conformational problem within the set of SOAz geometries can be actually reduced to a monodimensional one. In other words, the passage from SOAz(I) to SOAz(II B) can be then simulated upon rotation of the Az group linked to the sulphur, the four other wings staying unchanged, in first approximation at least.

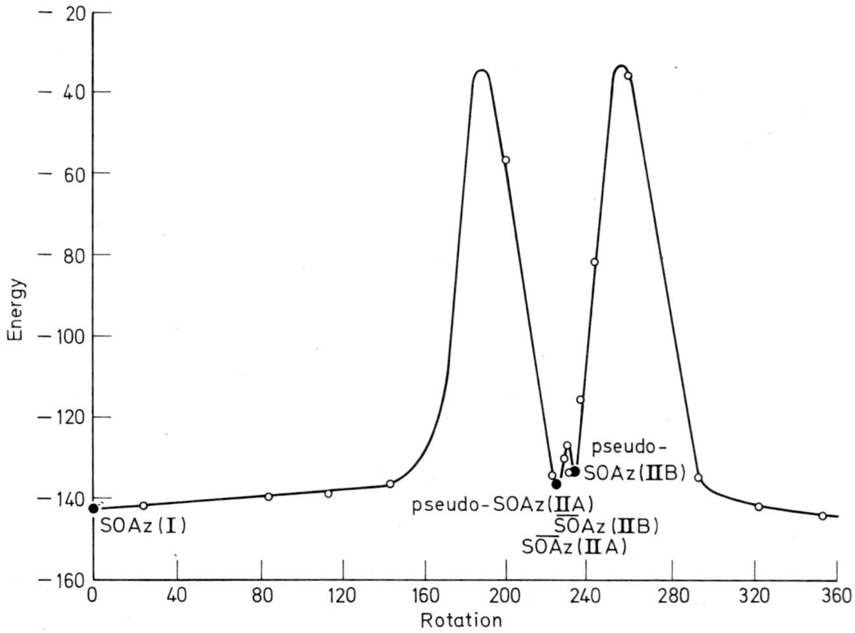

Fig. 65. SOAz(I) energy curve (of −140,000 kcal · mole^{-1}) upon rotation of the Az group linked to S

Consequently, such a reduced problem became amenable again to a common treatment by quantum mechanics, more especially as all the exocyclic N atoms keep *exactly* the same tetrahedral character whatever the molecule they belong to (Fig. 66).

When starting from the X-ray structure of SOAz(I), we computed the total energy upon the rotation through 360° of the Az group linked to S around the S—N_{exo} bond (Fig. 65) within CNDO approximation. This figure calls for the following remarks.

(i) The actual X-ray molecular structure of SOAz(I) is logically the stablest conformation on the basis of the minimal energy criterium (E(SOAz(I)) = —140143.35 kcal · $mole^{-1}$);

(ii) A second "mexican-hat" potential certainly does exist for rotation angles equal to 225° and 235° respectively which correspond closely to the conformations of the Az group linked to S both in SOAz(II A) and SOAz(II B). Thus, quantum calculations show that the three structures of SOAz as provided by X-ray crystallography are the three preferred conformations which could have been expected from the conformational analysis in the gas phase. However, the energies related to the two wells of the SoAz(II) area are in fact smaller than the real energies as computed from the SOAz(II A) and SOAz(II B) X-ray structures, owing to the fact that the conformations of the four Az groups linked to P atoms were assumed in the calculation to keep the situation they have in SOAz(I): the relative energies of the actual SOAz(II A) and SOAz(II B) and of their pseudo-structures are in fact the following:

E (SOAz(II A)) = —140147.48 kcal · $mole^{-1}$
E (pseudo-SOAz(II A)) = —140135.80 kcal · $mole^{-1}$
E (SOAz(II B)) = —140144.96 kcal · $mole^{-1}$
E (pseudo-SOAz(II B)) = —140133.24 kcal · $mole^{-1}$

Incidentally, the gaps of energy within the (SOAz(II A)/pseudo-SOAz(II A)) and (SOAz(II B)/pseudo-SOAz(II B)) pairs appear quite identical, i.e. 11.68 and 11.72 kcal · $mole^{-1}$ respectively;

(iii) Two huge rotational barriers (~100 kcal · $mole^{-1}$) are observed upon rotation around 190° and 260°. These particular angles correspond to forbidden conformations where one (N—C) bond of the rotating Az group eclipses one (S—N) endocyclic bond. Thus, if the three SOAz species may coexist in dilute solution or in gas phase, any conformational equilibrium between them cannot exist at all, except perhaps between SOAz(II A) and SOAz(II B). In other words, SOAz may

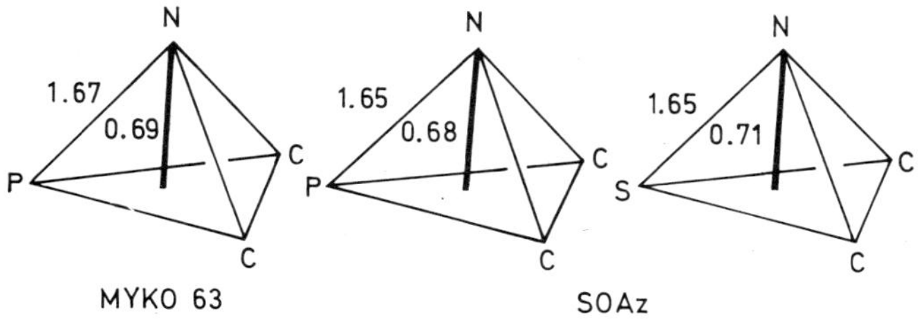

Fig. 66. Geometries of Aziridinyl groups linked to P and S

present either a rigid SOAz(I) conformation or a versatile behaviour around the (SOAz(II A), SOAz(II B)) position;

(iv) Anyhow, the relative stability of the three SOAz varieties decreases in the order

$$\text{SOAz(II A)} > \text{SOAz(II B)} > \text{SOAz(I)}$$

SOAz(II A) being the most "natural" form of SOAz (as MYKO-XYL is within the MYKO structures series).

These quantum results enable us to understand nicely why a given SOAz (or MYKO) structure is conformationally connected or not to another one. Coming back to the relationships we visualized in Figs. 57 to 61 and 62 to 64, the following concluding remarks may be made:

(i) There exists a direct structural link between *the two "natural" forms*, i.e. MYKO-XYL and SOAz(II A);

(ii) There also exists a close similarity between MYKO-C_6H_6 and SOAz(II B) which may be considered as perturbed forms of MYKO-XYL and SOAz(II A) respectively by D_{6h} and D_{3h} solvent fields (the D_{3h} perturbation being induced by the SOAz(II A) molecules).

(iii) Owing to the direct connection between SOAz(II A) and MYCLAz 632 on the one hand, and MYKO-XYL and SOPHi on the other hand, it may be concluded that the genuine MYCLAz 632 and SOPHi X-ray structures are the "natural" ones for these two inorganic ring systems;

(iv) Finally, we may understand thar MYKO-CCl_4 fits only MYKO-C_6H_6 (and not at all MYKO-XYL) simply by considering, as we pointed out previously, the ternary symmetry of both the solvent fields present in the unit cell (T_d for MYKO-CCl_4, D_{6h} for MYKO-C_6H_6).

In conclusion, a very simple use of a suitable microcomputing system enabled us to discriminate amongst a set of X-rays structures of anticancer drugs and determine

(i) what the "natural" preferred conformations (MYCLAz 632, SOAz(II A), SOPHi, MYKO-XYL) are and

(ii) what the direct relationships linking logically these various structures together are.

Such single-rotation-dependent relationships allowed a computational approach of the versatility or not of the SOAz molecule in dilute solution, and this is of considerable interest for a deep understanding of the wide antitumor activity of this drug, whatever the target may be [83].

9.5 Other Possibilities of the Microcomputing System Described here

The TRIDIM 80 software was used here in the case of dynamic conformational analysis. Many other applications of such software may be found. In the field of X-ray crystallography, the different steps of a structure refinement could be visualized for checking the quality of this refinement. In astronomy, views of the same portion of the sky at different times treated by TRIDIM 80 could be projected on to a TV screen at any angle so as to magnify all the movements which happen between them.

Actually, every time that two three-dimensional objects have to be compared, TRIDIM 80 is the simplest way — and a very effective one — to do it.

A copy of the complete program on diskette (Apple II Dos 3.2 Plus) is available from the author on request.

10 Concluding and Prospective Remarks

Thus, as demonstrated above, two generations of new anticancer drugs belonging to the inorganic ring system family were designed in a logical way from ten years of fundamental work within the field of quantum chemistry.

MYKO 63 (1st generation) was found to be significantly effective on *all the animal tumors* it was tested on and would have been really promising for clinical trials if it had not induced any cumulative toxicity when inoculated in daily polyinjections. In other words, the *molecular structure* of MYKO 63 was actually too stable to attack the tumor rapidly and to be excreted fast enough before a next injection would occur.

In order to destabilize this structure slightly, we arrived at the idea — on the basis again of previous quantum calculations — of substituting one phosphorus of the N_3P_3 ring by one sulfur atom. We knew for sure that the (P ... S) transannular binding interactions have a lower magnitude than (P ... P) ones.

The synthesis of the new anticancer drugs (2nd generation) was realized by J. C. van de Grampel and his team and antitumor tests proved immediately that we had been moving on the right track. SOAz and relatives did appear to be as effective as MYKO 63 on all animal tumors within a monoinjection protocol but they allow chronical daily treatment (polyinjections of large doses repeated daily) *without exhibiting any cumulative toxicity*. Furthermore, SOAz appeared to be not at all mutagenic, in contrast with the 30 drugs which are commonly used in clinics. Moreover, the pharmacological and toxicological data obtained by our industrial Japanese partner concerning SOAz are really in favour of its use for extensive clinical trials: In contrast with platinum and other drugs, SOAz does not in fact induce any nephrotoxicity, hepatotoxicity or cardiotoxicity.

Nevertheless, what can be done to design even better drugs, whatever the quality of the therapeutic properties already manifested by SOAz?

Two new generations of drugs are now currently being developed in our laboratory with the aim of further increasing the selectivity towards the malignant cells and of avoiding as a consequence any side-effect on healthy ones. For obvious reasons, we cannot disclose anything here about this up-to-date research, except that the driving idea is to link our drugs to some arrow which is known to have a deep affinity for vicious DNA. The future will tell us whether these human efforts have been blessed by the Creator or not . . .

11 References

1. Rosenberg, B. et al.: Nature (London) *222*, 385 (1969); Rosenberg, B., van Camp, L.: Cancer Res. *30*, 1799 (1970)
2. Ratz, R. et al.: Inorg. Chem. *3*, 757 (1964)

3. Geran, R. I. et al.: Cancer Chemother. Rep., part 3, *3*, 1 (1972)
4. Labarre, J-F. et al.: Europ. J. Cancer *15*, 637 (1979)
5. cf. for example: Nabi, S. N., Biddlestone, M., Shaw, R. A.: J. Chem. Soc. (Dalton) 2634 (1975) and references herein
6. Chou, F., Khan, A. H., Driscoll, J. S.: J. Med. Chem. *19*, 1302 (1976)
7. Connors, T. A. et al.: Biochem. Pharmacol. *24*, 1665 (1975)
8. Labarre, J-F., Levy, G., Sournies, F.: J. Mol. Struct. *63*, 127 (1980)
9. Monsarrat, B. et al.: Biomed. Mass Spectr. *7*, 405 (1980)
10. Millard, B. J.: Quantitative Mass Spectrometry, Heyden, London 1978
11. Sournies, F.: Thèse de Doctorat d'Université n° 350, Paul Sabatier Univ., Toulouse, France 1980
12. Labarre, J-F.: in Structure and Bonding, Vol. 35, (ed. Dunitz, J. D., Goodenough, J. B.) p. 1, Springer-Verlag 1978
13. Cameron, T. S., Labarre, J-F.: unpublished results
14. cf. for example: Rettig, S. J., Trotter, J.: Can. J. Chem. *51*, 1295 (1973) and references herein
15. Cameron, T. S. et al.: Z. Naturforsch. *35b*, 784 (1980)
16. Cameron, T. S., Labarre, J-F., Graffeuil, M.: Acta Crystallogr. Sect. B (1982), in the press
17. Galy, J., Enjalbert, R., Labarre, J-F.: ibid. Sect. B *36*, 392 (1980)
18. Bovin, J. O. et al.: J. Mol. Struct. *49*, 421 (1978)
19. Bovin, J. O., Labarre, J-F., Galy, J.: Acta Crystallogr. Sect. B *35*, 1182 (1979)
20. Scatchard, G.: Annal. N.Y. Acad. Sci. *51*, 660 (1949)
21. Cf. for example: Theophanides, T. et al.: J. Raman Spectrosc. *5*, 315 (1976); Bernard, L. et al.: Biochimie *60*, 1139 (1978)
22. Lowry, O. H. et al.: J. Biol. Chem. *193*, 265 (1951)
23. Defrance, P., Delesdain, N.: Pathol. Biol. *10*, 153 (1962)
24. Le Pecq, J-B.: Methods Biochem. Anal. *20*, 41 (1971)
25. Le Pecq, J-B., Paoletti, C.: J. Mol. Biol. *27*, 87 (1967)
26. Waring, M. J.: J. Mol. Biol. *13*, 269 (1965)
27. Le Pecq, J-B., Paoletti, C.: Annal. Biochem. *17*, 100 (1966)
28. Butour, J-L., Labarre, J-F., Sournies, F.: J. Mol. Struct. *65*, 51 (1980)
29. Manfait, M. et al.: J. Mol. Struct. *71*, 39 (1981)
30. Manfait, M., Beaudoin, J-L., Bernard, L.: in Advances in Raman Spectroscopy (ed., Mathieu, J-P), p. 76, Heyden 1972
31. Manfait, M., Lahana, R., Labarre, J.-F.: to be published
32. Adams, D. M., Fernando, W. S.: J. Chem. Soc. (Dalton) *22*, 2503 (1972)
33. Emsley, J. M.: J. Chem. Soc. A 109 (1970)
34. Mitchell, R. W., Burr, Jr., J. C., Merritt, J. A.: Spectrochim. Acta Sect. A, *23*, 195 (1967)
35. Alix, A. J. P., Manfait, M., Labarre, J-F.: J. Raman Spectrosc. (1982), in the press
36. Labarre, J.-F. et al.: J. Chim. Phys. Fr. *77*, 85 (1980)
37. Lord, R. C., Thomas Jr., G. J.: Spectrochim. Acta Sect. A, *23*, 2551 (1967)
38. Small, E. W., Peticolas, W. L.: Biopolymers *10*, 69 (1971)
39. Small, E. W., Peticolas, W. L.: ibid. *10*, 1377 (1971)
40. Erfurth, S. C., Peticolas, W. L.: ibid. *14*, 247 (1975)
41. Erfurth, S. C., Kiser, E. J., Peticolas, W. L.: Proc. Nat. Acad. Sci. USA *69*, 938 (1972)
42. Manzy, S., Engstrom, S. K., Peticolas, W. L.: Biochem. Biophys. Res. Commun. *68*, 1242 (1976)
43. Manzy, S., Peticolas, W. L.: Biochemistry *15*, 2650 (1976)
44. Tsuboi, M., Takahashi, S., Harada, I.: Physico-chemical Properties of Nucleic Acids (ed. Duchesne, M.), p. 91, Academic Press 1973
45. Tsuboi, M. et al.: J. Raman Spectrosc. *2*, 609 (1974)
46. Goodwin, D. A. et al.: Biochemistry *18*, 2057 (1979)
47. Theophanides, T., Berjot, M., Bernard, L.: J. Raman Spectrosc. *6*, 109 (1977)
48. Lawley, P. D., Brookes, P.: Biochem. J. *89*, 127 (1963)
49. Maxam, A. M., Gilbert, W.: Proc. Nat. Acad, Sci. USA *74*, 560 (1977)
50. Ames, B. N., Mac Cann, J., Yamasaki, E.: Mutation Res. *31*, 347 (1975)
51. Lecointe, P. et al.: ibid. *48*, 139 (1977)
52. Cf. for example: Lopez, M.: Thèse de Doctorat d'Etat n° 753, Paul Sabatier Univ., Toulouse, France 1977 and references herein

53. Faucher, J-P. et al.: J. Mol. Struct. *10*, 439 (1971)
54. Bruniquel, M. F. et al.: Phosphorus *3*, 83 (1973)
55. Faucher, J-P., Labarre, J-F.: ibid. *3*, 265 (1974)
56. Faucher, J-P. et al.: C.R. Acad. Sci. (Paris) Ser. C, *279*, 441 (1974)
57. Faucher, J-P., Labarre, J-F., Shaw, R. A.: J. Mol. Struct. *25*, 109 (1975)
58. Faucher, J-P., Labarre, J-F., Shaw, R. A.: Z. Naturforsch. *31b*, 677 (1976)
59. Faucher, J-P., Labarre, J-F.: Adv. Mol. Relax. Interact. Proc. *8*, 169 (1976)
60. Faucher, J-P. et al.: J. Chem. Research, S112, M1257 (1977)
61. Dewar, M. J. S., Lucken, E. A. C., Whitehead, M. A.: J. Chem. Soc. 2423 (1960)
62. Van de Grampel, J. C. et al.: Inorg. Chim. Acta *53*, L169 (1981)
63. Monsarrat, B. et al.: to be published
64. Labarre, J-F. et al.: Cancer Letters *12*, 245 (1981)
65. Labarre, J-F. et al.: French ANVAR patent n° 79-17336, July 4, 1979
66. Cf. for example: Le Pecq, J. B. et al.: Proc. Nat. Acad. Sci. USA *71*, 5078 (1974)
67. Galy, J. et al.: Acta Crystallogr., Sect. B *37*, 2205 (1981)
68. Cromer, D. T., Waber, J. T.: Tables for X-Ray Crystallogr., Kynoch Press, Vol. 4 (1974)
69. Cromer, D. T., Liberman, D.: J. Chem. Phys. *53*, 1891 (1970)
70. Germain, G., Main, P., Woolfson, M. M.: Acta Crystallogr. Sect. A, *27*, 368 (1971)
71. Dermer, O. C., Ham, G. E.: Ethyleneimine and other aziridines, Academic Press 1969
72. Johnson, C. K.: ORTEP, Report ORNL-3794, Oak Ridge Nat. Lab. 1965
73. Cameron, T. S. et al.: to be published
74. Manfait, M., Labarre, J-F.: J. Mol. Struct. *76*, 165 (1981)
75. Manfait, M. et al.: J. Raman Spectrosc. (1982), in the press
76. Huvenne, J-P., Vergoten, G., Legrand, P.: J. Mol. Struct. *63*, 47 (1980)
77. Brown, E. B., Peticolas, W. L.: Biopolymers *14*, 1259 (1975)
78. Lu, K. C., Prohofsky, W., Van Zandt, L. L.: ibid. *16*, 2491 (1977)
79. Orville-Thomas, W. J.: Int. rotation in Molecules, John Wiley and sons 1974
80. Cf. for example: Hargittai, I.: Az elektron diffrakcios atomtavolsag in A Kémia Ujabb Eredményei, Akadémiai kiado 1974
81. Pople, J. A., Segal, G. A.: J. Chem. Phys. *44*, 3289 (1966)
82. Galy, J. et al.: Z. Krist. (1982), in the press
83. Lahana, R., Labarre, J-F.: Theochem. (1982), in the press

Phosphorus(III)-Nitrogen Ring Compounds[1]

Rodney Keat

Department of Chemistry, University of Glasgow, Glasgow G12 8QQ, U.K.

Table of Contents

---

[1] Plenary lecture delivered at the 3rd IRIS Meeting held in Graz (Austria), August 17–22, 1981.

1 Introduction

This review is concerned with the chemistry of cyclic compounds where the ring contains alternating nitrogen and tervalent phosphorus atoms. Numerous heterocyclic compounds in which a tervalent phosphorus-nitrogen bond forms part of a ring containing other hetero-atoms are known, but these are outside the scope of this survey. A widely accepted form of nomenclature [1] classifies this group of inorganic ring compounds as cyclophosph(III)azanes and examples are given in Scheme 1.

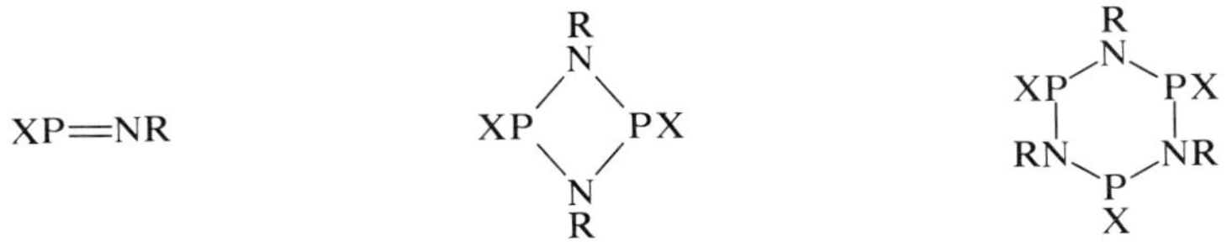

(*1*) (*2*) (*3*)

Scheme 1

Cyclodiphosph(III)azanes, (*2*), and cyclotriphosph(III)azanes, (*3*), are formally derived by oligomerisation of the phosph(III)azenes, (*1*), although few of the latter species are known. Nitrogen-substituents, R, are generally restricted to alkyl, aryl-or silyl-groups, but cyclophosph(III)azanes with a wide range of phosphorus substituents, X, are known. This review will also include a number of fused ring and cage compounds containing cyclophosph(III)azane rings, not strictly derived by oligomerisation of phosph(III)azanes, (*1*). Cyclophosph(III)azane chemistry has not formed the subject of a reasonably comprehensive survey since 1970 [2], although aspects of this topic form the subject matter of monographs [3–5] and reviews [6–9]. Here, because of limitations of space, emphasis will be placed on significant developments that have occurred in the last decade, particularly in the author's laboratory. However, an attempt has been made to provide a comprehensive list of references to work on cyclophosph(III)azane chemistry up to the end of August, 1981.

2 Conformations of Phosph(III)azanes

In recent years there has been an increasing awareness [10–12] of the importance of conformational effects on the chemistry of compounds containing P—O and P—N bonds. It is therefore useful to start by considering the conformational constraints imposed on the P(III)—N bond as a result of ring formation. In general, the preferred conformations adopted by acyclic hydrazines (*4*) (E = N) [13] and diphosphines (*4*) (E = P) [14] are those in which there is a gauche relationship between lone-pairs on adjacent nitrogen or phosphorus atoms. This is shown in the Newman projection (*5*) (Scheme 2), although if there are relatively electronegative or bulky groups on phosphorus other conformations may be preferred [14]. By contrast, structural studies of acyclic phosph(III)azanes (*6*) generally show that nitrogen has a planar, or near planar, distribution of bonds [15], and it is inferred that nitrogen and phosphorus lone-pairs prefer to adopt an orthogonal relationship (*7*). The reasons

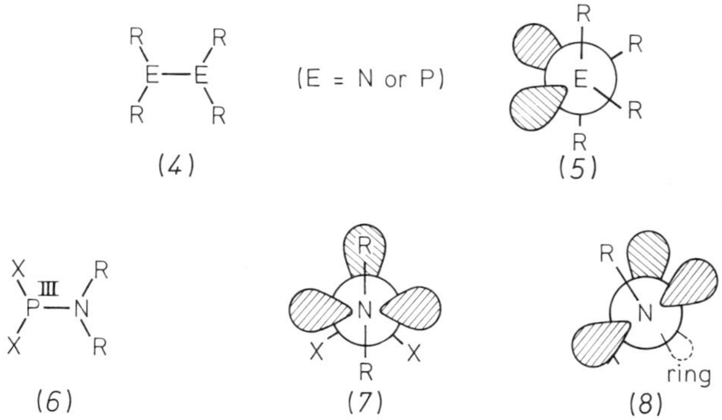

Scheme 2

for this have been studied, for example by *ab initio* molecular orbital calculations on the phosph(III)azane, H_2PNH_2 [16].

The result of cyclophosph(III)azane formation may therefore be to constrain the P—N bonds to conformations in which the nitrogen and phosphorus lone-pairs are no longer othogonal (*8*), and this increases the repulsive interaction between them [17]. This would result in a reduction in the P—N bond strength and will increase the Lewis base character of phosphorus. Two examples of the conformations adopted by the P—N bonds in cyclophosph(III)azane rings are shown (Fig. 1). The atomic co-ordinates for the *trans*-cyclodiphosph(III)azane (*9*) [18], and the cage-type molecule, $P_4(NMe)_6$ (*10*) [19] were used to obtain microcomputer drawn projec-

(*9*) (*10*)

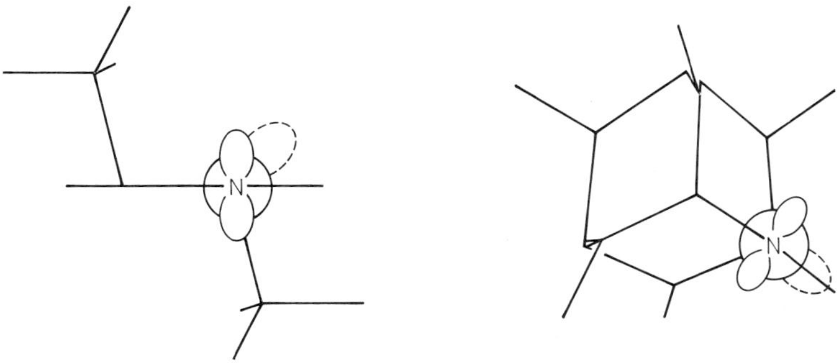

Fig. 1. Projections along the N—P bonds (arrowed) of (*9*) and (*10*), obtained using the atomic co-ordinates in Refs. [18] and [19] respectively (H-atoms omitted). Relative orientations of N- and P-lone-pairs are also shown

tions [20] along the N—P bonds (the six-membered ring fragment from the latter molecule was chosen because there are no crystal structure reports of cyclotriphosph(III)azanes). Although the position of the phosphorus lone-pair (broken line) in the cyclodiphosph(III)azane (*9*) is an approximation, it is clear that the nitrogen and phosphorus lone-pairs are not orthogonal. However, in the six-membered ring fragments (*10*), which form part of the P_4N_6 cage, the lone-pairs are approximately orthogonal when the phosphorus 'exo-substituent' occupies an axial position (see Sect. 4.2 for ring inversion effects). The cyclodiphosph(III)azanes in particular may therefore be expected to show significant differences in chemical properties compared to larger ring cyclophosph(III)azanes and acyclic phosph(III)azanes.

The conformational preferences of the P(V)—N bond in four co-ordinated phosphorus compounds, Scheme 3, (*11*) are less pronounced, and depend more on the type of the other phosphorus substituents. However, conformations close to (*12*) are found in a variety of crystal structures [21] of acyclic phosph(V)azanes.

(*11*) (*12*)

Scheme 3

3 Cyclodiphosph(III)azanes

3.1 Synthesis

The first synthesis of a cyclodiphosph(III)azane (*13*) (Scheme 4) was reported by Michaelis and Schroeter [22], who correctly identified (*13*) as a dimer by molecular weight determinations. These results have been confirmed by more recent work [23] which also shows that the reaction proceeds via an acyclic diphosphinoamine, $Cl_2PNPhPCl_2$. The synthesis of cyclodiphosph(III)azanes by the reaction of phosphorus trichloride with relatively bulky primary alkylamines [24,25], and related routes [26–30], has been widely studied. Examples of this work, leading to (*14*) and (*15*) are shown (Scheme 4). Cyclodiphosph(III)azanes with N-aryl [31–38] and N-sulphonyl [39] groups have been obtained similarly.

An alternative route [40–45] to closely related ring systems takes advantage of reactions which lead to the elimination of metal halides and trimethylsilyl halides (Scheme 5). The synthesis [42] of (*16*) is but one example of many cyclodiphosph(III)-azanes which have been obtained by this general route.

It is instructive to consider why the presence of relatively bulky N-substituents results in the formation of small rings. Of the reactions of phosphorus trichloride with amines, that with t-butylamine has been widely studied (Scheme 6) [24,27]. Intermediates (*17*) and (*18*) have been isolated, but not (*19*), which evidently undergoes a facile cyclisation step to give (*20*). Two effects can be identified, which are expected to favour the formation of small rings. Firstly, an n.m.r. study [46] of (*18*) shows that the preferred conformation in solution is that shown in (*21*) (X = Cl)

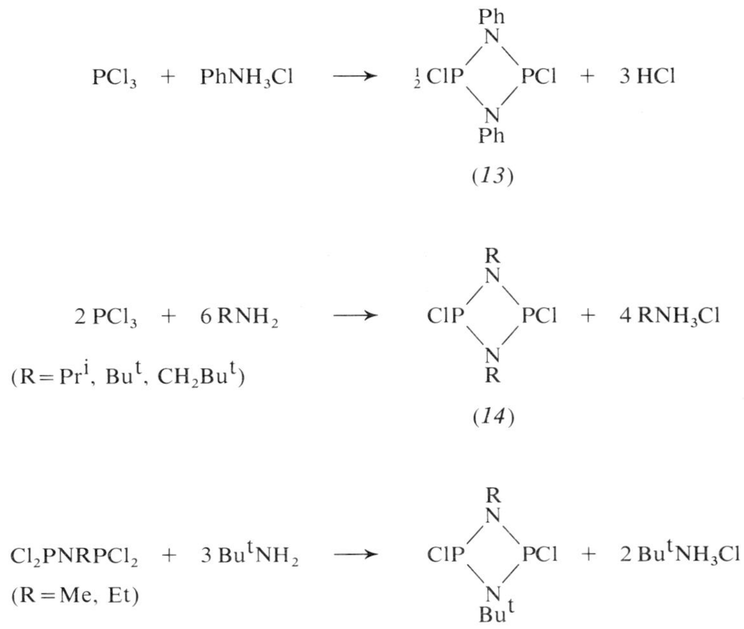

(15)

Scheme 4

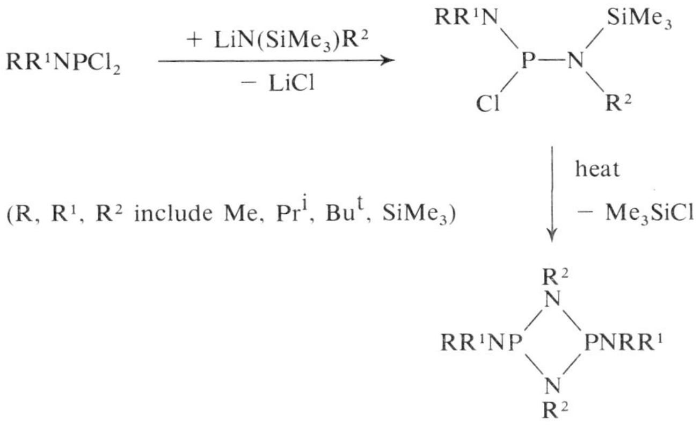

(16)

Scheme 5

(Scheme 7). This conformation is associated with a small negative *PNP* spin coupling (—20.1 Hz), whereas the analogous N-methyl compound adopts conformation (*22*) and is expected to have a large positive *PNP* spin coupling (>400 Hz) [46,47]. In both conformations an orthogonal relationship between adjacent nitrogen and phosphorus lone-pairs is retained. Thus the intermediate (*19*) may be expected to adopt conformation (*21*) (X = $NHBu^t$), particularly with a bulky X-group, where an intramolecular nucleophilic attack of nitrogen on the PCl_2-group

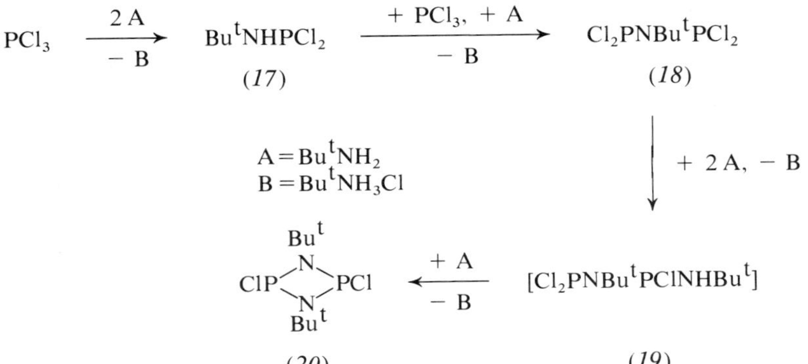

Scheme 6

(21) (22)

Scheme 7

would be facilitated over a conformation analogous to (*22*). It is worth noting that an intermediate analogous to (*19*), the diphosphinoamine, Cl(Bu^t)PNMeP(Bu^t)NHMe, has been isolated [41], and that the two rotameric forms have relatively large *PNP* couplings, [265 Hz (ca. 80%) and 190 Hz (ca. 20%)]. In this case it appears likely that relatively bulky phosphorus substituents are largely responsible for the cyclisation of the latter compound to a cyclodiphosph(III)azane (see below).

A second effect, widely believed to favour small ring formation [48], concerns the steric interactions between the P- and N-skeletal substituents (Scheme 8). In this case steric interference between the N-substituents is minimised in four membered rings (*23*), but greater in larger rings or polymers (*24*). This effect could equally well apply to phosphorus substituents.

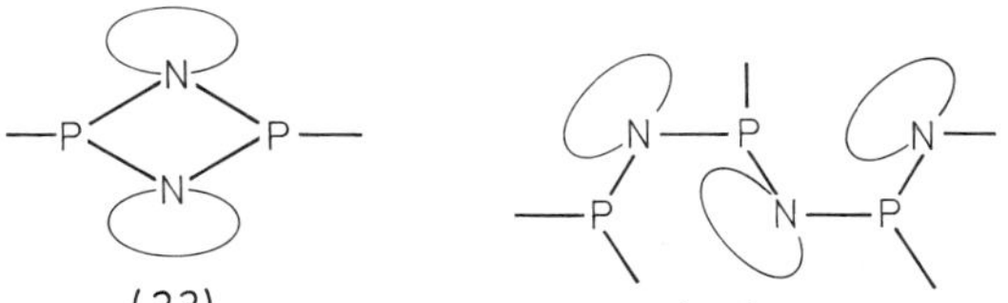

(23) (24)

Scheme 8

3.2 Relationships with Phosph(III)azenes

The presence of relatively bulky substituents on both phosphorus and nitrogen results in the formation of phosph(III)azenes, (*25*) [7,40,42,49–53], containing two co-ordinated tervalent phosphorus. Scheme 9 shows the final step in the synthesis [49,50] of (*25*) (R = R^1 = $SiMe_3$) from phosphorus(III) halides and the lithiated disilazane, $LiN(SiMe_3)_2$. These reactions are of interest here because in several cases it has been shown [42,54] that phosph(III)azenes (*25*) undergo a slow dimerisation process to give the corresponding cyclodiphosph(III)azanes. Phosph(III)-azenes may also be isolated in which the —P=N— bonds form part of a ring which

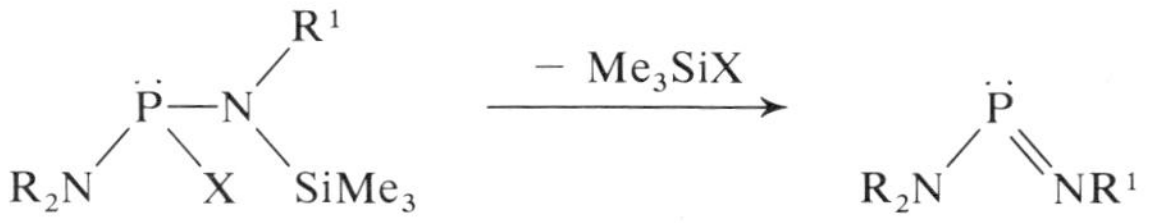

(*25*)

R, R¹ include Bu^t, $SiMe_3$
X=hal

Scheme 9

has a 6π-electron system. Examples are provided by the diaza- and triazaphospholes containing the ring systems, (*26*) [55] and (*27*) [56] respectively. Two reports of the reversible formation of phosph(III)azenes (*28*), (*29*) from cyclodiphosph(III)azanes [42] and cyclotetraphosph(III)azanes [38] respectively have appeared. Phosph(III)azene formation is favoured at elevated temperatures in the solution phase.

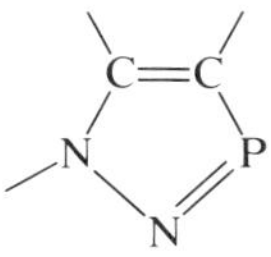

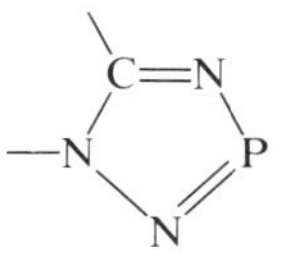

$R_2NNMeP{=}NR$
($R{=}SiMe_3$)

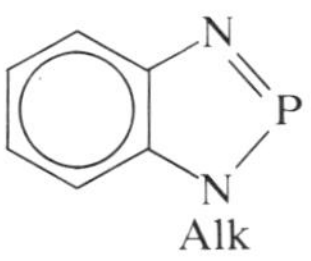

(*26*) (*27*) (*28*) (*29*)

Scheme 10

3.3 Properties of Cyclodiphosph(III)azanes

Some typical reactions of a chlorocyclodiphosph(III)azane are shown in Scheme 11. Chlorine-atom displacement by aminolysis [57,58], alcoholysis [47,59] or fluorination [60–62] of (*30*) all occur with retention of the four membered ring, but with interesting changes in its stereochemistry. Thus the dihalogen-derivatives (*30*) [63,64] and (*31*) [61] have a *cis*-arrangement of halogen atoms with respect to the ring. The products of aminolysis (*32*), or alcoholysis (*33*) are formed as a mixture of *cis*- and *trans*-isomers [65], but the *trans*-isomer of (*34*) is obtained stereospecifically [66]. Structures have not yet been assigned to (*35*) [24,67] and (*36*) [28]. Thermodynamic control of the products (*32*) and (*33*) can be achieved on standing, or more quickly, by heating, when *cis*-isomers predominate [65]. *Trans*-isomers of aminocyclodiphosph(III)azanes with relatively small endo N-alkyl- [58], or N-aryl- [57], or N-trimethylsilyl- [45,54] groups generally have thermodynamic stabilities at least comparable to the analogous *cis*-isomers. Numerous other alkyl- [28,41], amino- [22,23,43,57,58,68] alkoxy- [59,69–71] derivatives of cyclodiphosph(III)azanes have been synthesised, and in many cases geometrical isomers have been distinguished.

The oxidation of cyclodiphosph(III)azanes by chalcogens [24,28,34,41,45,47,59,67,72–74], methyl iodide [45,59,74], diborane [62], diketones [70,75], ketoimines [70,76], stannic chloride [77)], antimony pentachloride [77)], phosphorus pentachloride [77)], and transition metal compounds [44, 78–81)] has been studied.

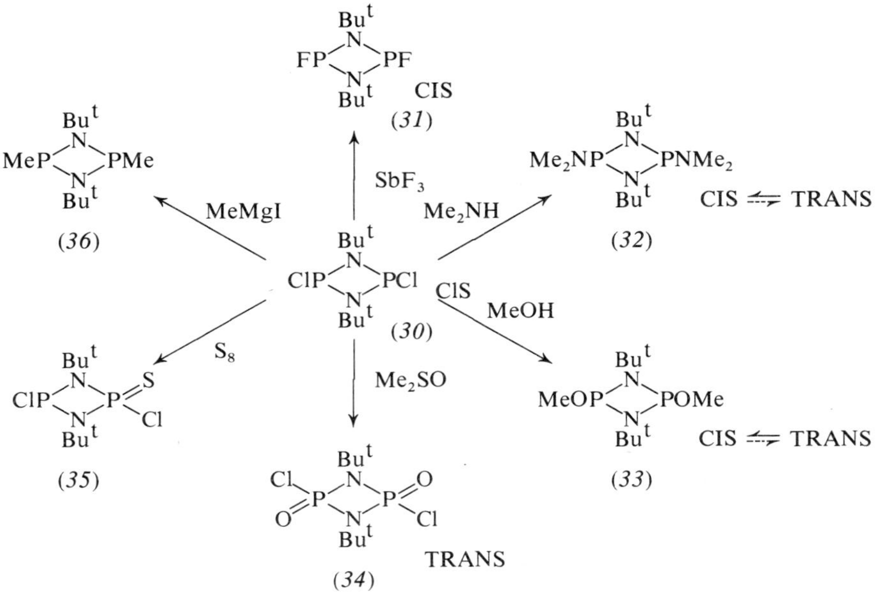

Scheme 11

Geometrical isomers of cyclodiphosph(III)azanes often have marked differences in chemical and physical properties and representative data are shown in Table 1. The ^{31}P chemical shift differences, typically 80–90 p.p.m., with the *trans*-isomer occurring at lower field, are unprecedented in inorganic ring systems. In cyclodiphosph(V)azanes [8)] and cyclophosph(V)azenes [82)], for example, ^{31}P chemical shift differences between geometrical isomers are generally less than 10 p.p.m. The photoelectron spectra [65)] also show that the ionisation of lone-pair electrons occurs at lower energy in the *trans*-isomers. This can be compared with the greater reactivity [58, 59)] of the *trans*-isomers to oxidation by sulphur or methyl iodide, and the stronger bonding of the *trans*-isomers to deuteriochloroforrm as measured by i.r. spectroscopy [65)]. These results are paralleled by the observation [83)] that only the *trans*-isomer of $(Pr^i_2NPNPr^i)_2$ undergoes electrochemical oxidation to give a stable radical cation. N.m.r. studies [65, 84)] have shown that *trans*-aminocyclodiphosph(III)-azanes also have significantly higher barriers to rotation about the exo-P—N bonds than the analogous *cis*-isomers (Difference *ca.* 6 kcal mol^{-1}) for the dimethylamino-derivatives (*37*) and (*38*). These results show that the conformation adopted by the amino-groups is that in which phosphorus and exo nitrogen lone-pairs are orthogonal (Table 1). The difference in rotational barriers has been attributed [65)] to a cross-ring interaction between the amino-groups destabilising the ground-state in the *cis*-isomer. There are several examples [43, 54, 84)] of n.m.r. effects that can be attributed to restricted rotation about the exo P—N bonds, and in a related ring system, $Me_2Si(NBu^t)_2PNBu^tSiMe_3$, the barrier is high enough to enable conformational isomers to be separated [85)].

Table 1. Properties of Geometrical Isomers (*37*) and (*38*)

Preferred Conformation	δ_P	p.e.s. bands/eV	reactivity $\gtrless P: \rightarrow \gtrless PS$	Δ(C—D) cm^{-1} Bonding $CDCl_3$	$\Delta G^{\neq}_{T_c}$ Kcal. mol^{-1} P—N rotation
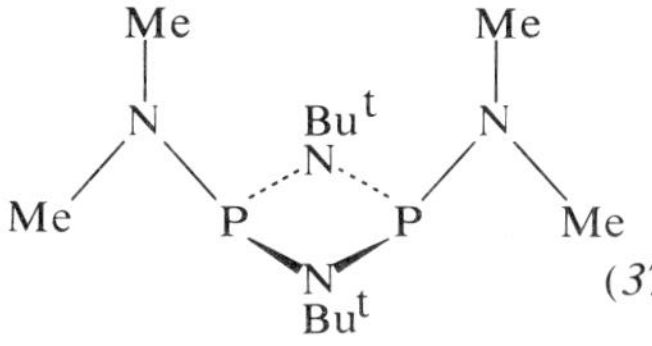 (*37*)	95.0	7.5, 8.2, 8.8	low	22	11.4_{215}
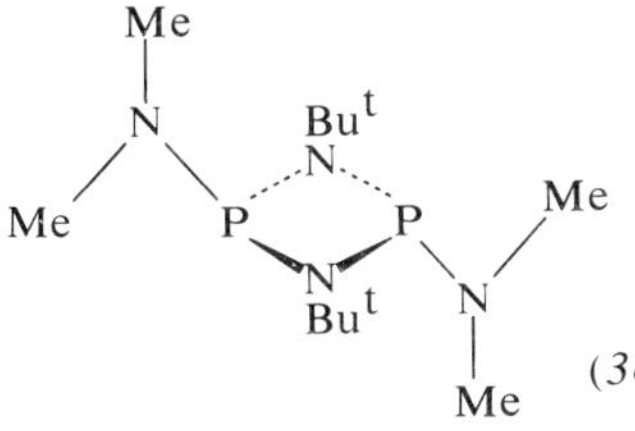 (*38*)	184.7	7.1, 7.5, 8.5	high	42	17.6_{327}

The reasons for the marked differences in chemical and physical properties that characterise the cyclodiphosph(III)azanes have not been clearly identified. However, some interesting comparisons can be made by examination of the conformations of cyclodiphosph(III)azane rings. The crystal structures of three *trans*-isomers of cyclophosph(III)azanes have been determined (Sect. 6) and these all show that the P_2N_2 ring is planar with the endo-nitrogen atoms having a planar distribution of bonds. By contrast, the P_2N_2 rings in the *cis*-isomers are puckered, and this is clearly seen in Fig. 2, which shows projections along the N—N (cross-ring) axis. Phosphorus substituents are omitted for the sake of clarity. The rings are puckered in the same way as cyclobutane and many of its derivatives; in the latter case the puckering of the ring arises largely as a result of eclipsing interactions between C—H bonds on adjacent carbon atoms [86]. A projection along the N—P bond of the *cis*-bispiperidino-derivative (*39*) (Fig. 2) [65] shows that an effect related to that in the cyclobutanes may also be operative, for the puckering of the ring relieves steric interactions between piperidino- and t-butyl-groups. Additionally the endo-nitrogen atoms have a non-planar distribution of bonds. This puckering effect may be expected to reduce the eclipsing of nitrogen and phosphorus lone-pairs (α ca. 40°) relative to the *trans*-isomers. Such effects would be expected to make the phosphorus less nucleophilic in the *cis*-isomers, as is observed [58]. Since it is not possible to define the axis of the phosphorus lone-pair with certainty it would be desirable to compare the crystal structures of geometrical isomers of the same compound, but this

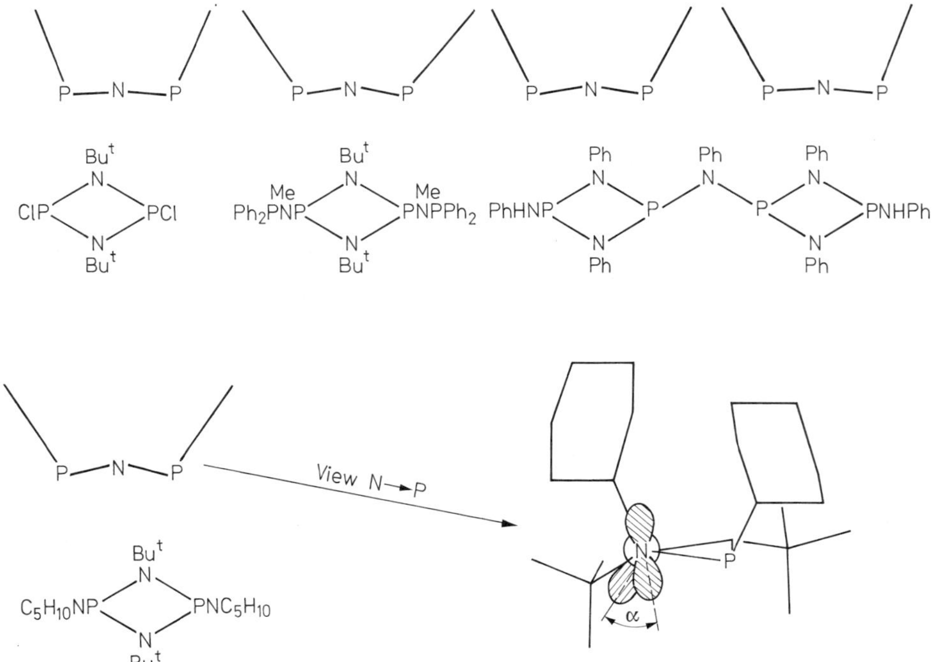

(39)

Fig. 2. Projections along the endo N—N and N—P axes in *cis*-cyclodiphosph(III)-azanes. Atomic co-ordinates were taken from the refs. in Table 2 (H-atoms omitted)

is not yet possible. Results supporting these suggestions come from observations on the reactivity of cyclodiphosph(III)azanes with less bulky endo N-substituents, where ring puckering effects should be reduced. Thus it is found that there is little difference in the reactivity of phosphorus towards sulphur in geometrical isomers of the N-methyl compound (*40*) [87]. The relative reactivity of isomers with endo N-phenyl substituents has not been compared, but it appears that *trans*-, rather than *cis*-isomers of (*41*) are thermodynamically favoured [57, 69].

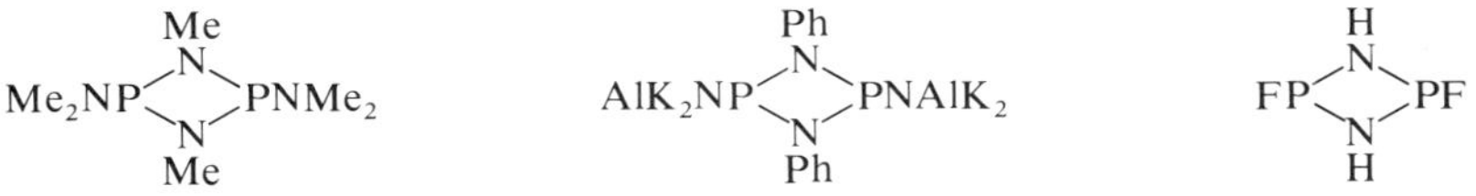

(*40*) (*41*) (*42*)

Scheme 12

Clearly these suggestions are empirical in nature, and a more accurate description of the underlying bonding effects is needed. Ideally this description would provide a rationale for the ^{31}P n.m.r. chemical shift differences which are still very large even for a given pair of isomers with endo N-substituents with relatively small steric demands. There has only been one report [88] of a study of the bonding in these ring systems and this is concerned with the isomers of the compound (*42*) (which has not been prepared) using the CNDO method. This predicted that the *trans*-isomer would be energetically favoured, a result consistent with the suggestions above, but an *ab initio* study would prove interesting, particularly if ring puckering effects could be included.

3.4 Cyclodiphosph(III)azanes with Extended P-N Skeletons

While the chemistry of the cyclodiphosph(III)azanes is now well founded, few compounds are known in which the P—N skeleton is extended exo to the P_2N_2-ring. Such compounds would provide examples of multidentate ligands, and present interesting comparisons with trimethylsilylamino-derivatives, where the bulky silyl-groups are known to kinetically stabilise phosph(III)azenes such as $(Me_3Si)_2NP=$

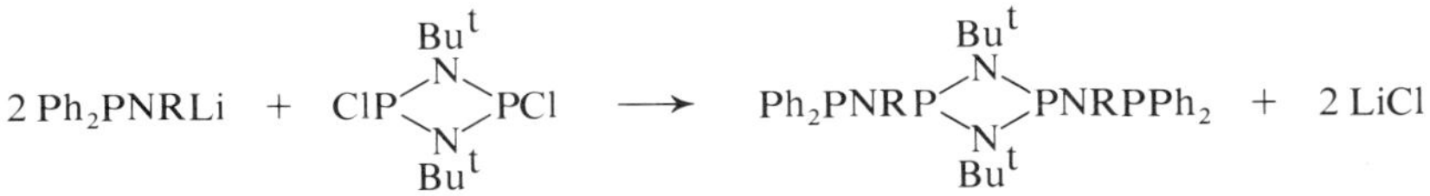

(R = Me, Et, Pr^i, Bu^t) (*43*)

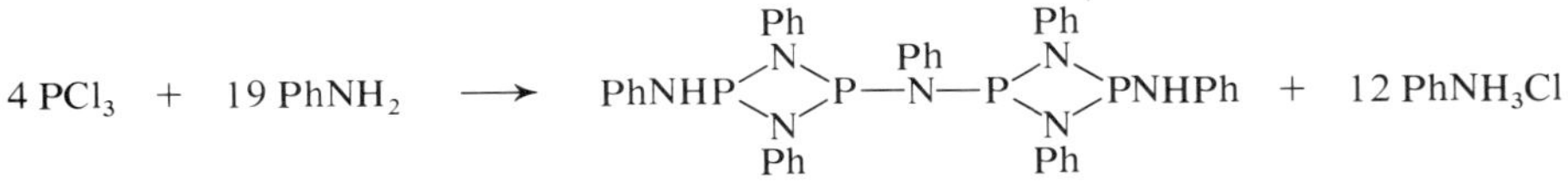

(*44*)

Scheme 13

$=NSiMe_3$ [49,50]. We have synthesised a series of cyclodiphosph(III)azanes containing extended P—N skeletal fragments by the reaction in Scheme 13, and studied their structures in some detail [89,90]. Even with relatively bulky groups on exo-nitrogen (R = Bu^t), no examples of phosph(III)azenes have been found. The n.m.r. spectra of these compounds give a good indication of the preferred conformations of the exo P—N—P skeletons in solution. Some of the most probable conformations of *cis*-isomers of (*43*) are shown in Fig. 3. Nitrogen atoms and their substituents, as well as phenyl-groups are omitted for clarity and the main constraints assumed for the exo-group conformations are that nitrogen has a planar distribution of bonds, and that adjacent nitrogen and phosphorus lone-pairs have an orthogonal relationship. Other conformations in which the phenyl-groups are directed towards the P_2N_2-ring, are considered sterically improbable. An analogous set of conformations is also possible for the corresponding *trans*-isomers.

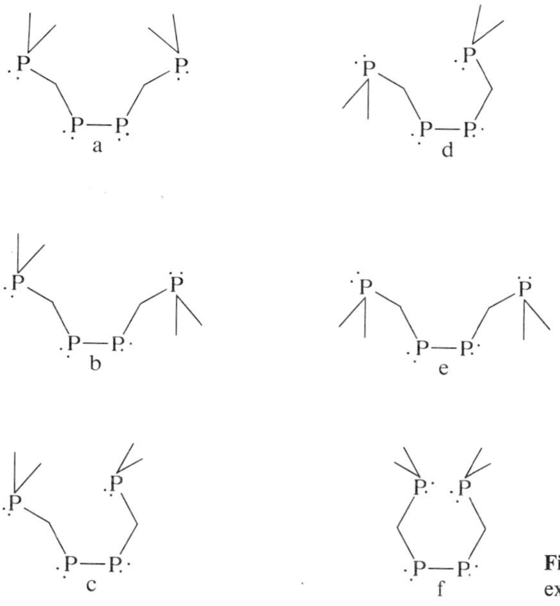

Fig. 3. Probable conformations of the exo-PNP skeletons in compounds (*43*)

In the case of (*43*) (R = Me [89] or Et [90]), *cis*- and *trans*-isomers are formed in comparable quantities, and these can be separated by fractional crystallisation. These compounds all have relatively large and positive exo-P—N—P spin-couplings (typically 350–400 Hz), and as shown in Scheme 7, this can be attributed to the preponderance of conformers a. The individual spin-couplings for the AA′XX′-type spin system exhibited by the $^{31}P\text{-}\{^1H\}$ n.m.r. spectra for the *cis*-isomer where R = Me are shown in Fig. 4, (i). The crystal structure of the same compound shows [89] that the molecule also adopts the same conformation in the solid state. When R = Pr^i or Bu^t, only *cis*-isomers of (*43*) are formed [90], a surprising result, which nevertheless may be related to the reduction of steric interactions by ring puckering as discussed above. At sub-ambient temperatures, the ^{31}P n.m.r. spectra of (*43*) (R = Pr^i) are difficult to interpret; possibly conformer d is present. Two conformers are present in the case of (*43*) (R = Bu^t), and the P ... P spin-coupling are shown

(i)

(ii)

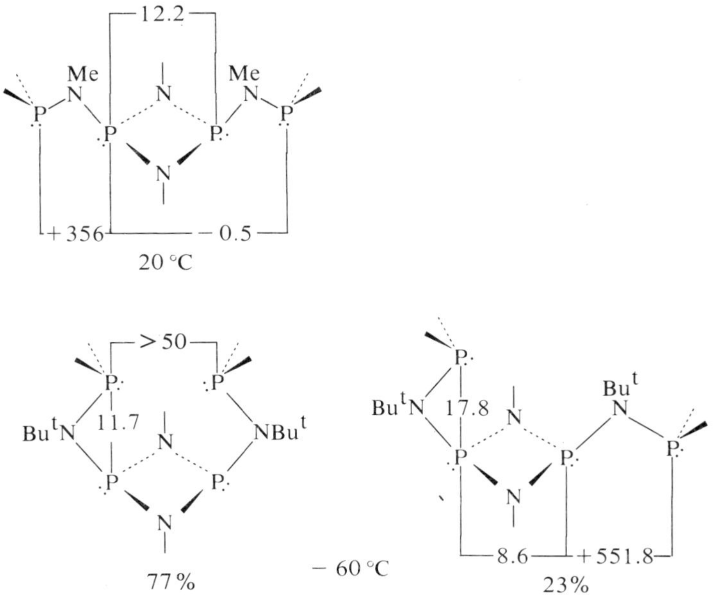

Fig. 4. *P ... P* spin-couplings (Hz.), obtained from the ^{31}P-{^{1}H} n.m.r. spectra of (*43*) (R = Me and Bu^t)

in Fig. 4, (ii). Single, double, and triple resonance experiments involving ^{1}H, ^{13}C and ^{31}P nuclei were used [90] for the identification of these conformers. The large exo P ... P coupling (>50 Hz) for (*43*) (R = Bu^t) (conformer f) suggests that a through-space spin-coupling effect is operative.

Two other examples of extended P—N skeletal species containing cyclodiphosph(III)azane rings have been reported. The first is a bicyclic structure (*44*), and this is formed [36] from the reaction of aniline with phosphorus trichloride in refluxing toluene solution (Scheme 14). This compound (as a methylene chloride adduct) formed the subject of an *X*-ray crystallographic study, which showed that both P_2N_2 rings have a *cis*-arrangement of phosphorus substituents, and that the bridging P—N—P skeleton has a conformation analogous to (*22*) (Scheme 7). Difficulty has been experienced in distinguishing (*44*) and $(PhNHPNPh)_2$, but the physical properties of each compound now seem well established [34,36,37].

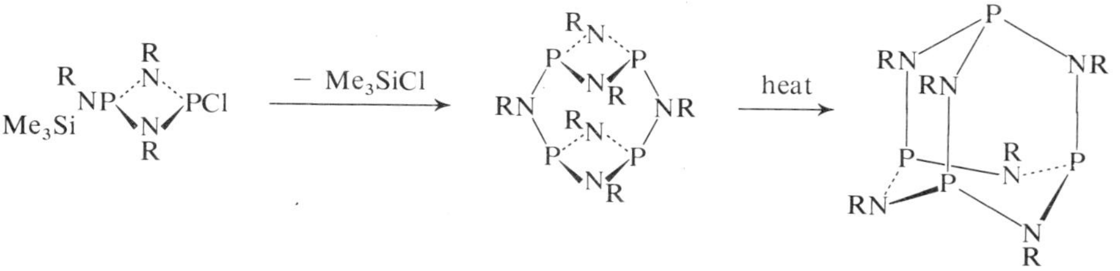

(*45*) (*46*) (*47*)

(R=Pr^i)

Scheme 14

The second example is provided by the novel tricyclic compound (*46*) which is obtained[91] by heating (*45*) in refluxing acetonitrile solution (Scheme 14). A feature of compound (*46*) is that it is thermodynamically unstable with respect to the adamantane-like compound (*47*); the latter is obtained on heating (*46*) over a period of several days at 156–158°.

4 Cyclotriphosph(III)azanes

4.1 Synthesis

In general, cyclotriphosph(III)azenes have proved more difficult to synthesize than cyclodiphosph(III)azanes, with the first well authenticated examples of the former appearing only two years ago[92]. The first report concerning a six-membered ring of this type (*49*) (R = Et, n = 3) arose from a study[93,94] of the reaction of the disilazane (*48*) (R = Et) with phosphorus trichloride (Scheme 15). It was also reported in this work that the cyclodiphosph(III)azane (*50*) could be obtained from an analogous reaction with heptamethyldisilazane, (*48*) (R = Me), but no n.m.r. or crystallographic data were available to support this contention. Subsequent attempts to repeat these reactions were unsuccessful[24,95], although the acyclic compounds

$$n\,PCl_3 \;+\; n\,Me_3SiNRSiMe_3 \longrightarrow (ClPNR)_n \;+\; n\,Me_3SiCl$$

(*48*) (*49*)

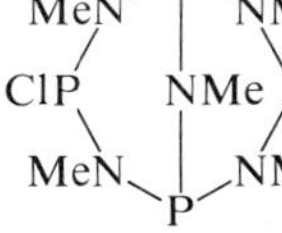

(*50*)

(*51*)

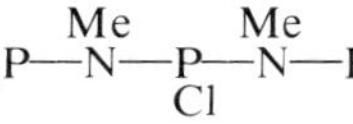

(*52*)

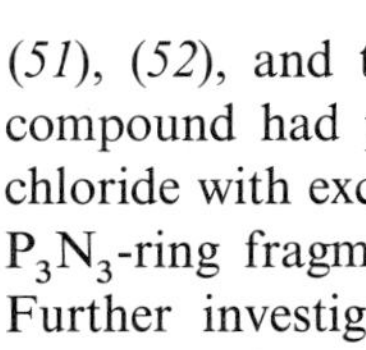

(*53*)

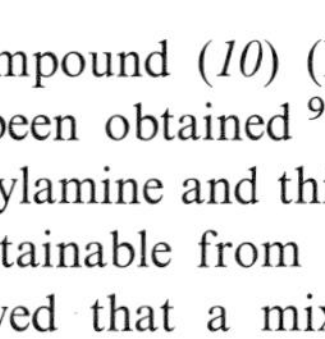

(*54*)

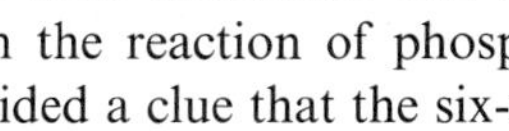

(*55*)

Scheme 15

(*51*), (*52*), and the cage compound (*10*) (Fig. 1) were identified. The latter cage compound had previously been obtained[96] from the reaction of phosphorus trichloride with excess of methylamine and this provided a clue that the six-membered P_3N_3-ring fragments are obtainable from amines with less-bulky N-alkyl groups. Further investigations showed that a mixture of the cage compound (*10*) and the fused ring compound (*53*) could be obtained from the reaction of the diphosphinoamine (*52*) with methylamine[27], or with the silazane, $(MeNH)_2SiMe_2$[97].

The first authentic cyclotriphosph(III)azane (*54*) was however obtained [92] by the reaction leading to (*49*) (R = Me) (Scheme 15). Yields of (*54*) were relatively low (30%), but the analogous bromide proved easier to obtain in a similar reaction with phosphorus tribromide [92]. Attempts to repeat the synthesis of (*54*) by the heptamethyldisilazane reaction also resulted in the formation of substantial quantities of acyclic compounds (*52*) and the triphosphorus compound (*55*) [98]. The isolation of chlorocyclotriphosph(III)azanes by this route is not straightforward.

It has been known for some time that alkyl- [99] or arylamine [23,24,100] hydrochlorides react with excess of phosphorus trichloride to give disphosphinoamines (*56*) (Scheme 16). However, we have recently found [98] that both the cyclotriphosph(III)-azane (*57*) and the fused-ring compound (*58*) are obtained when the stoichiometry is that shown (Scheme 16), and the reactions are carried out in reflexing *sym*-

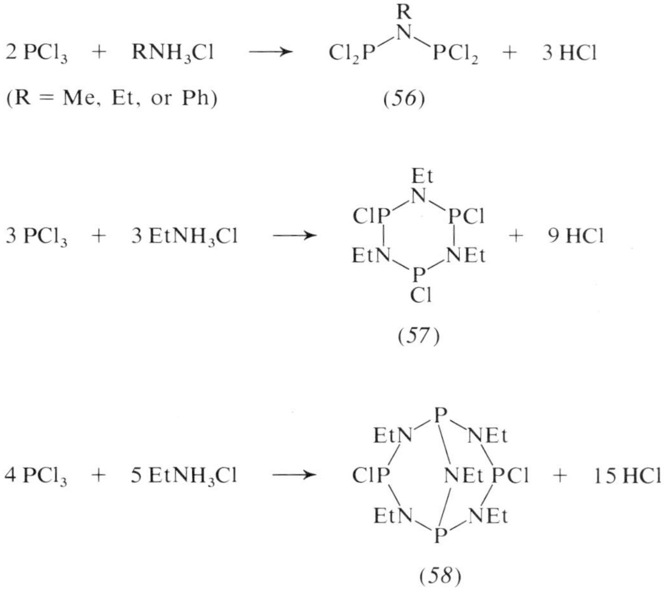

Scheme 16

terachloroethane solution. It was not possible to repeat these syntheses for the analogous N-methyl compounds, possibly because of the lower solubility of methylamine hydrochloride in *sym*-tetrachloroethane, but it may prove possible to overcome this difficulty with other solvents. Other claims [101–103] to have synthesised the cyclotriphosph(III)azane, $(PhPNPh)_3$, have yet to be substantiated.

4.2 Properties of Cyclotriphosph(III)azanes

Although no crystal structures of cyclotriphosph(III)azanes have yet been reported, it is reasonable to assume that the ring will adopt a chair conformation and that the nitrogen atoms will have a planar or near-planar distribution of bonds. This is consistent with the structures of the six-membered ring fragments found in the cage compound, $P_4(NMe)_6$ (Fig. 1) [19], and its monosulphide [104] and methyl iodide [105] adducts. This means that the identification of cyclotriphosph(III)azanes by n.m.r. methods is complicated not only by geometrical isomerism, but also by the axial/equatorial disposition of substituents (Fig. 5, where the nitrogen atoms and the ring

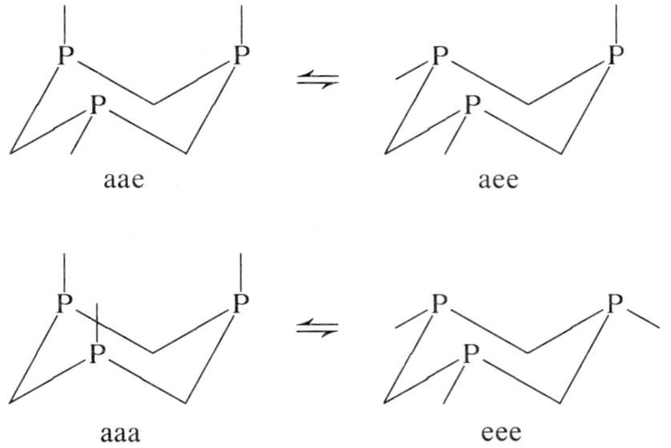

Fig. 5. Structures of cyclotriphosph(III)azanes (a = axial, e = equatorial phosphorus substituents)

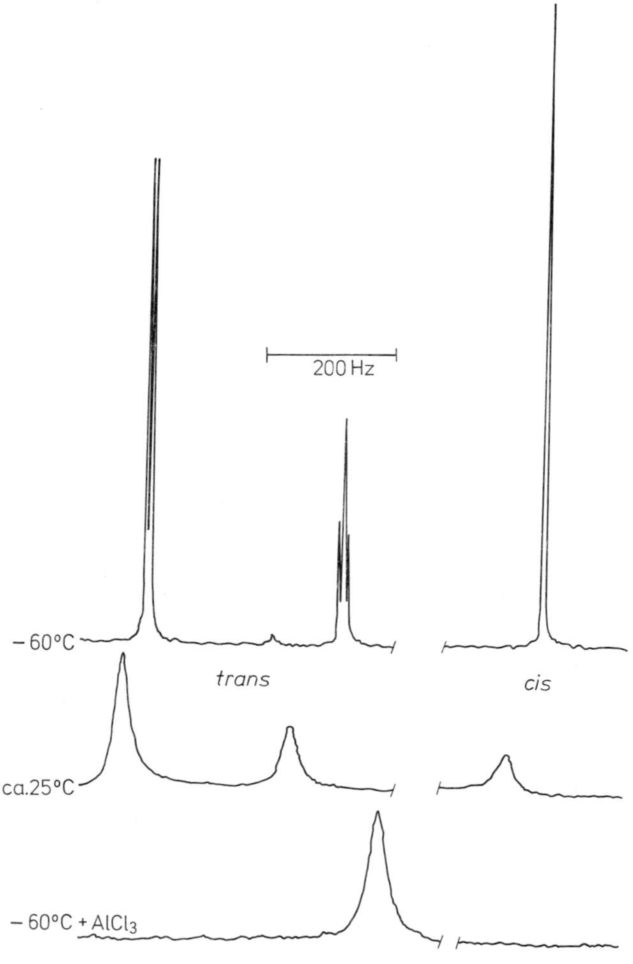

Fig. 6. ^{31}P-$\{^1H\}$ spectra of (*57*) ($CDCl_3$ solution)

substituents are omitted for clarity). Clearly, ^{31}P n.m.r. spectroscopy does not provide an unambiguous distinction between the aae or aee conformations of the *trans*-isomer (both of which would be expected to give two spin-coupled signals) or the aaa or eee conformations of the *cis*-isomer (both of which would give one signal only). Some examples of the ^{31}P-{^{1}H} n.m.r. spectra of (*57*) are shown (Fig. 6). At ambient temperatures there is a broadening effect in volving *cis*- and *trans*-isomers, which can be reduced at $-60°$ to resolve the *P*N*P* spin-coupling (7 Hz) in the *trans*-isomer. The population of the *cis*-isomer increases as the temperature is reduced, and at $-100°$, this isomer constitutes $>90\%$ of the mixture. The broadening process appears to be connected with chlorine-atom exchange, for the addition of anhydrous aluminium chloride (0.5 mole) collapses the sharp multiplet structure to a broad signal. Aluminium chloride is known to promote the ionisation of P(III)-Cl bonds and numerous cationic species derived from tervalent phosphorus compounds have been identified, including some cyclodiphosph(III)azanes [28,61,106]. It may be noted the ^{31}P chemical shifts between *cis*- and *trans*-isomers of cyclotriphosph(III)azanes are less than 30 p.p.m., much smaller than the shifts observed for cyclodiphosph(III)-azanes.

Some indication of the interactions between phosphorus and nitrogen lone-pairs can be obtained by consideration of the Newman-type projections along the N—P bonds (arrowed) in Fig. 7. When the phosphorus lone-pair adopts an equatorial site (*59*) the nitrogen and phosphorus lone-pairs are orthogonal, but a less favourable eclipsing interaction is apparent when the phosphorus lone-pair is axial (*60*). Similar arguments apply to the equatorial preference shown by phosphorus lone-pairs in dioxaphosphorinanes [107]. Since spin-coupling constants involving *P*N*P* [46] and *P*NC*H* [15] nuclei show a marked dependence on the conformation of the P—N bond it is possible that the axial/equatorial disposition of phosphorus substituents may be established on this basis. Observations on diphosphinoamines [46] suggest that the *P*N*P* spin-couplings will be small and negative in (*59*), and possibly positive in (*60*). Unfortunately, no signs for the relatively small *P*N*P* couplings in these rings have yet been determined. The magnitudes of *P*NC*H* couplings in N-methyl derivatives were found to cover a relatively narrow range (typically 13–16 Hz) [108] also making

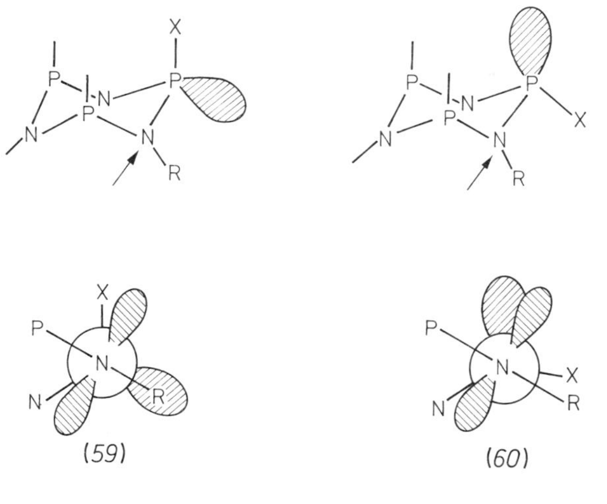

Fig. 7. Relative orientations of nitrogen and phosphorus lone-pairs in cyclotriphosph(III)azanes with (*59*) (axial) and (*60*) (equatorial) phosphorus substituents, X

a distinction between the two arrangements difficult. It is, of course, quite possible that the conformations shown (Fig. 7) are idealised, and that the ring may, in reality, be somewhat flattened offering a less clear-cut distinction between axial and equatorial sites than expected. Also, the relative populations of aae and aee conformers (Fig. 5) may be such as to make the apparent differences in *PNCH* spin-couplings small.

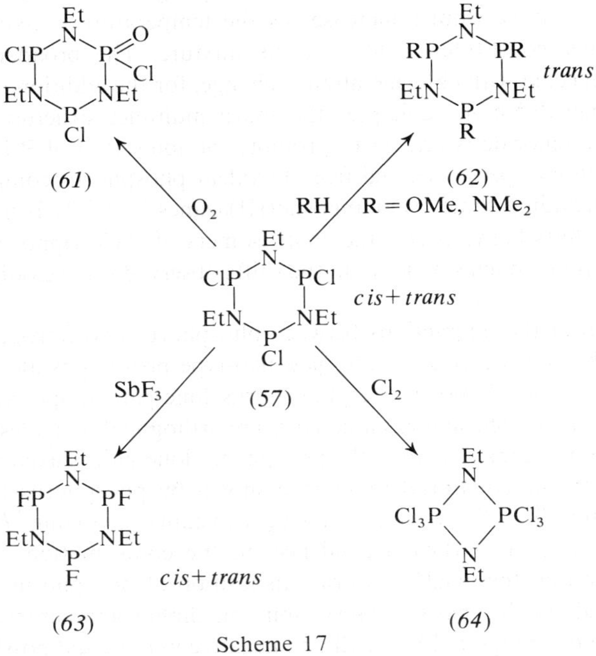

Scheme 17

Several halogen displacement [108–110] and oxidation [98,110,111] reactions have already been carried out on the cyclotriphosph(III)azenes and some of the results are shown in Scheme 17. Compounds *(61)*–*(63)* are formed with the ring still intact,

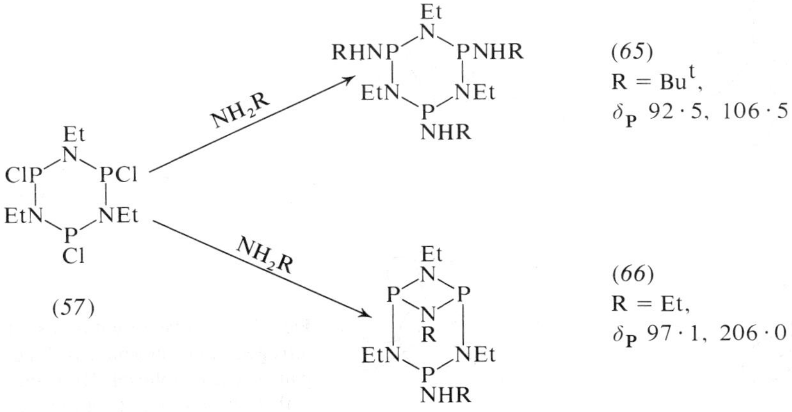

Scheme 18

but oxidation by chlorine [110] results in the formation of a well known cyclodiphosph(V)azane (*64*) [112], rather than a six-membered ring containing three five coordinated phosphorus atoms. Aminolysis by primary amines can result [110] in the preferential formation of a normal displacement product (*65*), or of the fused ring compound (*66*) (Scheme 18), depending on the amine used. It is not known whether the substituents on phosphorus adopt axial or equatorial positions, but clearly the result of bridging the two phosphorus atoms in (*66*) is to produce a very marked downfield ^{31}P shift compared with (*65*). The latter derivative can also be considered as a bridged cyclodiphosph(III)azane, and, as such, its ^{31}P shift is much closer to the analogous *cis*-dichloro-derivative, $(ClPNEt)_2$ (δ_p 227.3) [27], than to the *cis*-amino-derivative, $(Me_2NPNEt)_2$ (δ_p 107.9) [58]. An interesting reaction of an aminocyclotriphosph(III)azane, $(Et_2NPNMe)_3$, is its partial conversion [108] to the analogous cyclodiphosph(III)azane, $(Et_2NPNMe)_2$. Since this oligomerisation occurred during a distillation, it is not clear whether this is a reflection of the thermodynamic stability of P_3N_3 relative P_2N_2 rings. Sulphur [110,111] and selenium [110] induced oxidation reactions have been carried out on aminocyclotriphosph(III)azanes and, by contrast with the cyclodiphosph(III)azanes, or the cage compound (*10*) (Fig. 1) [113], it is easy to effect complete reaction at all of the phosphorus atoms.

4.3 Fused-Ring Cyclotriphosph(III)azanes

The direct synthesis [98] of the fused ring compound (*58*) from the reaction of phosphorus trichloride with ethylamine hydrochloride has already been described (Scheme 16). Its N-methyl analogue (*67*) can be obtained from the reaction of $(MeNPX)_3$ (X = Cl or Br) with heptamethyldisilazane [109] (Scheme 19). The identification of the structures of these fused-ring compounds, which are structurally related to the bicyclo(3.3.1)nonane carbocycles, pose difficult problems. An example of their

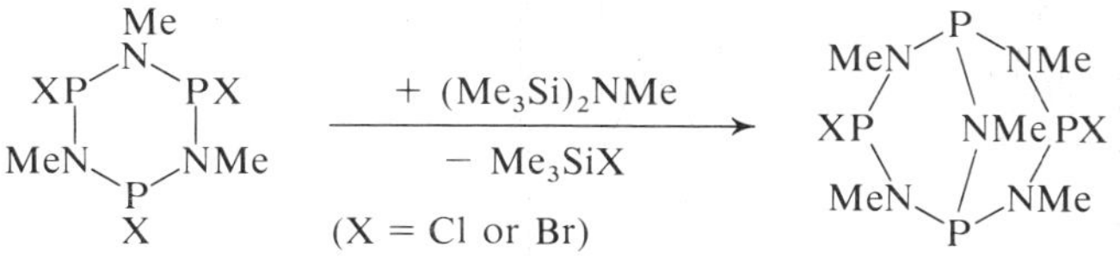

cf. (*54*) (*67*)

Scheme 19

^{31}P-$\{^1H\}$ n.m.r. spectra is shown in Fig. 8. Two isomers are apparent, which give rise to A_2B_2 and A_2MX spin-systems respectively, the former being favoured particularly at low temperatures. Only one isomer is evident with the N-methyl analogue (*67*) [109].

The possible conformations for these fused ring compounds are illustrated in Fig. 9. Any discussion of the relative stability of boat and chair forms depends on the axial/equatorial positions of the phosphorus substituents. It seems likely that ring inversion will be fast on the n.m.r. time-scale, so that the structures within a given group (i), (ii) or (iii) could all contribute (to varying degrees) to an observed spectrum.

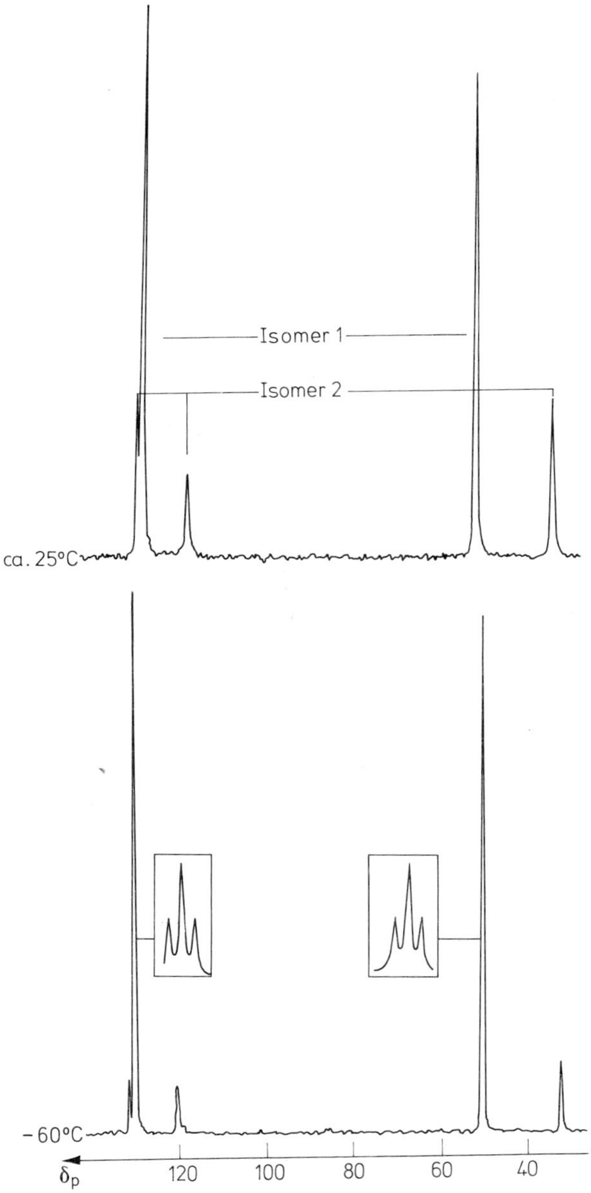

Fig. 8. ^{31}P-$\{^{1}H\}$ n.m.r. spectra of (*58*) ($CDCl_3$ solutions), with, inset, 10× expansion of the main signals

The reactions of the fused ring compound (*58*) with secondary amines, giving (*69*) as a mixture of isomers [110], and of (*67*) (X = Br) with t-butanol (and phenol) [109], have been studied. With primary amines, cage compounds such as (*68*) are the major products [110,111]. Cage-forming reactions of this general type were first studied with (*67*) (X = Br), and an interesting pair of tautomers (*70*), (*71*) identified (Scheme 20) [114]. See Ref. [113] for a summary of the chemistry of the cage compound (*10*) (Fig. 1).

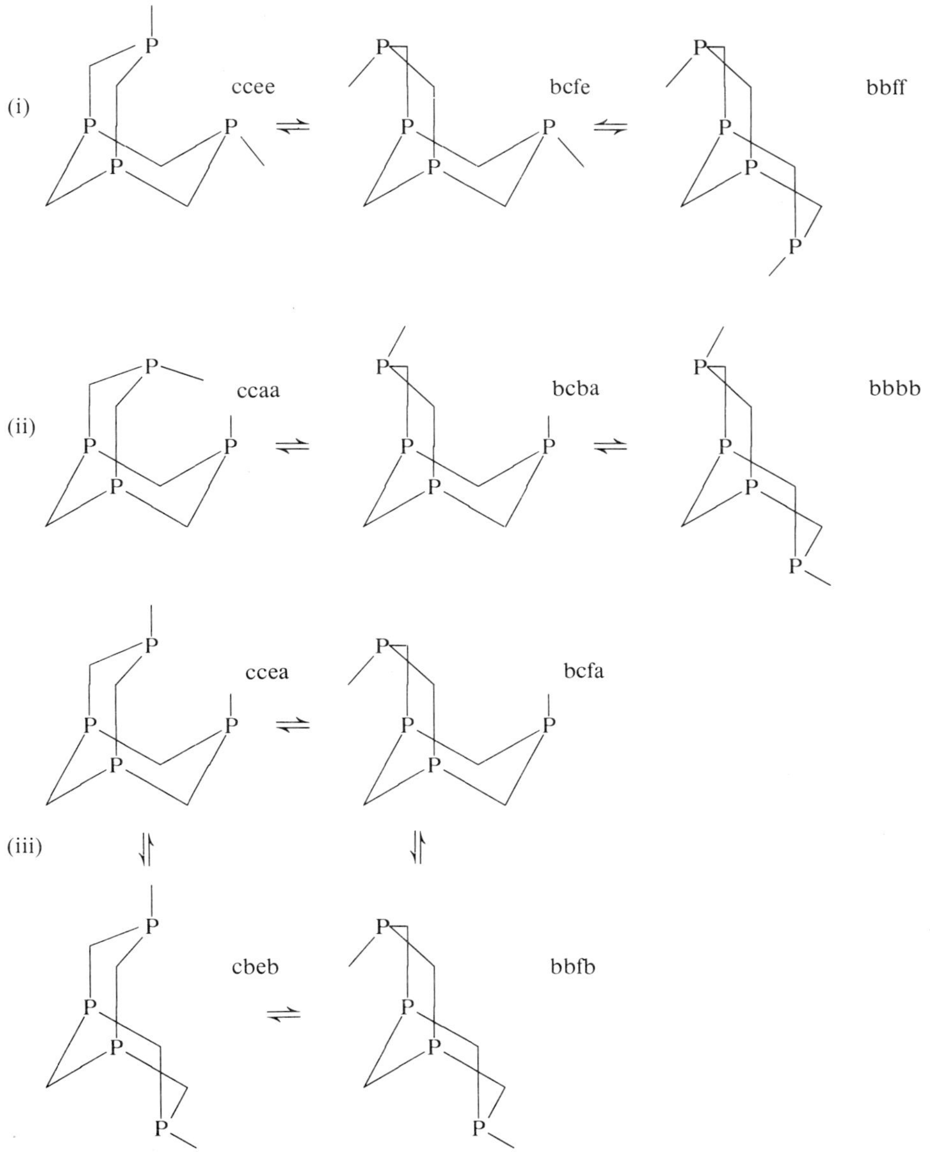

Fig. 9. Possible conformations of fused-ring compounds (e.g. (*58*)). Nitrogen atoms and ring substituents are omitted. The first two letters for each structure refer to the ring conformations (b = boat, c = chair) and the second two letters refer to the phosphorus-exo-substituent (a = axial, b = bowsprit, e = equatorial, f = flagpole)

The fused ring P—N skeleton in (*58*) can be extended in a manner analogous to that applied to the cyclodiphosph(III)azanes [115]. The ^{31}P-$\{^{1}H\}$ n. m. r. spectrum of one of these products (*72*) (Fig. 10) shows that only the two bridgehead phosphorus atoms (P_e) are chemically equivalent. The preferred conformations of the exo P—N—P units are analogous to those in (*21*) (Scheme 7) because the two J(*P*N*P*) (exo) couplings are large and positive (+342 and +352 Hz).

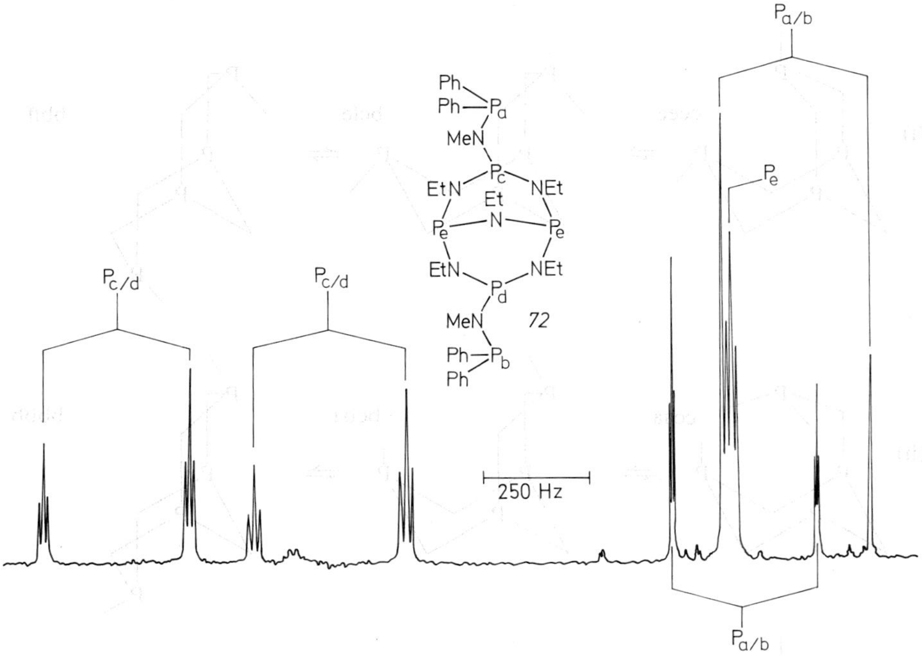

Fig. 10. ^{31}P-$\{^1H\}$ n.m.r. spectra of (*72*), obtained at $-60°$ in $CDCl_3$ solution

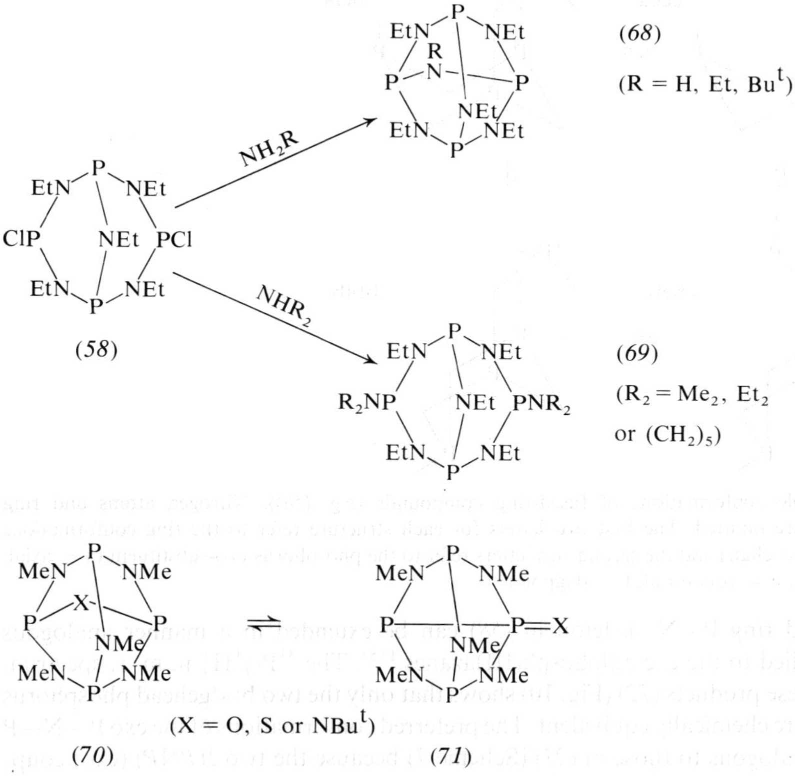

Scheme 20

5 Cyclotetraphosph(III)azanes

5.1 Synthesis and Properties

Very few examples of these P_4N_4 ring compounds are known, although the fused-ring compounds described above could also be considered as bridged cyclotetraphosph(III)azanes. An early claim [93,94] to have isolated $(EtNPCl)_4$ requires authentication, but two alkylated phosphorus derivatives containing the P_4N_4 ring have since been isolated (Scheme 21) [116]. The crystal structure [116] of the methyl-phosphorus compound (*73*) (R = Me) reveals that the ring has crown-type (C_{4v}) symmetry. Crystal structure data are also available for the trisulphide- [117], and mono- and bis(methyl iodide)-adducts [118] of (*73*) (R = Me). Adducts of the latter compound with acetylenes have been prepared [119].

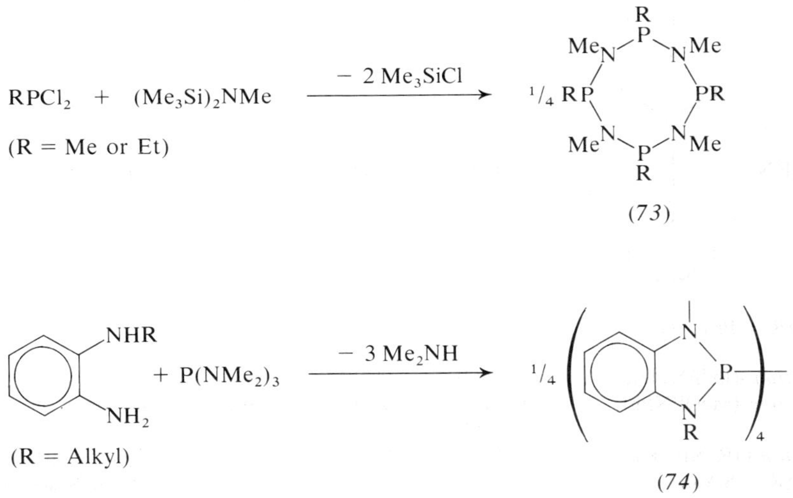

Scheme 21

The cyclophosph(III)azanes (*74*) (Scheme 21) exist mainly in the tetrameric form according to results obtained by field desorption mass spectrometry [38]. At ambient temperatures they have ^{31}P shifts in the range δ_p 81 to 89, but at elevated temperatures the shifts are in the range δ_p 225 to 236, the downfield shift being attributed to the formation of phosph(III)azenes (*29*) (Scheme 10).

6 Crystal Structure Data

A number of crystal structures of cyclophosph(III)azanes have now been reported and some of this data has been used earlier (Sect. 2) to illustrate the conformations adopted by the phosphorus-nitrogen rings. It appears that *X*-ray crystallographic methods will be crucial to an understanding of the conformations of six-membered ring and fused ring compounds, because n.m.r. correlations are not yet available to make definative structural assignments. Table 2 contains a summary of all the crystal structures reported to date.

Table 2. Crystal Structure Data for Cyclophosph(III)azanes[a]

Compound	P-N/Å[b]	< PNP/°[b]	< NPN/°[b]	Comment	Ref.
cis-$(ClPNBu^t)_2$	1.689(4)	97.3(5)	82.4(4)	almost C_s symmetry	64)
cis-$(C_5H_{10}NPNBu^t)_2$	1.720–1.750(2)	96.8(1)	80.3(1)	pronounced puckering of P_2N_2 ring	65)
cis-$(Ph_2PNMePNBu^t)_2$	1.723(11)	96.9(8)	80.1(6)	exo-PNP analogous to (22) (Scheme 7)	89)
cis-$[PhNHP(NPh)_2P]_2NPh/CH_2Cl_2$	1.723(8)	99.9(3)	79.4(3)	both P_2N_2 rings have *cis*-arrangements	36)
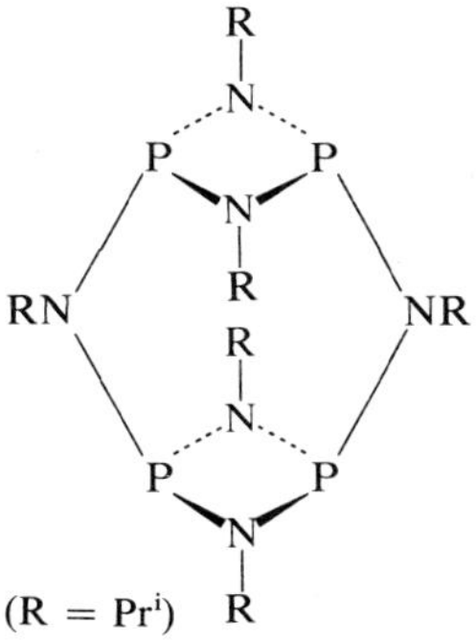 (R = Pr^i)	1.70–1.71(1)	99(1)	82(1)	disordered in the region of Pr^i-groups	91)
trans-$(Bu^tPNMe)_2$	1.716(3)	98.8(1)	81.2(1)	C_i symmetry	18)
trans-$(MeOPNPh)_2$	1.721–1.734(4)	99.9(2)	80.1(2)	C_i symmetry	120)
trans-$(R_2NPNR)_2$ (R = $SiMe_3$)	1.727	97.5	82.5	C_i symmetry, bonds at exo N distorted (< PNSi = 108.4 and 131.0°)	54)
$(MeNPMe)_4$	1.720, 1.723	111.8	107.9	C_{4v} symmetry crown-type P_4N_4 ring	116)
$P_4(NMe)_6$	1.695(13)	124.3(6)	101.1(7)	disordered methyl-groups-poor refinement	19)

a All *cis*-cyclodiphosph(III)azanes have puckered rings, whilst *trans*-isomers have planar P_2N_2 rings.
b Standard deviations in parentheses.

7 Summary and Conclusions

The P—N skeletal features of the cyclophosph(III)azanes and related cage compounds discussed here are summarised in Fig. 11. This shows that considerable advances have been made since 1960 when only one cyclophosph(III)azane ring system

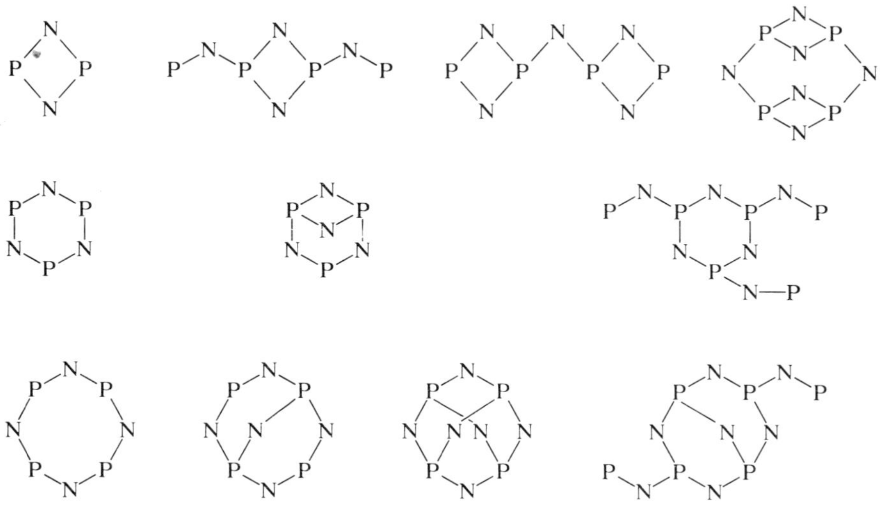

Fig. 11. P—N skeletons of cyclophosph(III)azanes and related compounds

(P_2N_2) had been identified. Compared with the phosph(V)azenes [1,3,8,82], only small ring systems have been obtained, and this may be largely due to the relatively small angles that prevail at tervalent phosphorus, and the presence of three rather than two coordinated nitrogen atoms. The cyclophosph(III)azanes already isolated are the precursors of some novel quinquivalent phosphorus compounds, and will make interesting multidentate ligands for metals. Work on the latter topic has been reported [44,78–81,121], but, so far, this is mainly concerned with the cyclodiphosph(III)azanes.

8 Acknowledgements

Thanks are due to Dr. W. Zeiss for communicating unpublished results, and to Professor J. G. Riess for providing the atomic coordinates of $P_4(NMe)_6$.

9 References

1. Shaw, R. A., Fitzsimmons, B. W., Smith, B. C.: Chem. Rev. *62*, 247 (1962)
2. Haiduc, I.: The Chemistry of Inorganic Ring Systems, Vol. II, p. 804, Wiley, Chichester 1970
3. Allcock, H. R.: Phosphorus-Nitrogen Compounds: Cyclic, Linear and High Polymeric Systems, Academic Press, New York, N.Y. 1972
4. Heal, H. G.: The Inorganic Heterocyclic Chemistry of Sulphur, Nitrogen and Phosphorus, Academic Press, London 1980
5. Lappert, M. F. et al.: Metal and Metalloid Amides, Ellis Horwood Ltd., Chichester 1980
6. Grapov, A. F., Mel'nikov, N. N., Razvodovskaya, L. V.: Russ. Chem. Rev. *39*, 20 (1970)
7. Niecke, E., Scherer, O. J.: Nachr. Chem. Techn. *23*, 395 (1975)

8. Shaw, R. A.: Phosphorus and Sulphur *4*, 101 (1978)
9. Gerrard, W., Hudson, H. R.: In Organic Phosphorus Compounds (Kosolapoff, G. M., Maier, L. (Eds.)) Vol. V, p. 21, Wiley, New York, N.Y. 1973
10. Verkade, J. G.: Phosphorus and Sulphur *2*, 251 (1976)
11. Hudson, R. F., Verkade, J. G.: Tetrahedron Lett. *1975*, 3231
12. Gorenstein, D. G., Luxon, B. A., Findlay, J. B.: J. Am. Chem. Soc. *101*, 5869 (1979)
13. Shvo, Y.: In The Chemistry of Hydrazo, Azo and Azoxy Groups (Patai, S. (Ed)) Pt. II, p. 1017, Wiley, New York, N.Y. 1975
14. Baxter, S. G. et al.: J. Am. Chem. Soc. *103*, 1699 (1981)
15. Bulloch, G., Keat, R., Rycroft, D. S.: J. Chem. Soc., Dalton Trans., *1978*, 764
16. Cowley, A. H. et al.: J. Am. Chem. Soc. *101*, 5224 (1979)
17. *c.f.* Hudson, R. F.: Angew. Chem. Internat. Edn. Egl. *12*, 36 (1973)
18. Pohl, S.: Z. Naturforsch. *34*, 256 (1979)
19. Cotton, F. A. et al.: Inorg. Chim. Acta *11*, L33 (1974)
20. Keat, R.: J. Chem. Educ., in press
21. Keat, R., Rycroft, D. S., Thompson, D. G.: J. Chem. Soc., Dalton Trans., *1980*, 1858
22. Michaelis, A., Schroeter, G.: Chem. Ber. *27*, 490 (1894)
23. Davies, A. R. et al.: J. Chem. Soc., Perkin Trans. 2, *1973*, 379
24. Jefferson, R. et al.: J. Chem. Soc., Dalton Trans., *1973*, 1414
25. Holmes, R. R., Forstner, J. A.: Inorg. Chem. *2*, 380 (1963)
26. Scherer, O. J., Klusman, P.: Angew. Chem. Internat. Edn. Engl. *8*, 752 (1969)
27. Bulloch, G., Keat, R.: J. Chem. Soc., Dalton Trans., *1974*, 2010
28. Scherer, O. J., Schnabl, G.: Chem. Ber. *109*, 2996 (1976)
29. Markovski, L. N. et al.: J. Gen. Chem. U.S.S.R. *50*, 273 (1980)
30. Barlow, M. G. et al.: J. Chem. Soc. (C) *1966*, 1592
31. Grimmel, H. W., Guenther, A., Morgan, J. F.: J. Am. Chem. Soc. *68*, 539 (1946)
32. Goldschmidt, S., Obermeier, F.: Justus Liebigs Ann. Chem. *588*, 24 (1954)
33. Peiffer, G., Guillemont, A., Traynard, J. C.: C.R. Hebd. Seances Acad. Sci. Ser. C *266*, 400 (1968)
34. Trishin, Yu. G. et al.: J. Org. Chem. U.S.S.R. *11*, 1747 (1975)
35. Trishin, Yu. G. et al.: ibid. *11*, 1750 (1975)
36. Thompson, M. L., Haltiwanger, R. C., Norman, A. D.: J. Chem. Soc., Chem. Commun., *1979*, 647
37. Norman, A. D., Haltiwanger, R. C., Tarassoli, A.: Abstract INOR 260, 178th. Am. Chem. Soc. Meet., Washington, D.C. 1979
38. Malavaud, C. et al.: Tetrahedron Lett. *1979*, 447
39. Bowden, F. L. et al.: J. Chem. Soc., Perkin Trans. 1, *1973*, 516
40. Scherer, O. J., Glässel, W.: Angew. Chem. Internat. Edn. Engl. *14*, 629 (1975)
41. Scherer, O. J., Schnabl, G.: ibid. *15*, 772 (1976)
42. Scherer, O. J., Glässel, W.: Chem. Ber. *110*, 3874 (1977)
43. Scherer, O. J., Andres, K.: Z. Naturforsch. B *33*, 467 (1978)
44. Zeiss, W., Feldt, Ch.: J. Organomet. Chem. *127*, C5 (1977)
45. Zeiss, W. et al.: Chem. Ber. *111*, 1180 (1978)
46. Keat, R. et al.: J. Chem. Soc. Dalton Trans., 1981, 2192
47. Colquhoun, I. J., McFarlane, W.: J. Chem. Soc. Dalton Trans., *1977*, 1674
48. Allcock, H. R.: Heteroatom Ring Systems and Polymers, Academic Press, New York 1967
49. Niecke, E., Flick, W.: Angew. Chem. Internat. Edn. Engl. *12*, 585 (1973)
50. Scherer, O. J., Kuhn, N.: Chem. Ber. *107*, 2123 (1974)
51. Scherer, O. J., Kuhn, N.: J. Organomet. Chem. *82*, C3 (1974)
52. Scherer, O. J., Kuhn, N.: Angew. Chem. Internat. Edn. Engl. *13*, 811 (1974)
53. Markovskii, L. N., Romanenko, V. D., Ruban, A. V.: J. Gen. Chem. U.S.S.R. *49*, 1681 (1979)
54. Niecke, E., Flick, W., Pohl, S.: Angew. Chem. Internat. Edn. Engl. *15*, 309 (1976)
55. Weinmaier, J. H. et al.: ibid. *18*, 412 (1979)
56. Barrans, J., Charbonnel, Y.: Tetrahedron *32*, 2039 (1976)
57. Bulloch, G., Keat, R., Thompson, D. G.: J. Chem. Soc. Dalton Trans., *1977*, 99
58. Keat, R., Rycroft, D. S., Thompson, D. G.: ibid. *1980*, 321

59. Keat, R., Rycroft, D. S., Thompson, D. G.: ibid. *1979*, 1224
60. Nixon, J. F., Wilkins, B.: Z. Naturforsch. B *25*, 649 (1970)
61. Keat, R., Thompson, D. G.: J. Chem. Soc. Dalton Trans., *1978*, 634
62. Jessup, J. S., Paine, R. T., Campana, C. F.: Phosphorus and Sulphur *9*, 279 (1981)
63. Muir, K. W., Nixon, J. F.: J. Chem. Soc., Chem. Commun., *1971*, 1406
64. Muir, K. W.: J. Chem. Soc., Dalton Trans., *1975*, 259
65. Keat, R. et al.: J. Chem. Soc., Chem. Commun., *1978*, 372
66. Keat, R., Manojlović-Muir, Lj., Muir, K. W.: Angew. Chem. Internat. Edn. Engl. *13*, 211 (1973)
67. Bulloch, G. et al.: Org. Magn. Reson. *12*, 708 (1979)
68. Keat, R., Thompson, D. G.: Angew. Chem. Internat. Edn. *16*, 797 (1977)
69. Zeiss, W., Weis, J.: Z. Naturforsch. B *32*, 485 (1977)
70. Kawashima, T., Inamoto, N.: Bull. Chem. Soc. Jpn. *49*, 1924 (1976)
71. Mitsunobu, O., Mukaiyama, T.: J. Org. Chem. *29*, 3005 (1964)
72. Scherer, O. J., Schnabl, G.: Angew. Chem. Internat. Edn. Engl. *16*, 486 (1977)
73. Keat, R., Thompson, D. G.: J. Organomet. Chem. *141*, C13 (1977)
74. Keat, R., Thompson, D. G.: J. Chem. Soc. Dalton Trans., *1980*, 928
75. Zeiss, W.: Angew. Chem. Internat. Edn. Engl. *15*, 555 (1976)
76. Schmidpeter, A., Weinmaier, J. H.: Chem. Ber. *111*, 2086 (1978)
77. Scherer, O. J., Schnabl, G.: Z. Naturforsch. B *31*, 1462 (1976)
78. Hawker, P. N., Jenkins, L. S., Willey, G. R.: J. Organomet. Chem. *118*, C44 (1976)
79. Jenkins, L. S., Willey, G. R.: J. Chem. Soc. Dalton Trans., *1979*, 777
80. Maisch, H.: Z. Naturforsch. B *34*, 784 (1979)
81. Burckett St. Laurent, J. C. T. R. et al.: Inorg. Chim. Acta *44*, L17 (1980)
82. Krishnamurthy, S. S., Sau, A. C., Woods, M.: In Adv. Inorg. Chem. and Radiochem. (Emeléus, H. J., Sharpe, A. G. (Eds.)) Vol. XXI p. 41, Academic Press, New York, N.Y. 1978
83. Diaz, A. F., Scherer, O. J., Andres, K.: J. Chem. Soc., Chem. Commun., *1980*, 982
84. Bulloch, G., Keat, R., Thompson, D. G.: J. Chem. Soc., Dalton Trans., *1977*, 1044
85. Scherer, O. J., Püttmann, M.: Angew. Chem. Internat. Edn. Engl. *18*, 679 (1979)
86. Moriarty, R. M.: In Topics in Stereochemistry, (Allinger, N. L., Eliel, E. L. (Eds.)) Vol. VIII, p. 271, John Wiley, New York, N.Y. 1974
87. Keat, R.: unpublished work
88. Perkins, P. G.: Phosphorus *5*, 31 (1974)
89. Harvey, D. A. et al.: Inorg. Chim. Acta *34*, L201 (1979)
90. Keat, R., Murray, L. H., Rycroft, D. S.: J. Chem. Soc., Dalton Trans., in press
91. Scherer, O. J. et al.: Angew. Chem. Internat. Edn. Engl. *19*, 571 (1980)
92. Zeiss, W., Barlos, K.: Z. Naturforsch. B *34*, 423 (1979)
93. Abel, E. W., Willey, G. R.: Proc. Chem. Soc. (London), *1962*, 308
94. Abel, E. W., Armitage, D. A., Willey, G. R.: J. Chem. Soc., *1965*, 57
95. Jefferson, R., Nixon, J. F., Painter, T. M.: J. Chem. Soc., Chem. Commun., *1969*, 622
96. Holmes, R. R.: J. Am. Chem. Soc. *82*, 5509 (1960)
97. Wannagat, U., Autzen, H.: Z. Anorg. Allg. Chem. *420*, 119 (1976)
98. Ahmed, K. Sh. et al.: unpublished work
99. Nixon, J. F.: J. Chem. Soc. (A)., *1968*, 2689
100. Goldschmidt, S., Krauss, H. L.: Justus Liebigs Ann. Chem. *595*, 193 (1955)
101. Johns, I. B., McElhill, E. A., Smith, J. O.: J. Chem. Eng. Data *7*, 277 (1962)
102. Johns, I. B., Wildi, B. S.: U.S. Pat. 3239561 (1966); C.A., *64*, 15924 (1966)
103. Johns, I. B., Wildi, B. S.: U.S. Pat. 3239785 (1966); C.A., *64*, 16811 (1966)
104. Cotton, F. A. et al.: Inorg. Chem. *17*, 3521 (1978)
105. Hunt, G. W., Cordes, A. W.: Inorg. Chem. *13*, 1688 (1974)
106. Cowley, A. H. et al.: Pure Appl. Chem. *52*, 789 (1980)
107. Maryanoff, B. E., Hutchins, R. O., Maryanoff, C. A.: In Topics in Stereochemistry (Allinger, N. L., Eliel, E. L. (Eds.)) Vol. XI, p. 187, John Wiley, New York 1979
108. Zeiss, W. et al.: Z. Anorg. Allg. Chem. *475*, 256 (1981)
109. Zeiss, W., Endrass, W.: Z. Naturforsch. B *34*, 678 (1979)
110. Harvey, D. A., Keat, R., Rycroft, D. S.: unpublished work

111. Zeiss, W., Klehr, H.: Z. Naturforsch. B *35*, 1179 (1980)
112. Bermann, M.: in Adv. Inorg. Chem. and Radiochem. (Emeléus, H. J., Sharpe, A. G. (Eds.)) Vol. XIV, p. 1, Academic Press, New York, N.Y. 1978
113. Riess, J. G. et al.: 1st. Int. Congr. on Phosphorus Compounds, Rabat 1977, abstracts p. 52
114. Zeiss, W., Endrass, W., Pointner, A.: IIIrd. Symp. on Inorganic Phosphorus Compounds, Halle 1979, abstracts p. 133, and Zeiss, W.: personal communication
115. Munro, I. T.: B. Sc. Thesis, Glasgow 1981
116. Zeiss, W., Schwarz, W., Hess, H.: Angew. Chem. Internat. Edn. Engl. *16*, 407 (1977)
117. Zeiss, W., Schwarz, W., Hess, H.: Z. Naturforsch. B *35*, 959 (1980)
118. Zeiss, W. et al.: Z. Naturforsch. B *36*, 561 (1981)
119. Zeiss, W., Henjes, H.: Chem. Ber., *111*, 1655 (1978)
120. Schwarz, W., Hess, H., Zeiss, W.: Z. Naturforsch. B *33*, 723 (1978)
121. Schmidpeter, A. et al.: Angew. Chem. Internat. Edn. Engl. *20*, 408 (1981)

Sulfur-Nitrogen Anions and Related Compounds[1]

Tristram Chivers and Richard T. Oakley

Department of Chemistry, University of Calgary, Calgary, Alberta, Canada T2N 1N4, Canada

Table of Contents

---

[1] Plenary lecture delivered at the 3rd IRIS Meeting held in Graz (Austria), August 17–22, 1981

1 Introduction — Scope of Review

In an excellent review of the chemistry of sulfur nitrides published in 1972, Heal drew attention to the slow development of the subject compared to more fashionable areas of inorganic chemistry [1]. Three years later, Banister contributed a comprehensive account of cyclic S—N compounds covering the 5 year period up to March 1973 [2]. He stated that "the subject is at an interesting stage at which general rationalizations of molecular structure and reactivity are just beginning to emerge" and predicted that important advances would be made by both theoreticians and experimentalists in the near future. The discovery, in 1975 [3], that polysulfur nitride is a superconductor at very low temperatures accelerated interest in S—N chemistry and the subsequent studies of this fascinating polymer have been thoroughly reviewed by Labes et al. [4]. In 1979 Roesky provided a major contribution to the review literature with the publication of a survey of the increasingly diverse range of cyclic S—N compounds, which were classified according to ring size and coordination numbers at sulfur and nitrogen [5]. Very recently, a book by Heal which surveys developments up to the end of 1979 has appeared [6].

Despite this extensive coverage of S—N chemistry, little attention has been paid to the behaviour of sulfur nitrides towards nucleophiles or the chemistry of S—N anions and related compounds which are the products of these reactions. Some of the earlier work on this topic, which was carried out without the advantages of modern physical techniques, has been shown to be incorrect and a number of significant discoveries have been made recently. The objective of this article is to provide an up-to-date account of this important branch of sulfur-nitrogen chemistry. The emphasis will be on the advances that have been made during the last 5–6 years in the preparation of S—N anions and their derivatives and in the application of physical techniques, e.g. X-ray diffraction, ^{15}N NMR spectroscopy, UV-visible, Raman and MCD spectroscopy, to our understanding of the molecular and electronic structures of these S—N compounds. The review is divided into two parts. The first part is concerned with the preparation, molecular structure, and reactions of sulfur-nitrogen anions and related compounds, while the second provides an account of our current understanding of electronic structure and bonding in these electron-rich compounds.

2 Binary Sulfur-Nitrogen Anions

2.1 General Comments

Sulfur and nitrogen exhibit versatile behaviour in the formation of binary cations, anions, and neutral molecules. Those species which have been structurally characterized by X-ray crystallography are listed in Table 1. In order to emphasise the recent progress of the subject of sulfur-nitrogen chemistry, the following developments should be noted: a) Seven of the eight known cations have been discovered and structurally characterized since 1970 [7–13], and the structure of the eighth ($S_4N_3^+$) has been redetermined recently [14], b) the first binary sulfur-nitrogen anion, $S_4N_5^-$ was isolated in 1975 [15], and its structure was reported a year later [16], c) the novel sulfur nitride, S_5N_6, was first identified in 1978 [17], and d) although tetrasulfur dinitride, S_4N_2, has been known since 1897, the structure was not unequivocally determined until 1981 [18].

Table 1. Binary Sulfur-Nitrogen Species[a]

Cations	Neutral Molecules[b]	Anions
SN^+	$(SN)_x$	S_3N^-
S_2N^+	S_2N_2	S_4N^-
$S_4N_3^+$	S_4N_2	$S_3N_3^-$
$S_4N_4^{2+}$	S_4N_4	$S_4N_5^-$
$S_4N_5^+$	S_5N_6	
$S_5N_5^+$		
$S_6N_4^{2+}$		
$S_6N_5^+$		

a Only those species which have been structurally characterized by X-ray crystallography are included
b Binary derivatives of cyclic sulfur imides e.g. $(S_7N)_2S_x$ (x = 2, 3 or 5) and the bicyclic compound $S_{11}N_2$, are also known, but the nitrogen atom is three coordinate in these compounds

2.2 The Tetrasulfur Pentanitride Anion, $S_4N_5^-$

2.2.1 Preparation and Molecular Structure

The $S_4N_5^-$ anion was discovered by Scherer and Wolmershäuser who reported the unexpected formation of the *tert*-butylammonium salt of this binary sulfur-nitrogen anion from the methanolysis of $(CH_3)_3CN{=}S{=}NSiMe_3$ [15]. The ammonium salt is formed from N,N′-bis-(trimethylsilyl)sulfur diimide in a similar way and it can readily be converted to tetraalkylammonium salts [16].

$$Me_3SiN{=}S{=}NSiMe_3 \xrightarrow{CH3OH} NH_4^+S_4N_5^- \xrightarrow{R4NOH} R_4N^+S_4N_5^-$$

An X-ray structural determination revealed that the structure of $S_4N_5^-$ (*2*) is closely related to that of S_4N_4 (*1*) [16], which can be considered as a slightly distorted tetrahedron of sulfur atoms (d(S—S) = 258(×2) and 269(×2) pm) with four of the edges bridged by nitrogen atoms. In $S_4N_5^-$, the fifth edge is bridged by the additional nitrogen leaving only one pair of sulfur atoms unbridged. Surprisingly, the tetrahedron of sulfur atoms in $S_4N_5^-$ is essentially regular with S—S distances in the narrow range 271–275 pm.

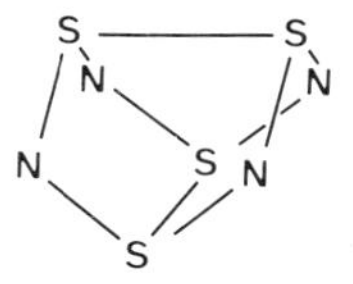

(1)

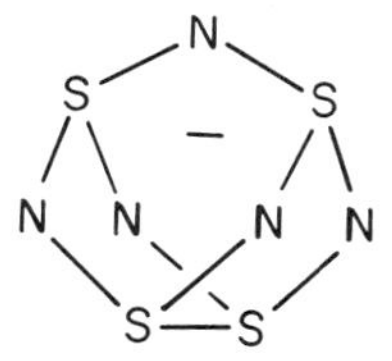

(2)

The structure proposed for $S_4N_5^-$ led Bojes and Chivers to investigate the preparation of this ion from S_4N_4 and alkali metal azides [19]. This approach was successful for azides of small cations (e.g. Li^+, Na^+, K^+), but the novel trisulfur trinitride anion, $S_3N_3^-$, was the predominant product when larger cations (e.g. Cs^+, R_4N^+, PPN^+) were used (see Sects. 2, 2.4, 2.4.1) [19]. Furthermore, when the investigation of the nucleophilic degradation of S_4N_4 was extended to include other anions, e.g. S^{2-}, CN^-, NH_2^-, it was found that $S_3N_3^-$ and/or $S_4N_5^-$ ions were invariably the products of these reactions [20]. The $S_3N_3^-$ ion is also the final product of the electrochemical reduction of S_4N_4 [21], which proceeds via the intermediate formation of the unstable radical anion, $S_4N_4^-$ [22]. The central role of these binary anions in the reductive or nucleophilic degradation of S_4N_4 is illustrated in Scheme 1.

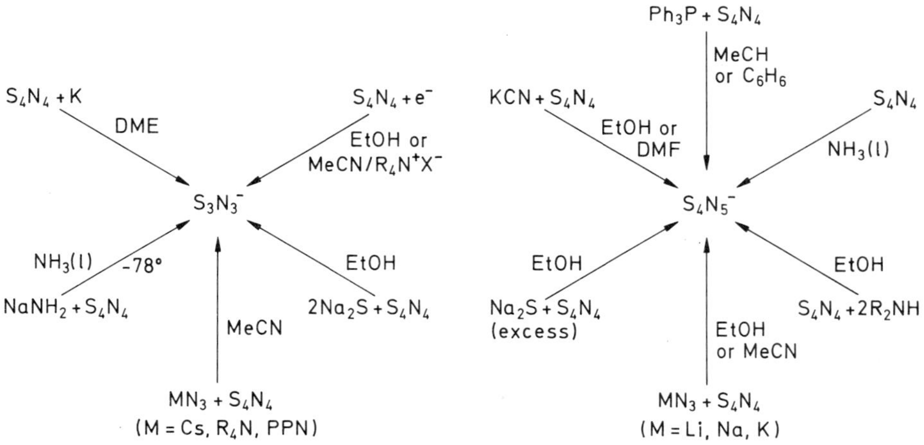

Scheme 1. Central Role of the $S_3N_3^-$ and $S_4N_5^-$ Ions in the Reductive or Nucleophilic Degradation of S_4N_4

To explain these results it has been proposed that nucleophilic attack at sulfur in S_4N_4 (or addition of an electron in the chemical or electrochemical reduction) causes ring opening to give a poly(sulfur-nitrogen) chain which cyclizes to give $S_3N_3^-$ [20]. In the presence of small cations or when an excess of S_4N_4 is present, $S_3N_3^-$ reacts with S_4N_4 to give $S_4N_5^-$.

It should be noted, however, that $S_4N_5^-$ is also formed in the reaction of ammonia with simple sulfur compounds, e.g. S_2Cl_2, SCl_2, or SCl_4 [23], and Scherer and Wolmershäuser have proposed the intermediate formation of the —N=S=NH group as a necessary step for the production of $S_4N_5^-$ in these reactions (and in the methanolysis of sulfur diimides).

2.2.2 Reactions with Halogens and Sulfur Oxychlorides: Preparation and Molecular Structure of S_4N_5Cl and S_5N_6

The reaction of $Na^+S_4N_5^-$ with chlorine in methylene chloride at −78 °C occurs rapidly to give S_4N_5Cl [17b], which, however, is more conveniently prepared by the reaction of $S_3N_3Cl_3$ with $Me_3SiNSNSiMe_3$ [10]. An X-ray crystal structure analysis of S_4N_5Cl showed that it was a polymeric, predominantly ionic structure in which bicyclic $S_4N_5^+$ cations (*6*) are symmetrically bridged by chloride ions [10]. It is interesting to compare the structure of the sulfur-nitrogen cage in this cation with that of the anion, $S_4N_5^-$, which contains an almost regular tetrahedron of sulfur atoms [d(S—S) = 271–275 pm] although only one of the S—S contacts represents a bonding interaction (see Sects. 3, 3.2). In the cation, five of the S—S separations are within the range 278–281 pm, but the sixth pair of sulfur atoms is 401 pm apart (i.e. non-bonding) consistent with the loss of 2 electrons in the oxidation of $S_4N_5^-$ to $S_4N_5^+$.

The oxidation of $S_4N_5^-$ with bromine (or iodine) occurs at 0 °C to give the new sulfur nitride, S_5N_6 [17a, b], which can be prepared by a variety of routes, the best being the reaction of $S_4N_4Cl_2$ with $Me_3SiNSNSiMe_3$ [24]. Pentasulfur hexanitride is an air-sensitive, explosive, yellow-orange solid which can be sublimed at 45 °C/10^{-2} Torr without significant decomposition. A X-ray crystal structure determination shows that it has a basket-like structure (*3*) in which an —N=S=N— unit (d(S—N) = 154 pm) bridges two sulfur atoms of an S_4N_4 cradle via S—N single bonds (170 pm) [17]. The introduction of this bridge widens one of the S—S transannualar separations in S_4N_4 to 394 pm while the other shortens to 243 pm. Thus the S_4N_4 unit in S_5N_6 can be viewed as two 5-membered rings sharing a S—S bond. In hot benzene, S_5N_6 decomposes to give S_4N_4, presumably via loss of the N_2S bridge.

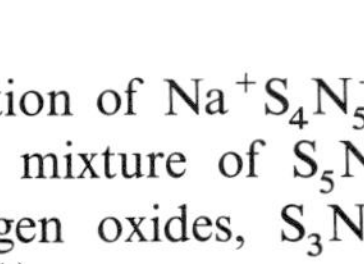

(*3*)

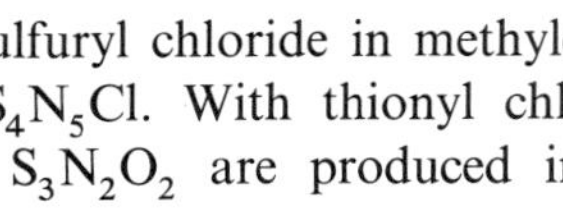

(*4*)

The reaction of $Na^+S_4N_5^-$ with sulfuryl chloride in methylene chloride at 0 °C produces a mixture of S_5N_6 and S_4N_5Cl. With thionyl chloride, however, the sulfur-nitrogen oxides, S_3N_2O and $S_3N_2O_2$ are produced in addition to S_4N_4 and S_5N_6 [17b].

2.2.3 Preparation and Molecular Structure of $S_4N_4X_2$ Derivatives

The nucleophilic degradation of S_4N_4 by phosphines, e.g. Ph_3P, in benzene at reflux was first studied by Krauss and Jung, who isolated $Ph_3P{=}N{-}S_3N_3$ in 11% yield [25].

A more recent and detailed investigation of this reaction has led to the identification of other products, the nature of which is dependent on solvent and temperature [26]. When the reaction is conducted in benzene at 23 °C, the yield of $Ph_3P{=}N{-}S_3N_3$ is 29% and a yellow salt $(Ph_3P{=}N)_3S^+S_4N_5^-$ (23%) was obtained. However, when Ph_3P reacts with S_4N_4 (2:1 molar ratio) in acetonitrile, the major product is 1,5-$(Ph_3P{=}N)_2S_4N_4$ (41%) in addition to $(Ph_3P{=}N)_3S^+S_4N_5^-$ (16%).

The structures of molecules of the type 1,5-$X_2S_4N_4$ (*4*, $X{=}Ph_3P{=}N$, [26], Cl [27], or Me_2N [28]) are similar to that observed for S_5N_6 with cross-ring S—S distances of 245 pm ($X{=}Ph_3P{=}N$ or Me_2N) and 248 pm (X = Cl), cf. 243 pm in S_5N_6. The exocyclic ligands are oriented in axial and equatorial directions and the different orientations are maintained in solution at low temperatures as indicated by ^{31}P ($X = Ph_3P{=}N$) and ^{15}N (X = Cl) NMR spectroscopy (see Fig. 1) [29]. By contrast, the radical addition of $(CF_3)_2NO$ to S_4N_4 gives 1,5-$S_4N_4[ON(CF_3)_2]_2$ in which the ligands are in a *cis*-configuration on the basis of ^{19}F NMR spectra [30].

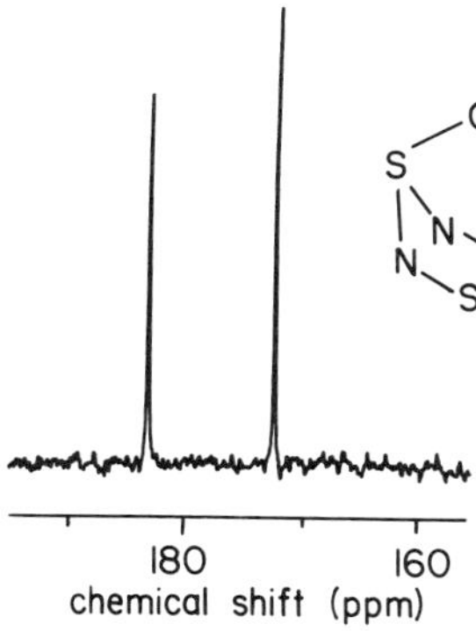

Fig. 1. Low temperature (—80 °C, CH_2Cl_2) ^{15}N NMR spectrum of 1,5-$Cl_2S_4^*N_4$ (*N = 99% ^{15}N)

2.2.4 Thermolysis

Salts of the $S_4N_5^-$ anion with small cations, e.g. alkali metals, exhibit a tendency to explode when subjected to mild pressure or heat [18–20]. In order to minimize the risk of explosions it has been necessary to use very large cations, e.g. Ph_4As^+ and $Ph_3P{=}N^+{=}PPh_3$ (PPN^+). With the latter counter-ion, both $S_4N_5^-$ and $S_3N_3^-$ can be handled safely and their thermal decompositions proceed smoothly. For example, when $PPN^+S_4N_5^-$ is heated to reflux in acetonitrile, a gradual color change from yellow through green to very deep blue takes place due to the following transformations [31].

$$PPN^+S_4N_5^- \xrightarrow{-N_2S} \underset{(\lambda_{max}\ 360\ nm)}{PPN^+S_3N_3^-} \xrightarrow{-N_2,\ +S^0} \underset{(\lambda_{max}\ 582\ nm)}{PPN^+S_4N^-}$$

The first step in this thermolysis apparently involves loss of an NSN unit (cf. $S_5N_6 \xrightarrow{-N_2S} S_4N_4$) to give $PPN^+S_3N_3^-$ which can be isolated if the reaction is stopped immediately after the initial formation of the blue color. Although there is no direct evidence for the appearance of the neutral molecule, N_2S [32], it is worth noting that *ab initio* Hartree-Fock-Slater SCF calculations for $S_4N_5^-$ show that introduction of an N^- bridge into the S_4N_4 framwork causes considerable weakening of the S—N framework bonds, thus providing a possible rationale for the facile loss of N_2S (see Sects. 3, 3.2) [10].

2.3 Preparation and Molecular Structure of the Tetrasulfur Oxide Anion, $S_4N_5O^-$

$NH_4^+S_4N_5O^-$ is obtained as yellow, water-soluble crystals on reaction of thionyl chloride with liquid ammonia (and subsequent hydrolysis of the reaction products) [33] or by air oxidation of a pyridine solution of the mixture obtained by dissolving S_4N_4 in liquid ammonia [34].

The structure of $S_4N_5O^-$ (*5*) is similar to that of $S_4N_5^-$ except that the exocyclic oxygen substituent introduces asymmetry into the SNS bridges of the cage. The unbridged S—S distance in $S_4N_5O^-$ is 263 pm, while the bridged S—S separations are in the range 266–274 pm [33b].

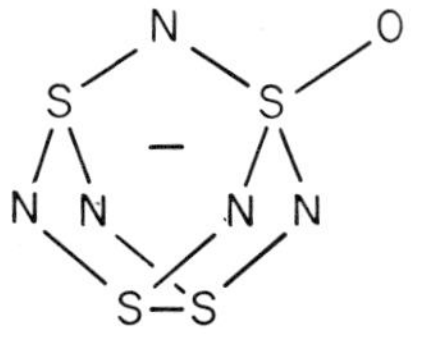

(5)

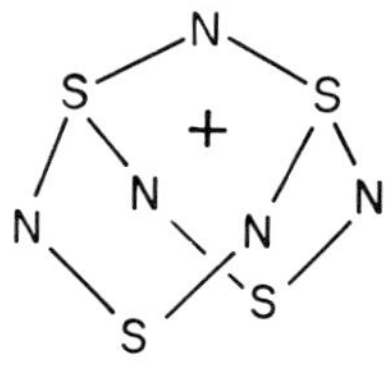

(6)

For molecules of the type S_4N_5X (where X is an exocyclic substituent) two isomers are possible. In one isomer the exocyclic substituent is attached to a sulfur atom which is bonded to three nitrogens (cf. $S_4N_5O^-$, *5*), whereas in the other isomer the sulfur atom bearing the substituent is connected to only two nitrogens (cf. S_4N_5Cl, *6*). Bartetzko and Gleiter have discussed intramolecular rearrangements in S_4N_5 cages [35], which could lead to interconversion of these isomers, but there is no experimental evidence in support of fluxional behaviour in sulfur-nitrogen cages. It is possible that the activation energies for such processes are greater than the energies required for breakdown of the cage e.g. by loss of N_2S.

2.4 The Trisulfur Trinitride Anion, $S_3N_3^-$, and Tetrasulfur Dinitride, S_4N_2

2.4.1 Preparation and Molecular Structure of $S_3N_3^-$ Salts

The preparation of the $S_3N_3^-$ ion by the reaction of S_4N_4 with azides of large cations, e.g. Cs^+, R_4N^+, PPN^+, has been mentioned in Section 2, 2.2, 2.2.1 and the central role of $S_3N_3^-$ in the nucleophilic and reductive degradation of S_4N_4 is summarized in Scheme 1 [19,20]. Surprisingly, $S_3N_3^-$ is also a major product of the deprotonation of $S_4N_4H_4$ with potassium hydride [20]. It seems likely that the deprotonation proceeds via intermedates of the type $S_4N_4H_{4-x}^{x-}$, but sulfur-nitrogen anions of this composition have not been conclusively identified [36]. The ion $S_2N_2H^-$ is, however, well known in complexes with metals such as nickel and cobalt (see Sects. 2, 2.10, 2.10.1).

X-ray structural determinations of the salts n-$Bu_4N^+S_3N_3^-$ [37] and $PPN^+S_3N_3^-$ [38] show that the anion is an essentially planar, 6-membered ring with bond angles at nitrogen of ca. 123° and at sulfur ca. 117° (7). The S—N bond distances fall

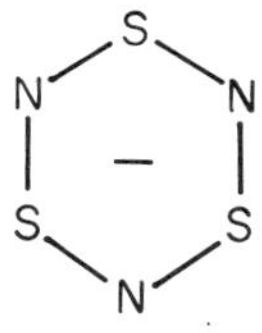

(7)

within the narrow range 158–163 pm. With one exception, the vibrational spectra of $S_3N_3^-$ salts of various cations are consistent with D_{3h} symmetry for the anion [37]. The exception is $Cs^+S_3N_3^-$ for which additional bands suggesting C_{3v} symmetry are observed in the infrared and Raman spectra. The existence of anion-cation interactions leading to distortions in the geometry of sulfur-nitrogen anions or cations is a common feature of the structural chemistry of sulfur-nitrogen ions e.g. S_4N^-, $S_4N_4^{2+}$, $S_5N_5^+$ (see Sects. 2, 2.7 and Ref. [4b, 11]). By contrast to the structure of $S_3N_3^-$, the isoelectronic neutral molecule S_3N_2O is thought to have a 5-membered ring structure with an exocyclic oxygen attached to one of the sulfur atoms of an S—S bond, but the crystal structure of this oily, red liquid has not been reported [39].

2.4.2 Molecular Structure of Tetrasulfur Dinitride, S_4N_2

On the basis of a variety of physical and spectroscopic measurements, Nelson and Heal concluded that S_4N_2 is a 6-membered ring with nitrogen atoms in the 1,3-positions [40]. However, because of its low melting point (23 °C) and thermal instability, the conformation of the ring has been difficult to ascertain. The discovery that S_4N_2 can be recrystallized from diethyl ether at —20 °C has enabled the low temperature (—100 °C) X-ray crystal structure to be determined [18]. As indicated

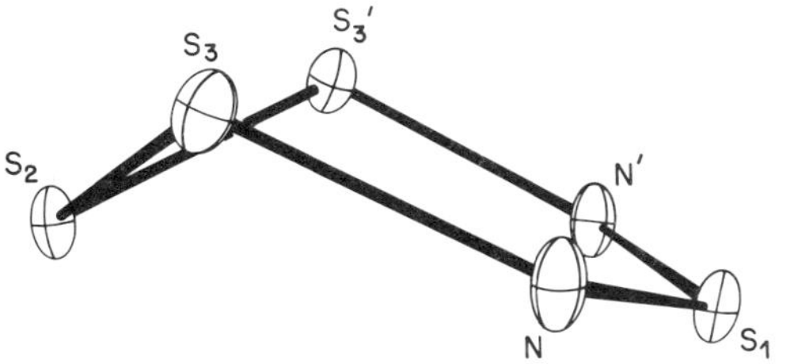

Fig. 2. ORTEP drawing of S_4N_2 (50% probability ellipsoids)

in Fig. 2, the 6-membered ring adopts a "half-chair" conformation in which the plane S(3)-S(2)-S(3)′ is tilted at an angle of 54.9° above the plane containing the 5 atom unit S(3)-N-S(1)-N′-S(3)′. The S—S bond lengths (206 pm) are typical of those found for homocyclic sulfur rings [41], but the S—N bond lengths deviate significantly from the values found for $S_3N_3^-$. The S—N bonds of the —N=S=N— unit are short (156 pm) compared with the S—N bonds connecting this group to the —SSS— unit (168 pm). The values approach those expected for double and single S—N bonds, respectively, in agreement with the π-bond orders calculated by Adkins and Turner [42]. These structural features are reminiscent of S_5N_6 (*3*) and suggest that the ready loss of the N_2S molecule may account for the thermal instability of S_4N_2.

S_4N_2 undergoes an irreversible one electron electrochemical reduction in S_4N_2 to give $S_4N_2^-$ (electrochemical evidence only) [43].

2.5 S_3N_3 Derivatives

2.5.1 Oxidation of $S_3N_3^-$ with Molecular Oxygen: Preparation of $S_3N_3O_x^-$ (x = 1, 2)

When a slow stream of dry O_2 is passed into a yellow solution of $PPN^+S_3N_3^-$ (λ_{max} 360 nm) the solution becomes red and crystals of $PPN^+S_3N_3O^-$ (λ_{max} 509 nm) can be isolated in 42% yield [44]. Further oxidation produces the purple $S_3N_3O_2^-$ ion (λ_{max} 562 nm), which has also been obtained by the reaction of $S_4N_4O_2$ with

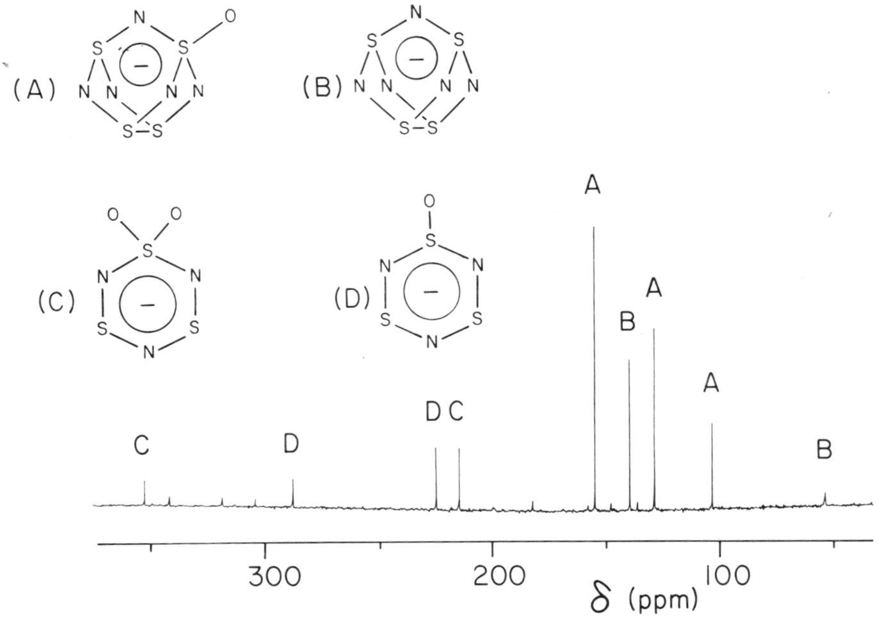

Fig. 3. ^{15}N NMR spectrum of the oxidation products of $PPN^+S_3^*N_3^-$ (*N = 30% ^{15}N, $CHCl_3$, 0.05 M $Cr(acac)_3$ added as PARR)

nucleophiles, e.g. NaN_3 or $Me_3Si—N{=}PPh_3$ [45]. The oxidation of ^{15}N-enriched $S_3N_3^-$ can be conveniently monitored by ^{15}N NMR spectroscopy which identifies the other products as $S_4N_5^-$ and $S_4N_5O^-$ (Fig. 3). The ^{15}N chemical shifts for $S_3N_3O^-$ are remarkably similar to those found for the structurally related molecules, $Ph_3E{=}N—S_3N_3$ (E=P, As; see Table 2 and following Sect.) [29].

Table 2. ^{15}N NMR Chemical Shifts for S_3N_3X Derivatives[a]

S_3N_3X	N_1 [b]	N_2, N_3 [b]
$S_3N_3O^-$	225.4	288.3
$S_3N_3{-}N{=}PPh_3$	203.5	282.5
$S_3N_3{-}N{=}AsPh_3$	209.6	282.6

a δ (in ppm) downfield of NH_3(l) at 25 °C
b N_1 is the unique nitrogen and N_2, N_3 are the equivalent nitrogen atoms in the S_3N_3 ring

2.5.2 Preparation and Molecular Structure of $Ph_3E{=}N{-}S_3N_3$ (E = P, As)

$Ph_3P{=}N{-}S_3N_3$ is preferably prepared by the reaction of S_4N_4 with $Ph_3P{=}{=}N{-}SiMe_3$ [46]. As indicated in Section 2, 2.2, 2.2.3, $Ph_3P{=}N{-}S_3N_3$ may also be prepared from S_4N_4 and Ph_3P in benzene [25,26,47]. The corresponding reaction with triphenylarsine proceeds much more slowly and very low yields of $Ph_3As{=}N{-}S_3N_3$ are obtained if the reaction is carried out in boiling benzene, as originally described [48], due to the thermal decomposition of the product (see Sects. 2, 2.5, 2.5.3) [47]. However, if Ph_3As and S_4N_4 are stirred together for 3 weeks in methylene chloride at 23 °C, an 81% yield of $Ph_3As{=}N{-}S_3N_3$ is obtained. The reaction of S_4N_4 with $Ph_3As{=}NH$ is more rapid and produces a 71% yield of $Ph_3As{=}N{-}S_3N_3$ after 24 h [47].

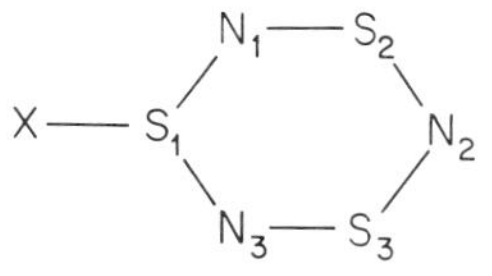

(8)

The structural parameters for S_3N_3X derivatives (*8*) (X = O^- [44], $Ph_3P{=}N$ [49], $Ph_3As{=}N$ [48]) are compared in Table 3. All three structures consist of a six-membered ring bearing an exocyclic substituent attached to one of the sulfur atoms which lies out of the plane containing the essentially planar 5 atom NSNSN unit. The dihedral angle is ca. 40° in all three derivatives. The trends observed for S—N bond lengths and bond angles are similar in all three rings. Thus the sulfur bearing the exocyclic

Table 3. Structural Parameters for S_3N_3X ($X{=}O^-$, $N{=}PPh_3$, $N{=}AsPh_3$)

X				X			
d(S—N) in pm	O^-	$N{=}PPh_3$	$N{=}AsPh_3$	$\hat{S}$ and $\hat{N}$ in deg.	O^-	$N{=}PPh_3$	$N{=}AsPh_3$
$S_1{-}N_1$	167	169	169	S_1	104	109	107
$N_1{-}S_2$	155	159	161	S_2	114	116	116
$S_2{-}N_2$	159	152	162	S_3	117	113	115
$N_2{-}S_3$	154	164	160	N_1	119	117	118
$S_3{-}N_3$	153	161	159	N_2	123	127	124
$N_3{-}S_1$	164	167	167	N_3	119	118	120

substituent is involved in the longest S—N bonds and has the smallest endocyclic bond angle. Similarly, the unique nitrogen has the largest bond angle (123–127°), while the bond angles at the two equivalent nitrogen atoms are close to 120° in all three rings. The average S—N bond distance (159 pm) in $S_3N_3O^-$ is shorter than that found for $Ph_3P{=}N{-}S_3N_3$ (d(S—N) = 162 pm) or $Ph_3As{=}N{-}S_3N_3$ (d(S—N) = 163 pm), probably due to the greater electronegativity of the exocyclic substituent in the oxyanion.

2.5.3 Thermolysis of $Ph_3E{=}N{-}S_3N_3$

During the investigation of the Ph_3P/S_4N_4 reaction a red compound, the yield of which increased with an increase in temperature, was observed in the chromatographic separation of products [26]. This red pigment is $Ph_3P{=}NSNSS$, a product of the thermolysis of $Ph_3P{=}N{-}S_3N_3$.

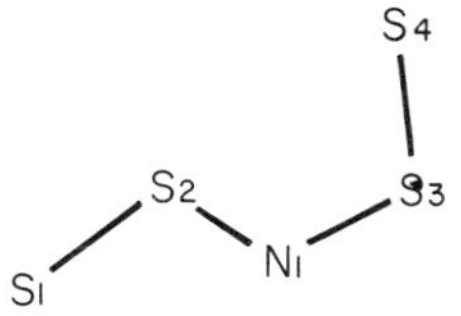

(*9*)

An X-ray crystal structure analysis reveals an open chain *cis*, *trans* structure in which the NSNSS unit is essentially planar (*9*, S_1 = Ph_3PN) [47]. The terminal S—S bond is short (191 pm) compared to an S—S single bond distance (ca. 206 pm) [41].

The thermal decomposition of $Ph_3As{=}N{-}S_3N_3$ proceeds in a significantly different manner from that of the phosphorus analog. In the solid state, thermolysis takes place smoothly at ca. 130°/10^{-3} Torr with the formation of disulfur dinitride, which can subsequently be polymerized to high quality $(SN)_x$ [47]. This route to S_2N_2 has a number of advantages over the synthesis involving thermolysis of S_4N_4 over a silver catalyst [50] viz. mild conditions are employed, no catalyst is required, and relatively pure S_2N_2 is obtained in good yield. In solution, the thermal degradation of $Ph_3As{=}N{-}S_3N_3$ is complete after 48 h in acetonitrile at reflux and S_4N_4 is formed, presumably via dimerization of S_2N_2. Thus the use of high temperatures in the synthesis of $Ph_3E{=}N{-}S_3N_3$ (E = P, As) should be avoided.

2.6 Cyclophosphadithiatriazenes, $R_2PS_2N_3$

2.6.1 Preparation, Molecular Structure and Formation of Adducts with Norbornadiene

The reaction of diphosphines, R_2PPR_2 (R = Me, Ph), with S_4N_4 in benzene at reflux produces intensely purple compounds of formula $R_2PS_2N_3$ (λ_{max} ~ 550 nm) [51]. The diphenyl derivative is an air and thermally stable crystalline solid whereas the dimethyl compound forms very volatile, low melting crystals which undergo slow decomposition at room temperature. An X-ray structure determination showed it (R = Pr) to consist of a 6-membered ring in which the phosphorus atom lies 28 pm out of the plane containing the NSNSN sequence (Fig. 4) [51]. This conformation is distinctly different from the highly puckered ring reported by Weiss for $(Me_3SiNH)_2PS_2N_3$ [52,53].

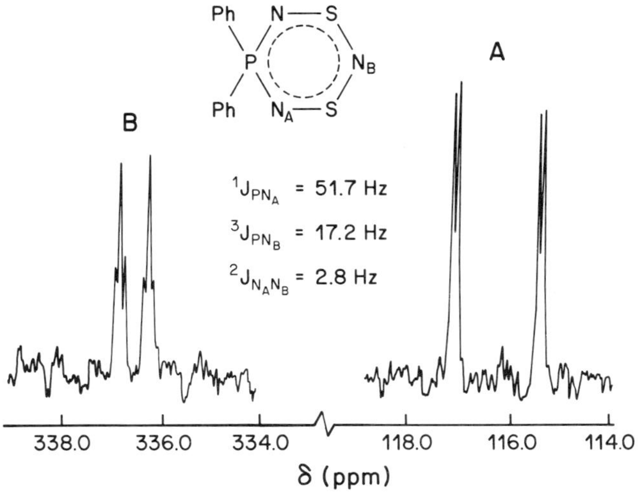

Fig. 4. ^{15}N NMR spectrum of $Ph_2PS_2^*N_3$ (*N = 99% ^{15}N, $CHCl_3$)

The phosphadithiatriazenes form white crystalline adducts with norbornadiene which facilitate the characterization of the less stable $R_2PS_2N_3$ compounds (R = Me, PhO) [51b]. An X-ray structural determination of $Ph_2PS_2N_3 \cdot C_7H_8$ has confirmed the prediction, based on NMR spectroscopic data, that the cyclo-addition occurs in a 1,3-S,S-fashion [51b].

The less stable cyclophosphadithiatriazenes have also been characterized by their ^{15}N NMR spectra which exhibit patterns very similar to the spectrum observed for $Ph_2PS_2N_3$ (Fig. 4) [51b].

2.6.2 Reaction of $Ph_2PS_2N_3$ with SO_2Cl_2: Preparation of $Ph_2PS_3N_5$

$Ph_2PS_2N_3$ is oxidized by SO_2Cl_2 to give the dichloro derivative (*10*) [54] which reacts with $Me_3SiNSNSiMe_3$ to form a bicyclic compound (*11*) of the type obtained

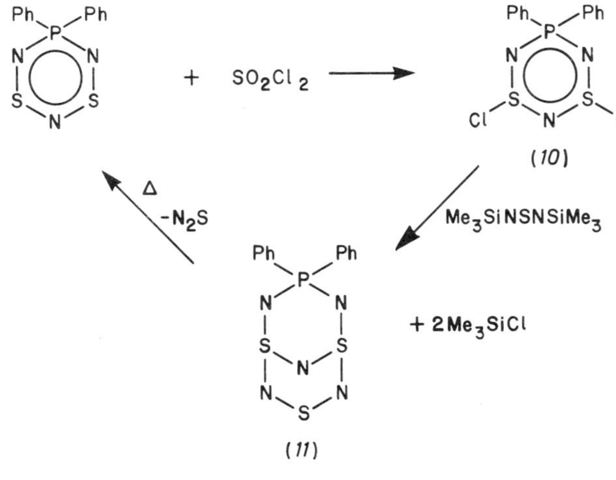

by Appel et al. from PF_5 (or $PhPF_4$) and $Me_3SiNSNSiMe_3$ [55,56]. Thermolysis of (*11*) regenerates the 6-membered ring, $Ph_2PS_2N_3$, via loss of N_2S as indicated above.

2.6.3 Preparation and Structure of $Me_2P(NSN)_2PMe_2$ and $RC(NSN)_2CR$ (R = Ph, Me_2N)

$Me_2PS_2N_3$ decomposes at ambient temperature to give $Me_2P(NSN)_2PMe_2$ which consists of an 8-membered ring with phosphorus atoms in the 1,5-positions and a cross-ring S—S contact of 255 pm (Fig. 5) [57]. The angle between the two intersecting planes of the ring is 114.9°. The variable temperature ^1H NMR spectrum of $Me_2P(NSN)_2PMe_2$ shows that the pairs of methyl groups remain non-equivalent up to at least 140 °C suggesting that the folded structure is maintained in solution.

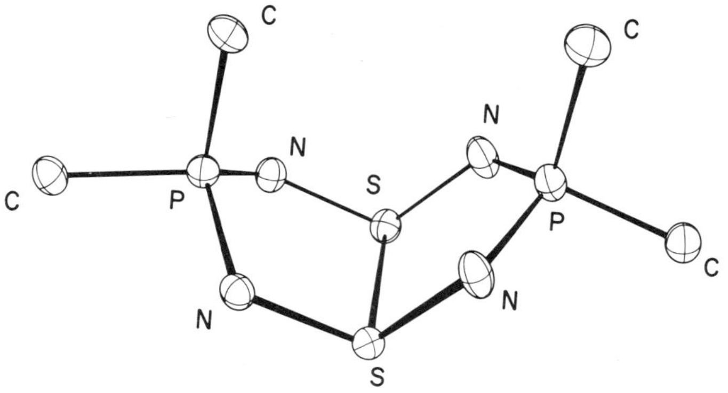

Fig. 5. ORTEP drawing of $Me_2P(NSN)_2PMe_2$

Isoelectronic carbon analogs of this 8-membered ring $RC(NSN)_2CR$ (R = Ph, Me_2N) have been reported recently and their structures are remarkably dependent on the nature of the exocyclic substituent [58]. Thus the phenyl derivative is completely planar whereas the Me_2N compound has a folded structure with a transannular S—S distance of 243 pm (see Sects. 3, 3.2, 3.2.2).

2.7 The S_4N^- Anion: Preparation and Molecular Structure

The deep blue chromophore (λ_{max} ca. 580 nm) formed by deprotonation of S_7NH was isolated as its tetra-n-butylammonium salt and identified as the S_4N^- anion in 1973 [59], but the structure of this unusual anion remained uncertain until an X-ray crystal structure determination of $PPN^+S_4N^-$, prepared by the thermal decomposition of $PPN^+S_3N_3^-$ in boiling acetonitrile, was reported in 1979 [31].

The structural data for $Ph_4As^+S_4N^-$ and $PPN^+S_4N^-$ are compared in Table 4 with those for the related molecule $Ph_3P{=}NSNSS$. In both salts the anion shows the same general structural features (*9*) viz. an essentially, planar *cis*, *trans* chain with nitrogen as the central atom and short, terminal S—S bonds (d(S—S) ~ 190 pm) [60].

The shortness of these bonds may be attributed, in part, to their polar nature as indicated by the strong intensity of the S—S stretching vibrations at ca. 590 and 565 cm^{-1} in the infrared spectrum of S_4N^- and supported by *ab initio* calculations of the atomic charge densities in the anion (see Sects. 3, 3.4)[31b]. In contrast to $Ph_3PNSNSS$, the S—N bonds of the central SNS unit in the two S_4N^- structures are unequal and the inequality is in opposite senses, presumably as a result of interactions with the cation.

Table 4. Bond Lengths (pm) and Bond Angles (deg) in S_4N^- Salts and $Ph_3PNSNSS$

	Ph_4As^+	PPN^+	$S_1 = Ph_3PN$
$S_1—S_2$	192	193	
$S_3—S_4$	193	191	191
$S_2—N_1$	162	157	159
$S_3—N_1$	156	162	159
$\hat{S}_2$	111.5	110.5	107.7
$\hat{N}_1$	120.1	120.5	120.9
$\hat{S}_3$	114.0	111.0	111.4

It has been known for many years that elemental sulfur dissolves in liquid ammonia to give colored solutions which are green or blue at room temperature and red at lower temperatures. The blue chromophores have recently been identified by Raman spectroscopy as S_4N^- and S_3^- ($\lambda_{max} \sim 610$ nm)[62]. The former ion is predominant except at low concentrations (ca. 10^{-2} M).

2.8 The S_3N^- Ion

The addition of more than one molar equivalent of triphenylphosphine to a solution of $PPN^+S_4N^-$ or $Ph_4As^+S_4N^-$ in acetonitrile causes an immediate color change from deep blue to orange due to the formation of the S_3N^- anion (λ_{max} 465 nm), which can be isolated in ca. 50% yield by addition of diethyl ether to the solution[63].

$$Ph_3P + S_4N^- \rightarrow Ph_3PS + S_3N^-$$

The S_3N^- ion is unstable with respect to the formation of S_4N^- in solution and in the solid state under the influence of heat or pressure. It appears that large counter-ions, e.g. Ph_4As^+, PPN^+, play an essential stabilizing role by keeping the S_3N^- ions well separated in the crystal lattice.

The vibrational spectra of $S_3^*N^-$ (*N = 30% ^{15}N) suggest an $SNSS^-$ or unbranched arrangement of atoms in the anion in contrast to the well established branched structure (D_{3h}) of the isoelectronic CS_3^{2-} ion. Thus only one band is observed in the S—S stretching region in both the infrared and Raman spectra (cf. S_4N^-) consistent with the cleavage of one S—S bond by triphenylphosphine[63].

Although accurate structural parameters for S_3N^- are not available due to problems with disorder in the X-ray structural determination, the preliminary data suggest a *cis* conformation for the ion [64].

2.9 Role of Sulfur-Nitrogen Anions in the Synthesis of Cyclic Sulfur Imides

The cyclic sulfur imides S_7NH and 1,3-, 1,4-, and 1,5-$S_6(NH)_2$ are obtained by acid hydrolysis of the deep blue intermediate formed from the reaction of S_2Cl_2 with ammonia in polar solvents, e.g. DMF [6], or by the reaction of sodium azide with elemental sulfur in HMPA [66]. The blue chromophore formed in both these reactions is S_4N^- and it has therefore been suggested that this anion is involved in equilibria with cyclic sulfur-nitrogen anions which are the precursors of the imides [66b].

$$S_7N^- \rightleftarrows S_4N^- + 3S^\circ \rightleftarrows 1/2S_6N_2^{2-} + 4S^\circ$$

The occurrence of such equilibria may also account for the fact that deprotonation of 1,3-$S_6(NH)_2$ with ethyl-lithium and reaction of the resulting anion with methyl iodide produces S_7NMe but no alkylated derivatives of the diimide [67].

More recently, it has been demonstrated that solutions of $PPN^+S_4N^-$ in liquid ammonia contain the S_3N^- ion which, on acid hydrolysis, produces low yields of the diimides [62]. Thus it seems likely that S_3N^- should be included in the equilibria shown above.

The reaction of S_7N^- (produced from S_7NH and n-butyl-lithium in THF) with chlorotrimethylsilane at —60 °C led to the exclusive formation of 1,4-$S_6N_2(SiMe_3)_2$. McGlinchey and co-workers have proposed the following mechanism involving the S_4N^- ion to explain this result [68].

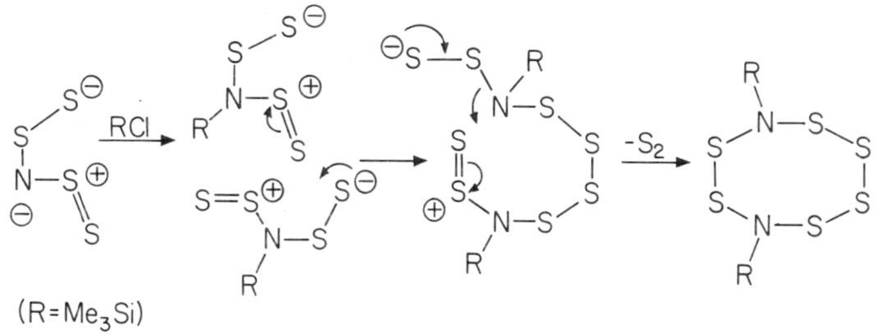

2.10 Metal Complexes of Sulfur-Nitrogen Anions

2.10.1 Complexes of the S_3N^- and $S_2N_2H^-$ Ligands

The formation of complexes of the type $M(S_3N)_2$ (M = Ni [69], Pd [70]) from the reaction of S_4N_4 with metal halides in methanol was reported more than 20 years ago, but the yields are very low despite recent improvements in the preparation and purification procedures [71]. In principle, the availability of salts of the S_3N^- ion should provide a more direct route to these complexes via metathetical reactions [63].

Indeed, $Ni(S_3N)_2$ was obtained in 30% yield [63] and several new complexes of copper(I) viz. $PPN^+[Cu(S_3N)_2]^-$, $PPN^+[ClCu(S_3N)]^-$, and $(Ph_3P)_2Cu(S_3N)$ are formed in good yields from the reaction of $PPN^+S_3N^-$ with the appropriate transition metal halides [72]. It should be noted, however, that complexes of the $S_2N_2H^-$ ligand, e.g. $M(S_3N)$ (S_2N_2H) and $M(S_2N_2H)_2$, (M = Co, Ni) which are the major products of the reactions of S_4N_4 with metal halides in methanol [70], are also formed in the reactions of S_3N^- with metal halides [73]. The complex $Co(NO)_2(S_3N)$ has been obtained by treating the initial product of the reaction of $Co_2(CO)_8$ and S_4N_4 with nitric oxide. The structure of the $Cu(S_3N)_2^-$ ion is shown in Fig. 6. Like $Pd(S_3N)_2$ [75], this complex contains *cis*-bidentate S_3N^- ligands coordinated through sulfur atoms to the metal.

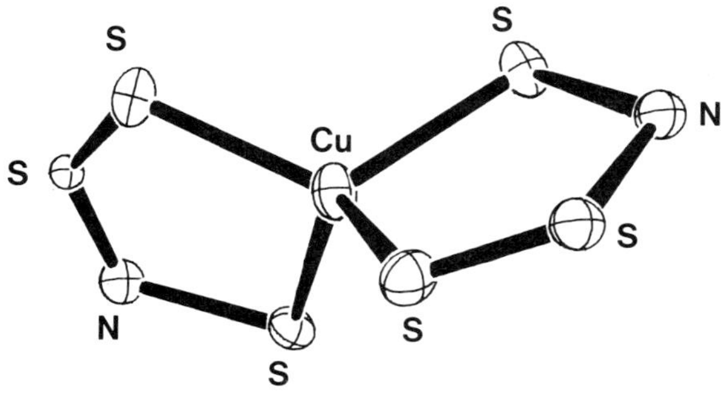

Fig. 6. ORTEP drawing of $Cu(S_3N)_2^-$

2.10.2 A Complex of the $S_4N_3^-$ Ligand

The preparation of the complex $ClPt(S_4N_3)$ from the reaction of S_4N_4 with *cis*-$Cl_2Pt(PhCN)_2$ has recently been reported [76]. An X-ray crystal structure analysis shows that the $S_4N_3^-$ ligand is bonded by two terminal sulfurs and the central nitrogen atom to platinum which is in a square planar configuration (*12*). As

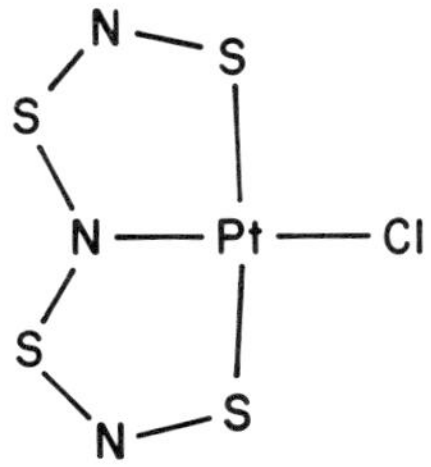

(*12*)

expected, the S—N distances involving the 3-coordinate nitrogen bonded to platinum are significantly longer (168 pm) than the other S—N distances (159 pm) in the ligand. Although the uncoordinated $S_4N_3^-$ ion has not been isolated, it is interesting to note that it may have an open chain structure similar to the intermediates proposed as precursors for $S_3N_3^-$ in the nucleophilic degradation of S_4N_4 (see Sects. 2, 2.2, 2.2.1).

3 Electronic Structures of Sulfur-Nitrogen Cages, Rings and Chains

3.1 Introduction

The rapid advances currently being made in the synthesis of molecules and ions containing conjugated sulfur-nitrogen units have led to a growing need for a simple and yet comprehensive theoretical framework within which the chemical and physical properties of these electron-rich molecules can be understood. The classical valence bond representations, which are still used extensively in the literature, provide a useful method of "book-keeping" electrons, but can give a misleading impression of ground state properties (e.g. charge densities, bond orders). Excited state properties (e.g. electronic spectra) are even less explicable in valence bond terminology. The empirical rules devised by Banister represent an improvement in the valence bond approach [77,78]. However, although this system allows a formal designation of the number of σ- and π-electrons in planar $S_xN_y^{z\pm}$ species, it fails to recognize the bonding or anti-bonding characteristics of the molecular orbitals that the π-symmetry electrons occupy; there is no differentiation between π, nπ and π*. Without such a distinction, the often invoked Hückel 4n + 2 rule is of little value in understanding thermodynamic stability and chemical reactivity.

In the last ten years, particularly since the discovery of superconductivity in $(SN)_x$ polymer [3], several molecules, especially S_2N_2 [42,79–89] and S_4N_4 [10b,80–82,90–95], have been the focus of many semi-empirical and *ab initio* molecular orbital calculations. More intuitive and flexible HMO descriptions have also been reported, and the agreement between the various levels of sophistication is encouraging. Two recent reviews, by Gleiter [96] and by Gimarc and Trinajstić [97], summarize much of the recent theoretical work on planar binary sulfur-nitrogen rings. It should be noted, however, that the latter two discussions, which rely heavily on the use of Hückel and extended Hückel methods, do not agree on an appropriate choice of Coulomb parameters for sulfur (α_S) and nitrogen (α_N). In the Gleiter approach, α_S is set more positive than α_N, in accord with the electronegativity scale of Pauling [98]. By contrast, Gimarc and Trinajstić assume α_S is less than α_N, in keeping with the orbital electronegativities prescribed by Streitweiser [99]. There are merits to both viewpoints, but the ordering and composition of the frontier-M. O.'s obtained from *ab initio* Hartree-Fock-Slater studies on several planar systems support the order $\alpha_S > \alpha_N$. Thus, for example, the B_{2u} HOMO of $S_4N_4^{2+}$ (*13*) is a sulfur-based lone pair, the corresponding nitrogen-centred orbital B_{1u} (*14*) lying some 1.02 e.v. lower [100]. In S_2N_2, the situation is less clearcut; the ordering of π-levels is very sensitive to the

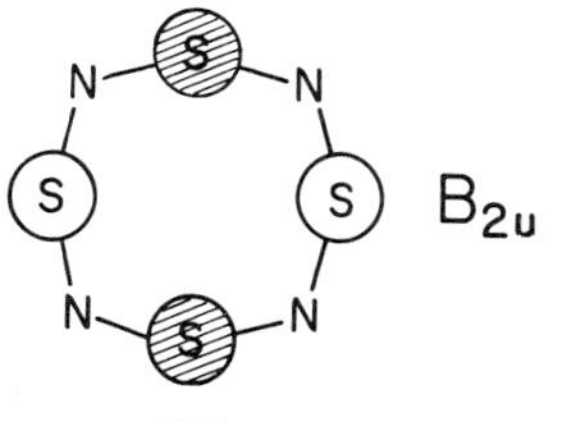

(*13*)

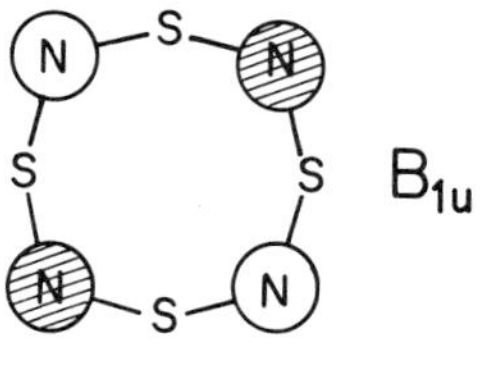

(*14*)

basis set employed. Indeed the P. E. spectrum of S_2N_2 demonstrates the near degeneracy of the B_{3g} and B_{2g} [(*15*) and (*16*)] orbitals [87,101]. Caution should therefore be exercised in the application of HMO theory to sulfur-nitrogen rings.

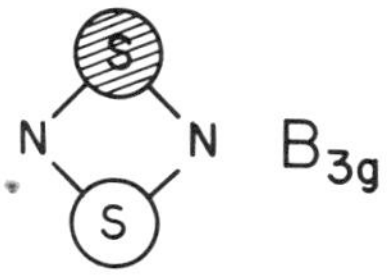

(*15*)

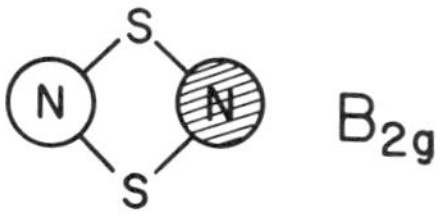

(*16*)

In the following paragraphs we present a generalized MO analysis of some of the compounds whose chemistry and structures were described in Section 2. Where possible, the results of simple HMO methods are compared with the findings of all electron calculations. The latter approach refines the former without contradicting its basic conclusions.

3.2 Cage Molecules

3.2.1 The $S_4N_4/S_4N_4^{2+}$ and the $S_4N_5^-/S_4N_5^+$ Systems

The unusual cage-like structure of S_4N_4 (*1*) has evoked continued debate over its electronic structure. The possible existence of bonding interactions between trans-annular sulfur atoms has received particular attention. Extended Hückel [90], CNDO

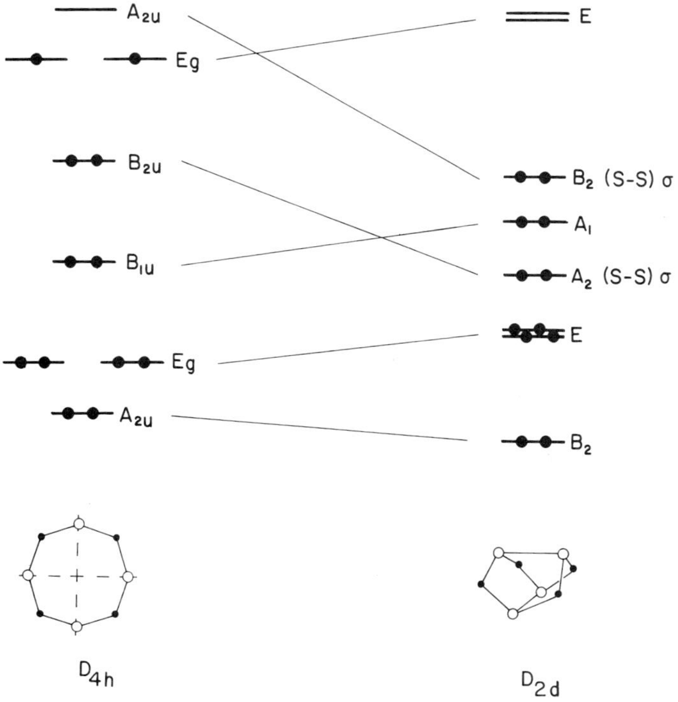

Fig. 7. A qualitative correlation diagram for the π MO's of a hypothetical planar (D_{4h}) S_4N_4 with the corresponding orbitals of the puckered (D_{2d}) form

[92,93)], $X_\alpha SW$ [82)], INDO-type ASMO-SCF [81)] and *ab initio* [10b,87)] calculations all illustrate the same fundamental features. There is a substantial bonding interaction between the transannular sulfurs, little or no N—N bonding, and polar S—N bonds. Electron density measurements are consistent with a concentration of charge between opposite sulfurs [102)]. In the E. H. analysis by Gleiter [90)], the M. O.'s of the D_{2d} conformation are correlated with the hypothetical π-levels of a planar S_4N_4 unit with 12 π-electrons. In D_{4h} symmetry S_4N_4 has an open shell configuration, and would be unstable with respect to Jahn-Teller distortion. In the D_{2d} structure the orbital degeneracy of the HOMO is lost, and 4 of the previously antibonding π* electrons are accommodated in S—S σ-bonds. Figure 7 illustrates an orbital correlation diagram for the π-MO's of a hypothetical planar S_4N_4 (D_{2d}) with the corresponding orbitals of the D_{2d} conformation. However, *ab initio* HFS calculations indicate that the HOMO of S_4N_4 is not the cross-ring σ—σ bond (the upper B_2 level in Fig. 7). It is, in fact, a nitrogen lone pair orbital. Thus it would appear that ionization of S_4N_4 to $S_4N_4^{2+}$ effects the removal of these latter electrons, the ensuing configuration being unstable with respect to the planar D_{4h} geometry, in which the B_2 level becomes virtual [10b,100)]. This concept is consistent with the observed structures of the Lewis acid adducts of S_4N_4; i.e. $S_4N_4 \cdot L$ (*17*, L = BF_3 [103)], SO_3 [104)], $SbCl_5$ [105)], $FeCl_3$ [106)], AsF_5 [107)]). In all these derivatives the coordination of a lone pair of electrons on nitrogen effects a structural change similar to that brought about by oxidation; the S_4N_4 cage is opened out with a complete loss of cross-ring sulfur-sulfur bonding.

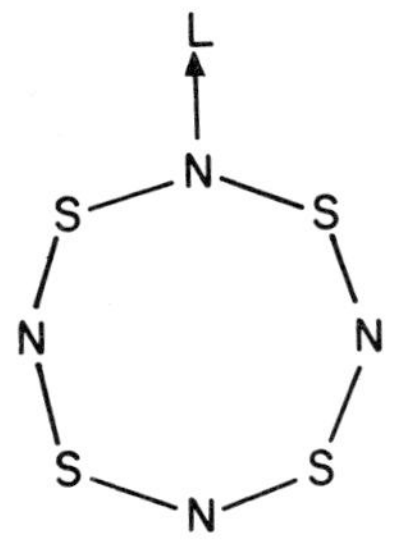

(*17*)

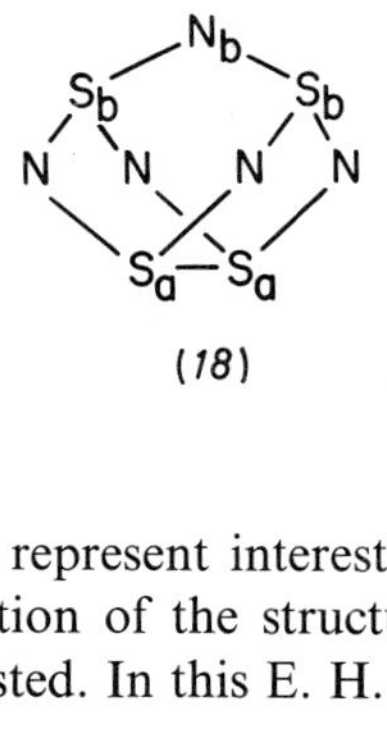

(*18*)

The cage ions $S_4N_5^-$ (*2*) and $S_4N_5^+$ (*6*) represent interesting modifications of the S_4N_4 molecule. A qualitative rationalization of the structural differences between these two species has recently been suggested. In this E. H. analysis, the framework

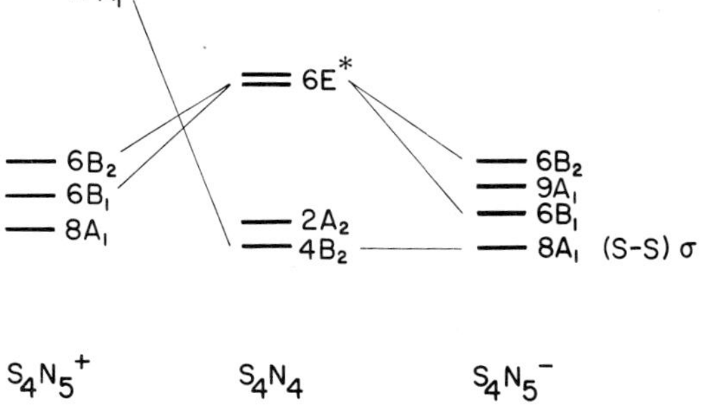

Fig. 8. A correlation diagram for the *ab initio* HFS frontier orbitals of $S_4N_5^+$, S_4N_4 and $S_4N_5^-$. The unoccupied levels are indicated by an asterisk

bonds are constructed from the interaction of an N_b^+ or N_b^- moiety (*18*) with two pseudo-allylic 5π-electron NS_aN fragments [35]. A more complete analysis of the problem, using the *ab initio* HFS method, reveals several important features. Figure 8 illustrates the frontier energy levels of S_4N_4 and how these change upon the formation of the $S_4N_5^{\pm}$ pair. The LUMO of S_4N_4 (6E) is lowered in both cation and anion (by interaction with the p_x and p_y orbitals on N_b), and the two orbitals so produced ($6B_1$, $6B_2$) are now filled. As a result of the occupancy of these two levels, which are antibonding with respect to the S_b—N bonds, the cage framework as a whole is weakened. The susceptibility of $S_4N_5^-$ to thermal decomposition via the loss of N_2S can thus be readily understood (see Sects. 2, 2.2, 2.2.4). The second point to emerge from the HFS calculations is that the $4B_2$ orbital of S_4N_4 (the cross-ring S—S σ-bond) remains largely unchanged in energy and composition in $S_4N_5^-$, where it now becomes the $8A_1$ level. As in S_4N_4 itself, this orbital is not the HOMO; it lies below $6B_1$ and $6B_2$ and also below a nitrogen lone pair on N_b ($9A_1$). In $S_4N_5^+$, a similar pattern of antibonding framework bonds ($6B_1$, $6B_2$) and an N_b lone pair ($8A_1$) is found near the Fermi surface. However, as a result of the opening of the $S_4N_5^+$ cage, the S_a—S_a bond is lost and the $4B_2$ orbital is now converted into the high lying virtual orbital $10A_1$. In conclusion, the $S_4N_5^-/S_4N_5^+$ ions form a redox pair similar to the $S_4N_4/S_4N_4^{2+}$ system. The net result of the oxidation is the rupture of the S_a—S_a bond. However, it would appear that the ionization process itself does not involve the electrons in this bond.

3.2.2 $RC(NSN)_2CR$ and $R_2P(NSN)_2PR_2$ Derivatives

Recently, a series of molecules of the type $E_2S_2N_4$ (E = CR, PR_2) have been prepared (see Sects. 2, 2.6, 2.6.3). Their molecular structures are of two types: a) those which are essentially planar 8-membered rings and b) those which exist as folded rings possessing a short cross-ring sulfur-sulfur contact. The reasons for

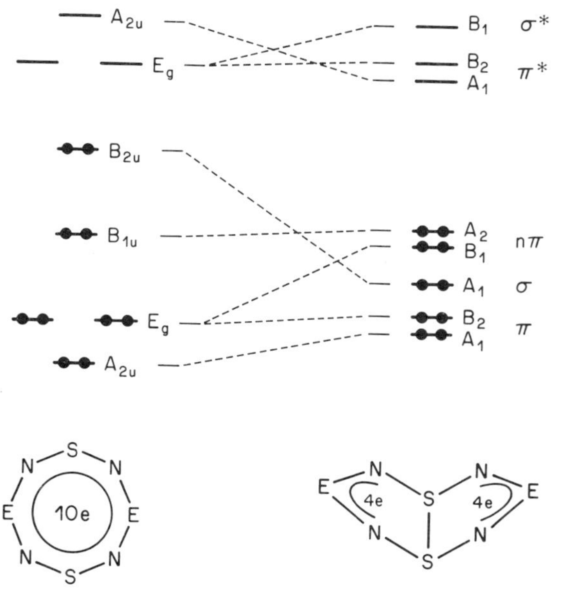

Fig. 9. A qualitative correlation diagram for the π MO's of a planar $E_2S_2N_4$ molecule (E here taken to be S^+) with the corresponding orbitals of the puckered (C_{2v}) form

this structural dichotomy can be understood by analysing the Hückel π-energies of the two modifications. Like $S_4N_4^{2+}$, molecules with a planar geometry (e.g. $PhC(NSN)_2CPh$ [58]) can be viewed as 10π-electron systems. In the folded geometry, two electrons are isolated in a sulfur-sulfur σ-bond (Fig. 9), while the 8 remaining π-electrons are distributed into 2 separated 3-center 4-electron —N=E=N— units. The π-energies of these models will of course depend on the relative magnitudes of the Coulomb parameters of sulfur (α_S), nitrogen (α_N) and E (α_E). Increasing the electronegativity of E stabilizes both systems, most especially the B_{2u} level of the planar conformation; the net effect is to favor the planar geometry. Thus at some point the π-energy of the planar structure will outweigh the combined σ-(S—S) and π-energies of the folded form, leading to an overall preference for this conformation.

These qualitative conclusions coincide well with the observed geometries of $E(NSN)_2E$ molecules. In $S_4N_4^{2+}$, where the electronegative perturbation is greatest, a planar geometry is found [9]. In molecules of the type $RC(NSN)_2CR$, the balance is sensitive to the nature of R. When R = Ph, the electronegativity of E is sufficient to favor the planar form. However, when a π-donor ligand, e.g. R = NMe_2, is present, the B_{2u} level of the planar model is destabilized, shifting the energy balance towards the folded form [58]. In phosphorus containing heterocycles (e.g. $Me_2P(NSN)_2PMe_2$, where an electropositive heteroatom is present, the observation of a folded conformation is expected [57].

3.3 Planar Six-Membered Ring Compounds

3.3.1 $S_3N_3^-$

Ab initio HFS calculations have shown that, consistent with the predictions of Banister [77,78], the $S_3N_3^-$ anion is a fully delocalized 10π-electron system [37,108]. The valence shell energy levels of the anion are reproduced in Fig. 10. Although the π-orbital pattern is reminiscent of that found for benzene, the occupancy of the

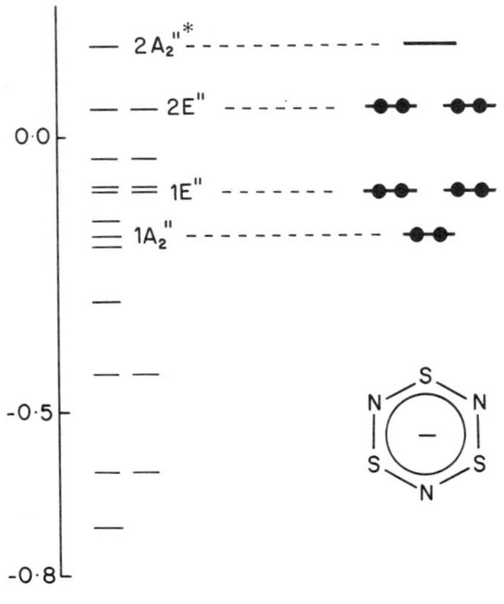

Fig. 10. *ab initio* HFS energy levels of the $S_3N_3^-$ ion. For purposes of clarity, only the π symmetry levels are labelled. The unoccupied level is indicated by an asterisk

π-orbitals is very different. The presence of four electrons in the 2E″ level in $S_3N_3^-$ dramatically weakens the S—N framework. In benzene the C—C Hückel π-bond order is 0.5, whereas in $S_3N_3^-$ the corresponding figure for the S—N bonds is 0.32. *Ab initio* overlap populations support this estimate [37].

The visible spectra of $S_3N_3^-$ salts display an intense absorption near 27,800 cm^{-1} (360 nm) (Fig. 11a), and on the basis of calculated transition moments and estimated energies this chromophore was assigned to a π*(2E″)→ π*(2A$_2''$) (HOMO-LUMO) excitation. This assignment has recently been confirmed by measurement of the magnetic circular dichroism (MCD) spectrum of the $S_3N_3^-$ anion [109]. The interpretation relies on the application of the Platt perimeter model [110], as developed by Michl [111]. Thus, the $S_3N_3^-$ anion can be viewed as a (4N + 2)-electron n-center (N = 2, n = 2N + 2) π-electron perimeter, which will have two π-holes. As such, the 2E″ → 2A$_2''$ transition, which is doubly degenerate, is predicted to generate a negative A-term in the MCD spectrum, as a result of the splitting of the excited state by the magnetic field (Scheme 2). As predicted, the observed spectrum does

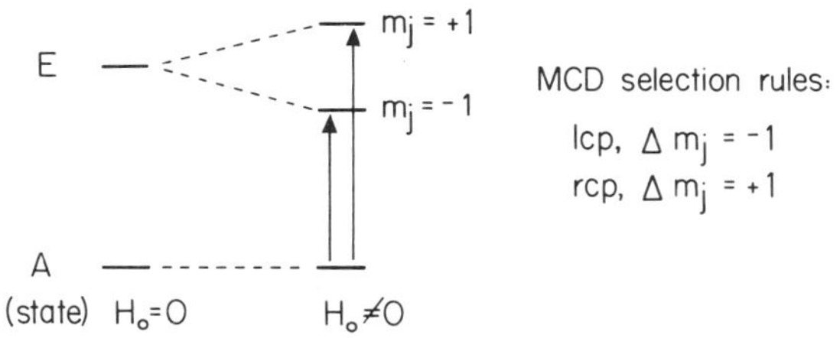

Scheme 2. Splitting of the Degenerate Excited State of $S_3N_3^-$ in the Presence of a Magnetic Field

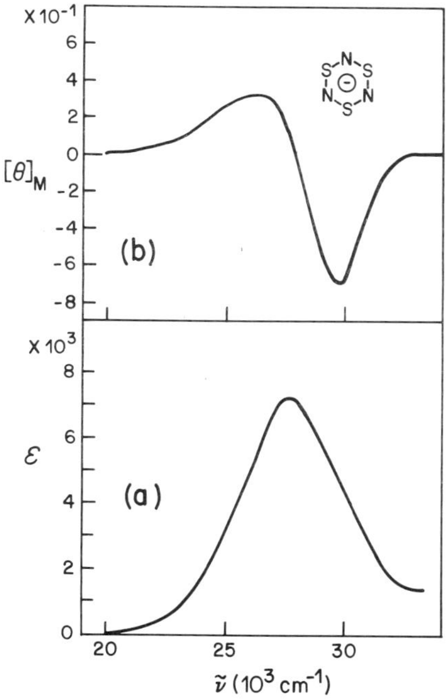

Fig. 11a and b. The visible absorption (**a**) and MCD (**b**) spectra of $S_3N_3^-$

indeed display the characteristic shape of a negative A-term (Fig. 11b), with the differential absorption $\varepsilon_{lcp} - \varepsilon_{rcp}$ being positive to low energy and negative to high energy of the normal electronic transition.

3.3.2 $S_3N_3O^-$ and $Ph_3E{=}N{-}S_3N_3$ (E = P, As) Derivatives

The molecular structures of all three of these species (*8*) (see Sects. 2, 2.5, 2.5.2) consist of a planar 5-atom N—S—N—S—N unit with the substituted sulfur displaced from this plane to produce a dihedral angle of *ca.* 40°. Because of the similarity of their molecular structures, the electronic structures of these molecules can justifiably be compared. Strictly speaking, the absence of a molecular plane of symmetry precludes a rigorous separation of σ- and π-electrons, but it is still useful for a qualitative discussion to relate the spectral properties of these molecules to a pseudo π-system generated from the HMO levels of the parent $S_3N_3^-$ ion. Substitution at one of the sulfur atoms in $S_3N_3^-$ will affect the ring π-system in two ways: a) the introduction of an electronegative ligand will increase the effective electronegativity of the remaining lone pair orbital on the substituted sulfur atom, and b) the π-system will be extended to include conjugation with the ligand (which is a strong π-donor). The effect of these two perturbations on the frontier MO's is illustrated in Fig. 12.

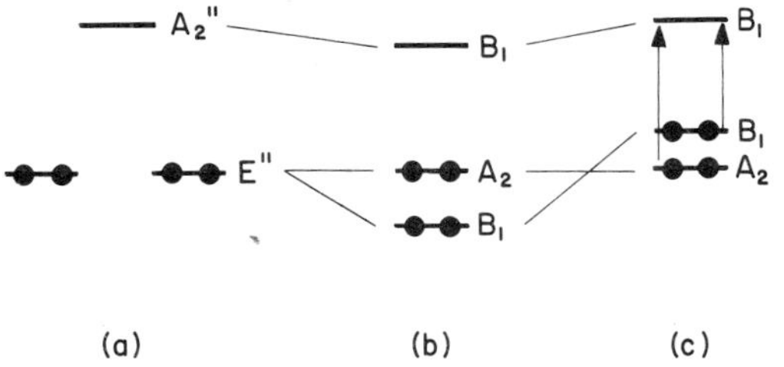

Fig. 12a—c. A schematic representation of the frontier π MO's of an S_3N_3X molecule, as derived from those of $S_3N_3^-$ (**a**). The diagram illustrates the effects of (**b**) electronegative perturbation at the tricoordinate sulfur atom and (**c**) conjugation with the π symmetry orbitals of the substituent

Regardless of which is the more important, the net result is a splitting of the formally degenerate HOMO's (2E'') of $S_3N_3^-$ (one of these is nodal at the substituted sulfur and remains unchanged) to produce a pair of high lying π* orbitals A_2 and B_1. The two visible absorptions observed for $X{-}S_3N_3$ molecules (Table 5) can then be assigned to excitations from these two orbitals to the LUMO B_1.

Table 5. Visible Spectra of $X{-}S_3N_3$ Derivatives

X	$\lambda_{max}(\varepsilon)$ nm ($M^{-1}\ cm^{-1}$)
O^-	509 (8×10^3), 340 (2×10^3)
$Ph_3P{=}N{-}$	478 (4×10^3), 330 (4×10^3)
$Ph_3As{=}N{-}$	488 (4×10^3), 336 (3×10^3)

The above interpretation of the $X{-}S_3N_3$ systems in terms of a 10π-electron perimeter has recently been supported by the analysis of the MCD spectra of $Ph_3E{=}N{-}S_3N_3$ (E = P, As), which show a negative B-term for the low energy transition and a positive B-term for the high energy absorption [112]. Linear dichroism studies suggest that the B_1 orbital is the HOMO, indicating that the effect of conjugative interactions outweighs the primary σ-inductive perturbation [112].

3.3.3 $S_3N_3O_2^-$ and $R_2PS_2N_3$ Derivatives

As in the case of X—S_3N_3 molecules, a discussion of the electronic structures of the $S_3N_3O_2^-$ ion [44b)] and the related $R_2PS_2N_3$ (R = Me, Ph, OPh [51)], NHSiMe$_3$ [52,53)]) derivatives is best approached by consideration of the changes in the π-orbital ordering and occupancy caused by perturbations on the $S_3N_3^-$ system. Although deviations from planarity are observed in the solid state, the ^{15}N NMR spectra of these molecules suggest that, in solution, planarity is achieved statistically. In our HMO analysis, ring planarity is assumed.

Removal of a sulfur atom from $S_3N_3^-$ effectively reduces the cyclic anion to an open chain N—S—N—S—N fragment containing 8π-electrons (Fig. 13b). Incorporation of a sulfone group (as in $S_3N_3O_2^-$) or a phosphonium cation (as in $R_2PS_2N_3$) reconstitutes the cyclic structure but, to a first approximation, the π-system can still be described in terms of a 5-atom 8π-electron unit. Such a description is analogous to the internal salt model used for λ_5-phosphorins [113)], except that in the present case there are eight π-electrons instead of six. Delocalization of π-charge onto the sulfone sulfur of $S_3N_3O_2^-$ or the phosphorus atom of $R_2PS_2N_3$ can then be considered as a perturbation of the ionic model. Such an interaction would probably require the participation of 3d-orbitals at the heteroatom position, but their inclusion would be unlikely to affect the ordering of levels.

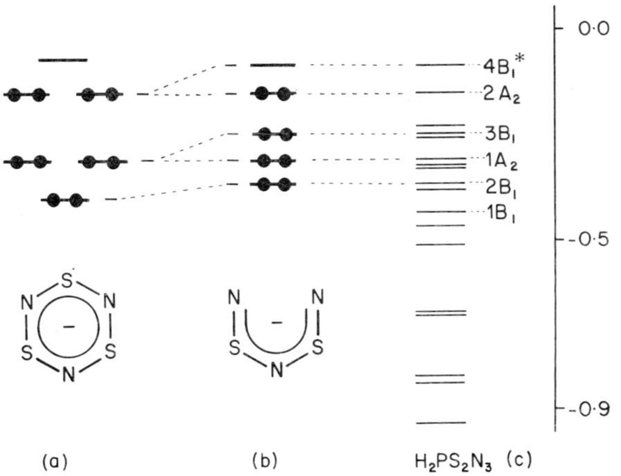

Fig. 13a—c. A correlation diagram for the Hückel π MO's of (**a**) $S_3N_3^-$ and (**b**) a hypothetical $S_2N_3^-$ ion. In (**c**), the *ab initio* HFS energy levels of a planar $H_2PS_2N_3$ molecule are shown. For purposes of clarity, only the π symmetry orbitals are labelled. Unoccupied orbitals are indicated by an asterisk

The above HMO analysis has been confirmed by *ab initio* HFS calculations on the hypothetical $H_2PS_2N_3$ molecule. Figure 13c illustrates the valence-shell energy levels for this model compound. There are in fact five occupied orbitals of π-symmetry, but one of these ($1B_1$) is an out-of-phase combination of P—H σ-bonds. The remaining orbitals correspond very closely in ordering and spacing to the HMO levels of Fig. 13a. The availability of phosphorus 3d-orbitals is an important feature in stabilizing the π-system, especially the π* levels. In fact, the Mulliken overlap populations for the P—N bond suggest a π-bond order approaching unity. Finally, the calculation of transition moments for all excitations >200 nm indicates that two transitions, the $3B_1 \rightarrow 4B_1$ and the $2A_2 \rightarrow 4B_1$ should be two orders of magnitude more intense than any other. On this basis, the *ca.* 550 nm band in the $R_2PS_2N_3$ derivatives and the $S_3N_3O_2^-$ ion (Table 6) can be assigned to a π* → π*

(HOMO-LUMO) transition. The higher energy absorption observed in $Me_2PS_2N_3$, $Ph_2PS_2N_3$ and $S_3N_3O_2^-$ (but obscured in the other derivatives) is assigned to a $3B_1 \rightarrow 4B_1$ (LUMO) excitation.

Table 6. UV-Visible Spectra of $S_3N_3O_2^-$ and $R_2PS_2N_3$ Derivatives

compound	$\lambda_{max}(\varepsilon)$ nm (M^{-1} cm^{-1})
$S_3N_3O_2^-$	562 (2×10^3), 362 (5×10^2)
$Me_2PS_2N_3$	543 (4×10^3), 295 (sh), 270 (5×10^2)
$Ph_2PS_2N_3$	550 (5×10^3), 301 (1×10^3)
$(PhO)_2PS_2N_3$	583[a, b]
$(HNSiMe_3)_2PS_2N_3$	570[c]

a Not measured
b Obscured by $\pi \rightarrow \pi^*$ transitions on phenyl rings
c From Ref. [53)]

Having established the validity of the Hückel analysis of these molecules, it is useful to extend the method to the study of the cycloaddition reaction of the $R_2PS_2N_3$ molecules with norbornadiene. Thus the addition of a 2e-olefin to the $R_2PS_2N_3$ system can be regarded as a novel example of a thermally allowed $8_s + 2_s$ reaction. As in the addition of olefins to S_4N_4 [114)], the preference for S,S-addition over N,N-addition can be understood in terms of the size and spatial distribution of the HOMO ($2A_2$, *19*) and LUMO ($4B_1$, *20*). Both orbitals are primarily sulfur-based, so that efficient overlap in the transition state is best achieved via addition at sulfur as opposed to nitrogen.

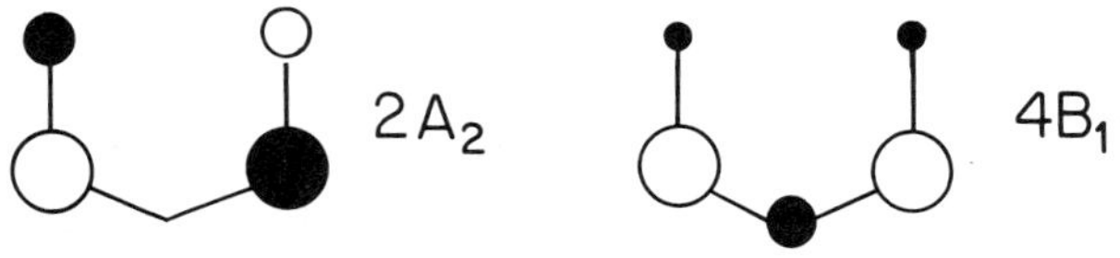

(*19*) (*20*)

3.4 Open Chain Anions: S_4N^- and S_3N^-

The two sulfur-rich anions S_4N^- and S_3N^- represent a novel class of binary sulfur-nitrogen compounds. Although they can be considered as anions of the hypothetical sulfur imides S_4NH and S_3NH, and indeed their intermediacy in the preparation of cyclic sulfur imides is well recognized (see Sects. 2, 2.9), they do not possess cyclic structures. The X-ray crystallographic analysis of S_4N^- and the vibrational analysis of S_3N^- indicate open chain structures for both ions (See Sects. 2, 2.7 and 2.8). *Ab initio* HFS calculations for various conformations of S_4N^- show that the statistical energy of the *cis,trans* conformation (*21*) is not substantially lower than that of the *trans,trans* form (*22*) [31 b)]. However, these calculations also show that the terminal sulfur atoms carry a significant negative charge while the internal sulfur atoms have a small but positive charge. Thus an electrostatic interaction between terminal and internal sulfur atoms, which is possible in (*21*) but not in (*22*), may account for the observed conformation [31b,61]. The S_4N^- ion is an 8π-electron system, and on the basis of calculated transition moments, the intensive visible absorption near 580 nm in S_4N^- salts has been assigned to a $\pi^* \rightarrow \pi^*$ (HOMO-LUMO) excitation (calculated value 610 nm).

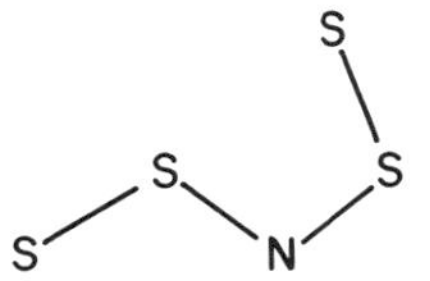

(*21*)

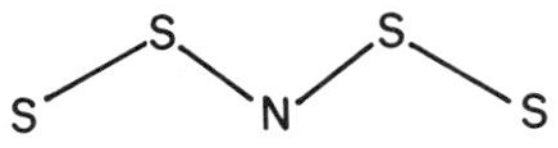

(*22*)

In the case of the S_3N^- ion, *ab initio* HFS calculations suggest that the *cis* conformation (*23*) is favored over the *trans* form (*24*) by approximately 20 kcal/mole. In both conformers there are 6π-electrons distributed over the chain, and the most

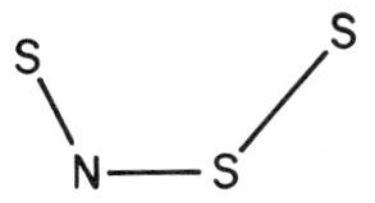

(*23*)

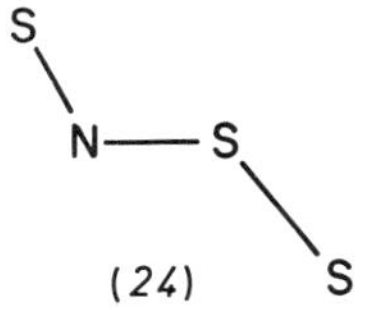

(*24*)

intense calculated electronic transition is again a $\pi^* \rightarrow \pi^*$ (HOMO-LUMO) excitation. The calculated energy for this transition in the two forms is 468 nm (*23*) and 566 nm (*24*). In view of the fact that S_3N^- salts show a strong visible absorption at 465 nm, it seems reasonable to assign the *cis* conformation to the S_3N^- ion [115] by [73].

4 Summary

The molecular orbital description of binary sulfur-nitrogen anions and related molecules provides a simple and flexible method for understanding many of their ground and excited state properties. Most molecules exhibit a strong tendency to adopt planar or near-planar geometries. The electron-rich character of the π-system of these structures sets them apart from organic π-systems, and accounts for the weakness of their skeletal framework and hence their thermodynamic instability. The intense low energy absorptions in the electronic spectra of all of these molecules also originate from their π-systems. *Ab initio* calculations and MCD spectroscopy confirm that the electronic transitions involved are $\pi^* \rightarrow \pi^*$ excitations.

In eight-atom perimeters such as S_4N_4, a planar structure would give rise to an open shell configuration for the π-system. Under these circumstances the molecule adopts a cage-like structure, the otherwise anti-bonding π^*-electrons being accommodated in cross-ring S—S σ-bonds. A similar structural phenomenon is observed for 10π-electron 8 atom rings $E(NSN)_2E$.

Although cyclic structures are unlikely for the sulfur-rich anions, S_4N^- and S_3N^-, because of ring strain, it should also be noted that the π and π^* levels would be fully occupied in planar rings. In fact, these anions adopt open chain structures in which each terminal sulfur atom accommodates two lone pairs of electrons and the anti-bonding levels of the delocalised π-system are only partially occupied.

5 Conclusions and Future Developments

The binary sulfur-nitrogen anions $S_4N_5^-$, $S_3N_3^-$, S_4N^- and S_3N^-, have been prepared and characterized. As representatives of S—N cages ($S_4N_5^-$), planar rings ($S_3N_3^-$), and open chain species (S_4N^- and S_3N^-), the elucidation of the molecular and electronic structures of these anions and related compounds has greatly enhanced our understanding of the behaviour of these electron-rich molecules. Investigations of the chemical reactions of these molecules have already led to the characterization of novel S—N molecules and ions, e.g. S_5N_6 and $S_3N_3O^-$, and have provided some insights into the formation of cyclic sulfur imides. It seems likely that a systematic study of the chemistry of S—N anions will provide further unexpected developments and the synthesis of transition metal derivatives appears to be an area worthy of particular attention.

The technique of X-ray crystallography has been, and will remain, indispensable for the determination of the unusual structures of S—N compounds. A more recent development is the application of ^{15}N NMR spectroscopy in S—N chemistry. Despite the necessity to employ ^{15}N-enriched materials for these studies, the judicious application of this technique in both structural determinations and in monitoring the progress of reactions will undoubtedly accelerate the progress of the subject. The advent of MCD spectroscopy and the use of the perimeter model have also enhanced our understanding of the electronic structures of cyclic S—N molecules. Rapid advances in this area are to be expected.

6 Acknowledgements

The authors wish to thank the Natural Sciences and Engineering Research Council (NSERC) of Canada for financial support. The collaborative efforts of Professors A. W. Cordes (X-ray crystallography), W. G. Laidlaw and M. Trcis (MO calculations), and J. Michl (MCD spectra) are also gratefully acknowledged.

7 References

1. Heal, H. G.: Adv. Inorg. Chem. Radiochem. *15*, 375 (1972)
2. Banister, A. G.: MTP Int. Rev. of Science, Inorg. Chem., Series 2, Vol. 3, p. 41 (1975)
3. Greene, R. L., Street, G. B., Suter, L. J.: Phys. Rev. Lett. *34*, 577 (1975)
4. Labes, M. M., Love, P., Nichols, L. F.: Chem. Rev. *79*, 1 (1979)
5. Roesky, H. W.: Adv. Inorg. Chem. Radiochem. *22*, 239 (1979)
6. Heal, H. G.: The Inorganic Heterocyclic Chemistry of Sulfur, Nitrogen, and Phosphorus, Academic Press, New York 1981
7. SN^+: Clegg, W. et al.: Acta Cryst. *B37*, 548 (1981)
8. S_2N^+: Faggiani, R. et al.: Inorg. Chem. *17*, 2975 (1975)
9. $S_4N_4^{2+}$ (a) Gillespie, R. J., Slim, D. R., Tyrer, J. D.: J. Chem. Soc., Chem. Commun. 253 (1977)
   (b) Gillespie, R. J. et al.: Inorg. Chem., *20*, 3799 (1981)
   (c) Sharma, R. D., Aubke, F., Paddock, N. L.: Can. J. Chem., *59*, 3157 (1981)
10. $S_4N_5^+$: (a) Chivers, T., Fielding, L.: J. Chem. Soc., Chem. Commun. 212 (1978)
    (b) Chivers, T. et al.: Inorg. Chem. *18*, 3379 (1979)

11. $S_5N_5^+$: (a) Roesky, H. W. et al.: J. Chem. Soc., Chem. Commun. 735 (1975)
(b) Banister, A. J. et al.: J. Chem. Soc., Dalton Trans. 928 (1976)
12. $S_6N_4^{2+}$: (a) Krebs, B. et al.: Chem. Ber. *113*, 226 (1980)
(b) Gillespie, R. J., Kent, J. P., Sawyer, J. F.: Inorg. Chem., *20*, 3784 (1981)
13. $S_6N_5^+$: Passmore, J.: private communication
14. (a) Moss, G., Guru Row, T. N., Coppens, P.: Inorg. Chem. *19*, 2396 (1980)
(b) Mayerle, J. J., Wolmershäuser, G., Street, G. B.: *18*, 1161 (1979)
15. Scherer, O. J., Wolmershäuser, G.: Angew. Chem. Int. Ed. Engl. *14*, 485 (1975)
16. Flues, W. et al.: ibid. *15*, 379 (1976)
17. (a) Chivers, T., Proctor, J.: J. Chem. Soc., Chem. Commun. 642 (1978),
(b) Chivers, T., Proctor, J.: Can. J. Chem. *57*, 1286 (1979),
(c) Roesky, H. W. et al.: Chem. Ber. *112*, 3531 (1979)
18. Chivers, T., Codding, P. W., Oakley, R. T.: J. Chem. Soc., Chem. Commun. 584 (1981)
19. (a) Bojes, J., Boorman, P. M., Chivers, T.: Inorg. Nucl. Chem. Lett. *12*, 551 (1976),
(b) Bojes, J., Chivers, T.: J. Chem. Soc., Chem. Commun. 453 (1977),
(c) Bojes, J., Chivers, T.: Inorg. Chem. *17*, 318 (1978)
20. Bojes, J. et al.: Inorg. Chem. *17*, 3668 (1978)
21. Hojo, M.: Bull. Chem. Soc., Jpn., *53*, 2852 (1980)
22. (a) Meinzer, R. A., Pratt, D. W., Myers, R. J.: J. Am. Chem. Soc. *91*, 6623 (1969),
(b) Williford, J. D. et al.: J. Electrochem. Soc. *120*, 1498 (1973)
23. Scherer, O. J., Wolmershäuser, G.: Chem. Ber. *110*, 3241 (1977)
24. Sheldrick, W. S., Rao, M. N. S., Roesky, H. W.: Inorg. Chem. *19*, 538 (1980)
25. Krauss, H. L., Jung, H.: Z. Naturforsch. *16B*, 624 (1961)
26. (a) Bojes, J. et al.: Can. J. Chem. *57*, 3171 (1979),
(b) Bojes, J. et al.: Inorg. Chem. *20*, 16 (1981)
27. Zak, Z.: Acta Cryst. *B37*, 23 (1981)
28. Roesky, H. W. et al.: Angew. Chem. Int. Ed. Engl. *20*, 592 (1981)
29. Chivers, T. et al.: Inorg. Chem. *20*, 914 (1981)
30. Mews, R.: J. Fluorine Chem. *18*, 155 (1981)
31. (a) Chivers, T., Oakley, R. T.: J. Chem. Soc., Chem. Commun., 752 (1979),
(b) Chivers, T. et al.: J. Amer. Chem. Soc. *102*, 5773 (1980)
32. Laidlaw, W. G., Trsic, M.: Inorg. Chem. *20*, 1792 (1981)
33. (a) Steudel, R.: Z. Naturforsch. *243*, 934 (1969),
(b) Steudel, R., Luger, P., Bradaczek, H.: Angew. Chem. Int. Ed. Engl. *12*, 316 (1973)
34. Becke-Goehring, M., Erhardt, K.: Naturwissenschaften *56*, 415 (1969)
35. Bartetzko, R., Gleiter, R.: Chem. Ber. *113*, 1138 (1980)
36. Banister, A. J., Younger, D.: J. Inorg. Nucl. Chem. *32*, 3766 (1970)
37. Bojes, J. et al.: J. Amer. Chem. Soc. *101*, 4517 (1979)
38. Cordes, A. W., Swepston, P. N.: private communication
39. Roesky, H. W., Wiezer, H.: Angew. Chem. Int. Ed. Engl. *14*, 258 (1975)
40. Nelson, J., Heal, H. G.: J. Chem. Soc., Dalton Trans. 136 (1971)
41. Steudel, R.: Angew. Chem. Int. Ed. Engl. *14*, 655 (1975)
42. Adkins, R. R., Turner, A. G.: J. Amer. Chem. Soc. *100*, 1383 (1978)
43. Tweh, J. W., Turner, A. G.: Inorg. Chim. Acta *47*, 121 (1981)
44. (a) Chivers, T. et al.: J. Chem. Soc. Chem. Commun., 1214 (1981)
(b) Chivers, T. et al.: to be published
45. Witt, M., Aramaki, M., Roesky, H. W.: paper presented at the 2nd Int. Symp. on Inorganic Ring Systems, Göttingen, W. Germany, Aug. 21–24, 1978
46. Ruppert, I., Bastian, V., Appel, R.: Chem. Ber. *107*, 3426 (1974)
47. (a) Chivers, T. et al.: Inorg. Chem. *20*, 2376 (1981),
(b) Chivers, T. et al.: J. Chem. Soc., Chem. Commun. 35 (1980)
48. Holt, E. M., Watson, K. J.: J. Chem. Soc., Dalton Trans. 514 (1977)
49. Holt, E. M., Holt, S. L., Watson, K. J.: ibid. 1357 (1974)
50. Mikulski, C. M. et al.: J. Amer. Chem. Soc. *97*, 6358 (1975)
51. (a) Burford, N. et al.: J. Chem. Soc., Chem. Commun. 1204 (1980),
(b) Burford, N. et al.: J. Amer. Chem. Soc., in press
52. Weiss, J.: Acta Cryst. *B33*, 2272 (1977)

53. Appel, R., Halstenberg, M.: Angew. Chem. Int. Ed. Engl. *15*, 696 (1976)
54. Burford, N., Chivers, T., Oakley, R. T.: unpublished data
55. Appel, R. et al.: Chem. Ber. *107*, 380 (1974)
56. Weiss, J., Ruppert, I., Appel, R.: Z. Anorg. Allg. Chem. *406*, 329 (1974)
57. Burford, N. et al.: Inorg. Chem., in press
58. Ernest, I. et al.: J. Amer. Chem. Soc. *103*, 1540 (1981)
59. (a) Chivers, T., Drummond, I.: J. Chem. Soc., Chem. Commun. 734 (1973), (b) Chivers, T., Drummond, I.: Inorg. Chem. *13*, 1222 (1974)
60. Burford, N. et al.: Inorg. Chem., *20*, 4430 (1981)
61. Gleiter, R., Bartetzko, R.: Z. Naturforsch. *36B*, 492 (1981)
62. Chivers, T., Lau, C.: Inorg. Chem., *21*, 453 (1982)
63. Bojes, J., Chivers, T.: J. Chem. Soc., Chem. Commun. 1023 (1980)
64. Cordes, A. W., Swepston, P. N.: private communication
65. Heal, H. G., Kane, J.: Inorg. Synth. *11*, 192 (1968)
66. (a) Bojes, J., Chivers, T.: Inorg. Nucl. Chem. Letters *10*, 735 (1974), (b) Bojes, J., Chivers, T.: J. Chem. Soc., Dalton Trans. 1715 (1975)
67. Tingle, E. M., Olsen, F. P.: Inorg. Chem. *8*, 1741 (1969)
68. Bruce, R. B., Stephan, D. W., McGlinchey, M. J.: Inorg. Chim. Acta *53*, L19 (1981)
69. Piper, T. S.: J. Amer. Chem. Soc. *80*, 30 (1958)
70. Weiss, J., Thewalt, U.: Z. Anorg. Allg. Chem. *346*, 234 (1966)
71. Woollins, J. D. et al.: J. Chem. Soc., Dalton Trans. 1910 (1980)
72. Bojes, J., Chivers, T., Codding, P. W.: J. Chem. Soc., Chem. Commun., 1171 (1981)
73. Bojes, J. et al.: J. Amer. Chem. Soc. in press
74. Herberhold, M., Haumaier, L., Schubert, U.: Inorg. Chim. Acta *49*, 21 (1981)
75. Weiss, J., Neubert, H. S.: Z. Naturforsch. *21B*, 286 (1966)
76. Endres, H., Galantai, E.: Ang. Chem. Int. Ed. Engl. *19*, 653 (1980)
77. Banister, A. J.: Nature Phys. Sci. *237*, 92 (1972)
78. Banister, A. J.: ibid. *239*, 69 (1972)
79. Kertesz, M. et al.: Chem. Phys. Lett. *44*, 53 (1976)
80. Salanek, W. R. et al.: Phys. Rev. *B13*, 4517 (1976)
81. Tanaka, K. et al.: J. Phys. Chem. *82*, 2121 (1978)
82. Salahub, D. R., Messmer, R. P.: J. Chem. Phys. *64*, 2039 (1976)
83. Deutsch, P. W., Curtiss, L. A.: Chem. Phys. Lett. *51*, 125 (1977)
84. Collins, M. P. S., Duke, B. J.: J. Chem. Soc. Chem. Commun. 701 (1976)
85. Jafri, J. A. et al.: J. Chem. Phys. *66*, 5167 (1977)
86. Fujimoto, H., Yokoyama, T.: Bull. Chem. Soc., Jpn., *53*, 800 (1980)
87. Findlay, R. H. et al.: Inorg. Chem. *19*, 1307 (1980)
88. (a) Millefiori, S., Millefiori, A.: Inorg. Chem. Acta *45*, L19 (1980), (b) Millefiori, S., Millefiori, A.: J. Chem. Research (S) 244 (1980)
89. Haddon, R. C. et al.: J. Am. Chem. Soc. *102*, 6687 (1980)
90. Gleiter, R.: J. Chem. Soc. A 3174 (1970)
91. Cassoux, P. et al.: J. Mol. Struct. *13*, 405 (1972)
92. Gopinathan, M. S., Whitehead, M. A.: Can. J. Chem. *53*, 1343 (1974)
93. Turner, A. G., Mortimer, F. S.: Inorg. Chem. *5*, 906 (1966)
94. Cartwright, H. M., Bossomaier, T. R. J., Grinter, R.: Theoret. Chim. Acta *44*, 265 (1978)
95. Braterman, F. S.: J. Chem. Soc. A 2297 (1965)
96. Gleiter, R.: Angew. Chem. Int. Engl. Ed. *20*, 444 (1981)
97. Gimarc, B. M., Trinajstić, N.: Pure Appl. Chem. *52*, 1443 (1980)
98. Pauling, L.: The Nature of the Chemical Bond, Cornell University Press, Ithaca Press 1960
99. Streitwieser, A.: Molecular Orbital Theory for Organic Chemists, John Wiley, New York 1961
100. Laidlaw, W. G., Trsic, M., Oakley, R. T.: unpublished results
101. Frost, D. C. et al.: J. Chem. Soc., Chem. Commun. 217 (1977)
102. DeLucia, M. L., Coppens, P.: Inorg. Chem. *17*, 2336 (1978)
103. Drew, M. G. B., Templeton, D. H., Zalkin, A.: ibid. *6*, 1906 (1967)
104. Gieren, A., Dederer, B.: Z. Anorg. Allg. Chem. *440*, 119 (1978)
105. Neubauer, D., Weiss, J.: ibid. *303*, 28 (1960)
106. Thewalt, U.: Z. Naturforsch. *835*, 855 (1980)

107. Gillespie, R. J., Kent, J. P., Sawyer, J. F.: Acta Crystallogr. *B36*, 655 (1980)
108. Chivers, T. et al.: Inorg. Chim. Acta *53*, L189 (1981)
109. Waluk, J. W., Michl, J.: Inorg. Chem. *20*, 963 (1981)
110. Platt, J. R.: J. Chem. Phys. *17*, 484 (1949)
111. Michl, J. R.: J. Am. Chem. Soc. *100*, 6801 (1978)
112. Waluk, J. W. et al.: Inorg. Chem., in press
113. Schafer, W. et al.: J. Am. Chem. Soc. *48*, 4410 (1978)
114. Yamabe, T. et al.: J. Phys. Chem. *83*, 767 (1979)

Homocyclic Sulfur Molecules[1]

Ralf Steudel

Institut für Anorganische und Analytische Chemie der Technischen Universität Berlin, Sekr. C2, D-1000 Berlin 12, FRG

Table of Contents

---

[1] Plenary lecture delivered at the 3rd IRIS Meeting held in Graz (Austria), August 17–22, 1981.

1 Introduction

Several types of compounds containing homocyclic structures of sulfur atoms are known (see Table 1). The neutral "naked" rings of type S_n are well known and form a long series of homologous species since n can assume all integer values from 6 on upwards. Eight of these rings have been prepared in a pure crystalline state. Oxidation of rings S_n formally yields either ions of types S_n^+ or S_n^{2+}, respectively, or molecular oxides of types S_nO and S_nO_2, respectively. In addition, at least two ionic homocyclic sulfur iodides exist, e.g. S_7I^+ and $S_{14}I_3^{3+}$. In the present review, due to the lack of space, only neutral species will be dealt with; for reviews of the ionic molecules see Ref. [1-5].

The chemistry of elemental sulfur has been reviewed before [6,7], but since that time considerable progress has been made. New and simple preparative methods for several allotropes have been developed, the molecular composition of liquid sulfur has been elucidated, X-ray structural analyses of several species have been published for the first time and new analytical techniques to detect and determine molecules S_n in mixtures have been worked out. In addition, several new reactions of sulfur rings have been discovered. These recent developments will be reviewed in the present paper.

The chemistry of the homocyclic sulfur oxides given in Table 1 has recently been covered elsewhere [8]. Therefore, these compounds will be mentioned only briefly as the oxidation products of the corresponding molecules S_n.

Table 1. Presently known compounds containing homocyclic sulfur rings

Elemental sulfur	Sulfur cations	Sulfur oxides	Sulfur halide cations
S_n	S_5^+	S_nO	S_7Br^+
n = 6–26[a]	S_4^{2+}, S_8^{2+}, S_{19}^{2+}	n = 5–10 S_7O_2, $S_{12}O_2$	S_7I^+, $S_{14}I_3^{3+}$

a S_5 which has been observed in sulfur vapor only so far may also be cyclic [9]

2 Preparation of Crystalline Homocyclic Sulfur Allotropes

The following homocyclic allotropes of elemental sulfur have been prepared in a pure, crystalline state:

$$S_6 \quad S_7(\alpha - \delta) \quad S_8(\alpha - \gamma) \quad S_9(\alpha, \beta) \quad S_{10} \quad S_6 \cdot S_{10}$$
$$S_{12} \quad \alpha\text{-}S_{18} \quad \beta\text{-}S_{18} \quad S_{20}$$

The greek letters given in brackets indicate the different crystal lattices observed for one and the same molecule. α and β-S_{18}, on the other hand, consist of different molecules since the conformations of the two rings are different. The molecular addition compound $S_6 \cdot S_{10}$ is a unique case among the many allotropes of the non-metallic elements in so far as it consists of two different molecules of the same element.

In addition to the above-mentioned species the solvate $S_{12} \cdot CS_2$ has been obtained as single crystals.

In the following, the best preparative methods of these compounds will be described, except β-S_{18} [10] whose preparation has not been published yet.

Preparation of S_6

S_6 was first obtained by acid decomposition of aqueous sodium thiosulfate [11,12] which according to recent results [13] yields a mixture of S_6, S_7 and S_8:

$$Na_2S_2O_3 + 2\,HCl\,(aq) \rightarrow \frac{1}{n}\,S_n + SO_2 + 2\,NaCl + H_2O$$

The S_6 can easily be obtained from the sulfur mixture by recrystallization from toluene, but the evolution of large quantities of SO_2 makes this preparation unpleasant if not dangerous.

A more convenient method uses the thermal decomposition of di-iododisulfane which can be generated in solution from commercial chemicals [14]:

$$S_2Cl_2 + 2KI \xrightarrow[-2\,KCl]{H_2O/CS_2} S_2I_2 \xrightarrow{20\,°C} \frac{1}{n}\,S_{2n} + I_2$$

The sulfur formed in this reaction mainly consists of S_6 and S_8 with small amounts of larger even-membered rings [15] of which S_{12}, α-S_{18} and S_{20} could be isolated [14]. The iodine can be recycled by reaction with stoichiometric amounts of sodium thiosulfate, and the sulfur mixture is separated by precipitation with pentane and recrystallization from CS_2; yield of S_6: 36% [14]. S_6 forms orange-yellow crystals which decompose at 20° within a few days [16].

The formation of S_6 from S_2I_2 is likely to proceed via higher iodosulfanes like S_4I_2 and S_6I_2 rather than via a polymerization of S_2 which would have to be formed from S_2I_2 in a first order reaction.

Preparation of S_7, S_{12}, α-S_{18} and S_{20}

These allotropes are best prepared from liquid sulfur (molten S_8) which after equilibration contains rings of all sizes (see below). The simple preparation of the title compounds is based on their high (S_7) or low (S_{12}, S_{18}, S_{20}) solubility in CS_2, respectively, compared with the main constituent of the melt, S_8 [16–20].

S_{12}: Commercial sulfur (S_8), whose purity is not of crucial importance, is heated electrically to about 200 °C and is then allowed to cool to 140–160 °C. As soon as the melt has become less viscous, it is poured, in as thin a stream as possible, into liquid nitrogen in order to quench the equilibrium. The liquid nitrogen is decanted off the yellow powder-like sulfur, which is then dissolved in CS_2 at 20 °C. The yellow solution after filtration is cooled to −78 °C, whereupon a mixture of much S_8 (large yellow crystals) and a little $S_{12} \cdot CS_2$ (small, almost colorless crystals) crystallizes out and can be separated by flotation in CS_2 yielding pure $S_{12} \cdot CS_2$ in 0.2% yield compared with the starting amount of S_8 [16,18]. On standing in air $S_{12} \cdot CS_2$ converts to S_{12}, single crystals of which can be obtained by recrystallization from benzene. S_{12} forms yellow, needle-like crystals which are metastable at 20 °C; m.p. 146–148 °C (dec.).

S_7: The above CS_2 solution from which S_{12} and most of the S_8 has been crystallized at —78 °C is used for the preparation of S_7, S_{18} and S_{20}. Stirring of this solution at —78 °C after addition of some glass powder (or seed crystals of S_7) for about 2 hours results in the precipitation of finely powdered sulfur which is isolated by removing the solution by means of an immersion filter frit. The residue is extracted three times with small amounts of toluene (leaving an organe residue) from which S_7 crystallizes on cooling to —78 °C. On recrystallization from CS_2 pure S_7 is obtained in 0.7% yield; m.p. 39 °C [16,17].

Depending on the crystallization conditions S_7 is obtained as either α, β, γ or δ allotrope all of which contain the same kind of molecules [21,22]. Solid S_7 decomposes completely at 20 °C within 10 days but can be stored at —78 °C for longer periods of time without decomposition. The first signs of the decomposition products S_8 and polymeric sulfur (S_μ) can be detected after 30 min at 20 °C [21]. In CS_2 solution S_7 is quite stable (see below).

S_{18} and S_{20}: The amorphous organe residue obtained during the S_7 preparation consists of a complex mixture of sulfur rings, S_x, with x possibly ranging up to 50 or more. The mean molecular weight shows that the average value of x amounts to 25 [16,18], and the rings upt to S_{26} have been detected chromatographically [15] (see below). S_x is stable in CS_2 solution only; on standing of the concentrated solution at 20° for 2 to 3 days small crystals of α-S_{18} (intense yellow orthorhombic plates) and S_{20} (light yellow rods) are precipitated. This mixture can be separated by flotation in a $CHCl_3/CHBr_3$ mixture since the density of α-S_{18}, is slightly higher than that of S_{20}. Yields: 0.02% α-S_{18}, 0.01% S_{20}. Both allotropes melt at 120–130 °C with decomposition [16].

Preparation of S_9, S_{10} and $S_6 \cdot S_{10}$

S_9 is the most difficult of the unstable sulfur allotropes to prepare. It decomposes in the solid state at 20 °C within several weeks to S_8 and polymeric sulfur.

The synthesis starts from dichlorotetrasulfane, S_4Cl_2, and the heterocyclic titanocene pentasulfide, $(C_5H_5)_2TiS_5$, which are prepared according to the following equations [23,24]:

$$H_2S_2 + 2\,SCl_2 \rightarrow S_4Cl_2 + 2\,HCl$$

$$2\,NH_3 + H_2S + \frac{1}{2}S_8 \rightarrow (NH_4)_2S_5$$

$$(C_5H_5)_2TiCl_2 + (NH_4)_2S_5 \rightarrow (C_5H_5)_2TiS_5 + 2\,NH_4Cl$$

S_9 is obtained in an elegant reaction in 30% yield [25] according to:

$$(C_5H_5)_2TiS_5 + S_4Cl_2 \rightarrow S_9 + (C_5H_5)_2TiCl_2$$

S_9 forms intense yellow needle-shaped crystals of no sharp melting point which depending on the conditions crystallize as either α or β-S_9 whose Raman spectra are similar but not identical [26,27]. Despite considerable effort [26] single crystals have never been obtained and, therefore, the molecular structure is unknown.

S_{10} can be prepared according to different methods. If larger amounts (several grams) are needed, Schmidt's synthesis [28] is the most convenient one:

$$2\,(C_5H_5)_2TiS_5 + 2\,SO_2Cl_2 \rightarrow S_{10} + 2\,(C_5H_5)_2TiCl_2 + 2\,SO_2$$

The reaction takes place in CS_2 at −78 °C (yield 35%). S_{10} forms intense yellow crystals which polymerize at about 60 °C [28]. If only small amounts of S_{10} are needed and S_6 and S_7 are available, the oxidation of either one with trifluoroperoxyacetic acid provides S_{10} in a reaction of unknown mechanism. The homocyclic oxides S_6O_2 and S_7O decompose at 5 °C in CS_2 or CH_2Cl_2 solution within several days to give S_{10}, insoluble sulfur and SO_2:

$$S_6 + 2\,CF_3CO_3H \longrightarrow S_6O_2$$
$$S_7 + CF_3CO_3H \longrightarrow S_7O$$
$$S_6O_2,\ S_7O \xrightarrow{5^\circ C} S_{10} + S_\mu + SO_2$$

Since the oxides do not have to be isolated, the sulfur solution after addition of the peroxyacid solution is simply kept in the refrigerator until S_{10} has formed which is then isolated by cooling and recrystallization [29].

When both S_6 and S_{10} are dissolved in CS_2 and the solution is cooled, then, under special concentration conditions, a new sulfur allotrope crystallizes out as orange-yellow opaque crystals of m.p. 92 °C. This compound has been shown by vibrational spectroscopy [29] and X-ray structural analysis [30] to consist of equal amounts of S_6 and S_{10} molecules in their usual conformations. In solution the mean molecular weight of 258 corresponding to 8 atoms per molecule indicates complete dissociation [29]. This is the first example of an allotrope of a chemical element consisting of molecules of different sizes.

3 Molecular and Crystal Structures of Homocyclic Sulfur Allotropes

13 homocyclic sulfur allotropes have been investigated by X-ray diffraction on single crystals, and the most recent data obtained are listed in Table 2. All neutral sulfur rings form puckered structures as can be seen from Fig. 1.

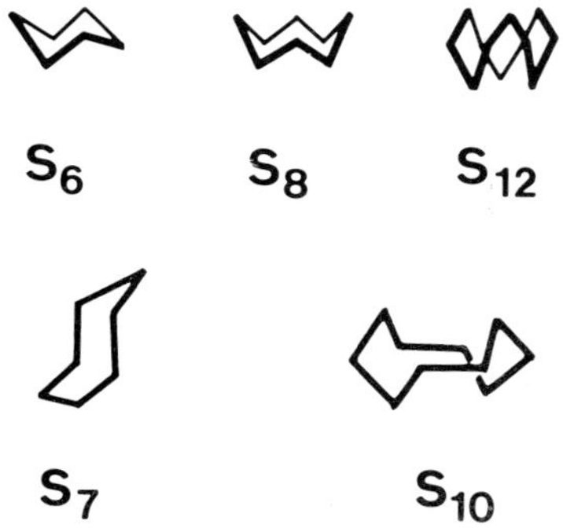

Fig. 1. Molecular structures of the homocyclic molecules S_6 (D_{3d}), S_8 (D_{4d}), S_{12} (D_{3d}), S_7 (C_s) and S_{10} (D_2). The molecular point groups of the free molecules are given in brackets

Table 2. Molecular parameters of solid homocyclic sulfur molecules as determined by X-ray crystallography

	Space group	Site symmetry	SS bond lengths (pm)	Bond angles (°)	Torsional angles (°)	Ref.
S_6	$R\bar{3}$	D_{3d}	206.8	102.6	73.8	[31]
γ-S_7	$P2_1/c$	C_1	199.8–217.5	101.9–107.4	0.4–108.8	[22]
δ-S_7	$P2_1/n$	C_1	199.5–218.2	101.5–107.5	0.3–108.0	[22]
α-S_8	Fddd	C_2	204.6–205.2	107.3–109.0	98.5	[32]
β-S_8	$P2_1/c$	C_1	204.7–205.7	105.8–108.3	96.4–101.3	[33]
γ-S_8	P2/c	C_2	202.3–206.0	106.8–108.5	97.9–100.1	[34]
S_{10}	C2/c	C_2	203.3–207.8	103.3–110.2	75.4–123.7	[35]
S_{12}	Pnnm	C_{2h}	204.8–205.7	105.4–107.4	86.0–89.4	[36]
α-S_{18}	$P2_12_12_1$	C_1	204.4–206.7	103.8–108.3	79.5–89.0	[37]
β-S_{18}		C_i	205.3–210.3	104.2–109.3	66.5–87.8	[10]
S_{20}	Pbcn	C_2	202.3–210.4	104.6–107.7	66.3–89.9	[37]
$S_6 \cdot S_{10}$	I2a	$S_6:C_i$	206.1–206.4	102.1–103.1	73.5–74.2	[30]
		$S_{10}:C_2$	203.9–208.0	103.7–111.4	73.1–123.2	
$S_{12} \cdot CS_2$	$R\bar{3}m$	$S_{12}:D_{3d}$	205.4	105.8; 106.7	87.2	[36]

The symmetry of these molecules can be quite high (S_8: D_{4d}) or very low (S_7: C_s); the site symmetry in the crystal lattices is, however, generally lower than the molecular symmetry of the free (gaseous) molecule.

Despite the fact that no substituents are present, the molecular parameters of the homonuclear sulfur bonds very tremendously with ring size and symmetry. While the rings of high symmetry (S_6, S_8, S_{12}) show "normal" bond distances (d = 205 ± 2 pm), bond angles (α = 105 ± 3°) and torsional angles (τ = 85 ± 15°) which are similar to those of the simplest compound containing a sulfur-sulfur single bond, H_2S_2 (d = 205.5 pm, τ = 90.6° [38]), the rings of lower symmetry (S_7, S_{10}, S_{18}, S_{20}) show considerable deviations of all bond properties. The structure of the S_7 ring is most exciting. The molecule does not exhibit the C_2 symmetry predicted by simple mechanical "force field" models [39] and expected by analogy to the structures of cycloheptane and cyclohexathiepane [40], but instead is of C_s symmetry both in the solid and solution states [21]. This means that four neighboring atoms are located in a plane and that one torsional angle is close to 0°. Since this is an energetically unfavorable situation for atoms of the VIth main group [38], the corresponding bond is the longest in the molecule (218 pm). Surprisingly, the two neighboring bonds (200 pm) are the shortest ones in the molecule and clearly of higher bond order than 1. This can be explained by a pair of coordinate π bonds within the plane of the four atoms:

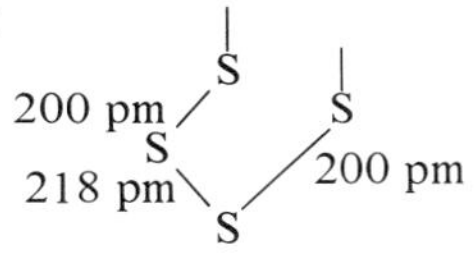

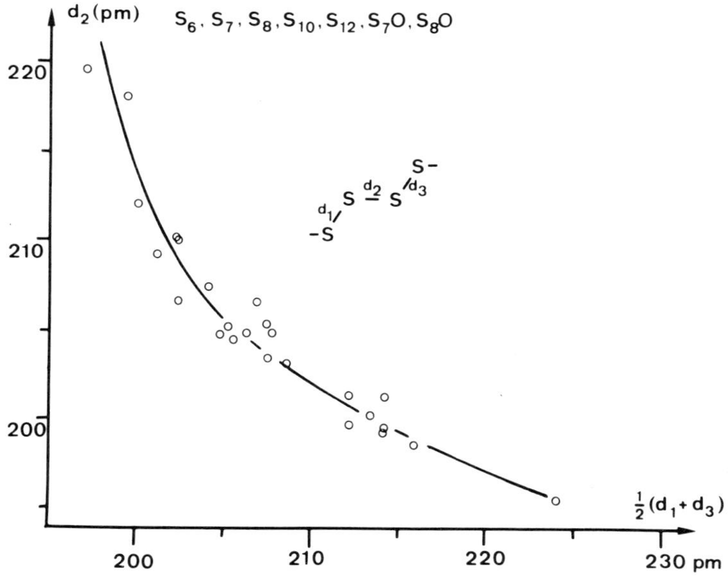

Fig. 2. Relationship between the bond lengths of neighboring bonds in sulfur rings indicating strong bond-bond interaction. The distance, d_2, of a bond between two-coordinated atoms is a function of the arithmetic mean of the lengths of the two neighboring bonds, $^1/_2(d_1 + d_3)$. The values were taken from the compounds indicated in the Figure. The curve ends on the right side at d_2 = 189 pm, the bond length of the diatomic molecule S_2

In this way the repulsive π electron density at the bond with $\tau = 0°$ is reduced and the C_s symmetry becomes more stable than the C_2 conformation [21].

It is general observation that the longest bonds in sulfur rings are those whose torsional angles show the largest deviation from the optimum value of 80–100° [38]. Furthermore, the shortest bonds are always neighboring to the longest ones as can be seen from Fig. 2 [41].

Since CS_2 is by far the best solvent for S_n rings, a special interaction between the two kinds of molecules has long been suspected. From the temperature dependence of the CS_2 solubility in liquid sulfur at 94–158 °C an interaction enthalpy of 27 kJ/mol has been calculated [42]. S_{12} is the only sulfur molecule which crystallizes from CS_2 as an adduct; its crystal structure shows, however, that there is no stronger than van-der-Waals interaction between the two components. The solvent molecules just occupy the empty spaces of the S_{12} layer lattice [36].

The conformation of sulfur rings can be described in terms of the so-called motif which is the order of the signs (+ or —) of the torsional angles around the ring. If 0 stands for $\tau = 0°$, then the motifs are:

S_6: + — + — + —

S_7: + — + 0 — + —

S_8: + — + — + — + —

S_{10}: — + — — + — + — — +

S_{12}: + + — — + + — — + + — —

α-S_{18}: + — — + + — + + + — + + — — + — — —

β-S_{18}: + + — + — — + + + — — + — + + — — —

S_{20}: + + + + — + + + + — + + + + — + + + + —

A regular helical chain molecule would have torsional angles of identical signs (left-handed chain: —; right-handed: +). The variation of the signs allows the formation of almost unstrained rings of all sizes above n = 5. The question of conformational changes of rings S_n is unsettled. According to CNDO/2 molecular orbital calculations the chair conformation of S_6 is by 16 kJ/mol more stable then the boat form with an interconversion barrier of 90 kJ/mol [43]. α-S_{18} does not convert to β-S_{18} on refluxing in CS_2 solution [18]. In the case of S_7, however, pseudorotation in the vapor phase has been postulated on the basis of entropy calculations [21].

The bond properties of SS bonds in various inorganic and organic compounds have been previously reviewed [38].

4 Molecular Spectra of Sulfur Rings

Molecular spectroscopy provides a means of detecting single compounds sometimes even in complex mixtures and of deducing information about the molecular structure from the spectra of pure compounds. In the case of homocyclic sulfur molecules, however, only a few spectroscopic techniques can be successfully applied.

UV spectra of S_6 [44], S_8 [44] and S_{12} [45] have been measured. The absorbance is very intense but rather uncharacteristic and therefore of little value. ^{33}S NMR spectra of sulfur rings have never been observed since even a saturated solution of S_8 in CS_2 (30% by weight S_8) shows the CS_2 signal only [46,47]. Mass spectra of S_6 [48,49], S_7 [50], S_8 [48,49], S_9 [25], S_{10} [50], and S_{12} [51] have been recorded and molecular ions have been observed, but extensive fragmentation and thermal decomposition of the larger rings on even gentle heating make accurate analyses complicated if not impossible when mixtures are to be investigated. Furthermore, the volatility of the different rings is quite different and rapidly decreases with increasing ring size. However, under special conditions ions up to S_{23}^+ have been observed [52].

Of all spectroscopic techniques vibrational spectroscopy is the only suitable method for detecting and determining single S_n rings in mixtures with other molecules of this homologous series. The number of vibrational degrees of freedom depends of the number of atoms in the molecule, and the high symmetry of some sulfur rings leads to a degeneracy of certain vibrations. As a consequence of these two facts, the infrared and Raman spectra of S_6 [29,53,54], S_7 [21], S_8 [55], S_9 [26,27], S_{10} [26,29], S_{12} [56,57], α-S_{18} [16], and S_{20} [16] are quite different and usually allow the detection of single members in mixtures with others (see Fig. 3).

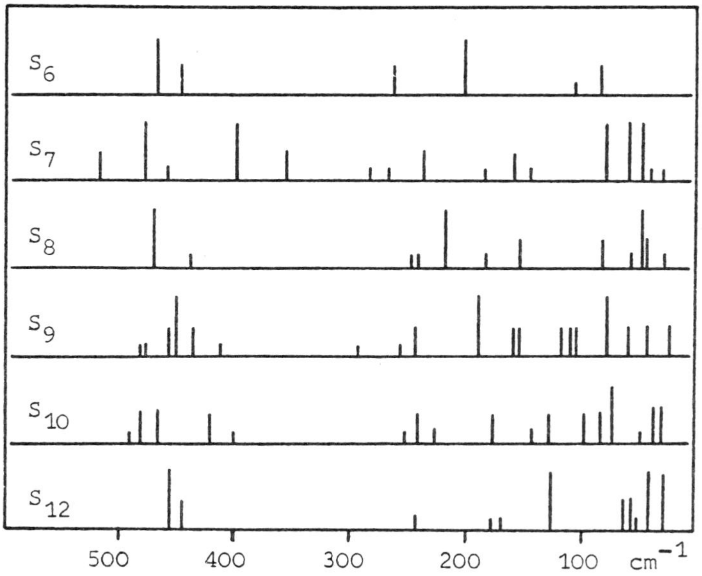

Fig. 3. Schematic representation of the Raman spectra of the cyclic molecules S_6, S_7, S_8, S_9, S_{10} and S_{12}. For each compound characteristic strong lines can be found which allow its detection in mixtures

Generally, i.r. absorptions of cyclic molecules S_n are weak since these species are of low or no polarity (rings belonging to the molecular point groups C_1, C_s, C_n or C_{nv} with $n \geqq 2$ may have a small dipole moment). On the other hand, Raman scattering from crystalline sulfur allotropes is very intense and leads to excellent spectra with low noise levels at good resolution conditions. However, since most compounds of this type are thermally unstable and all are light-sensitive, a sample temperature of −100 °C or below and the use of a red laser line is recommended to avoid decomposition.

Vibrational spectra of solid samples are also influenced by the packing of the molecules in the crystal lattice. For instance, the spectra of orthorhombic α-S_8 and monoclinic β-S_8 are somewhat different [55]. Thus it was found by Raman spectroscopy that S_7 crystallizes as four [21] and S_9 as two [26,27] allotropes which consist of identical molecules but must have different crystal structures.

Solution Raman and i.r. spectra have been used to determine the concentrations and concentration ratios, respectively, of S_6, S_7 and S_8 in CS_2 [20].

The vibrational spectra of S_6 [54,58], S_7 [21], S_8 [57,59] and S_{12} [57] have been thoroughly analyzed by assigning the observed wavenumbers to the normal vibrations of the molecules and by calculation of force constants. From this work it follows that SS stretching vibrations of S_n molecules are found in the region 520–350 cm^{-1}, while bending, torsional and lattice modes occur below 320 cm^{-1}. The wavenumber of the SS stretching vibrations (averaged for all equivalent bonds within the molecule) depends on the bond length [38,60] as is shown in Fig. 4. Furthermore, the stretching force constant f_r linearly depends on the SS bond length [38,60]; see Fig. 5. These relationships are useful for the assignment of vibrational spectra of new sulfur rings, e.g. as in homocyclic sulfur oxides.

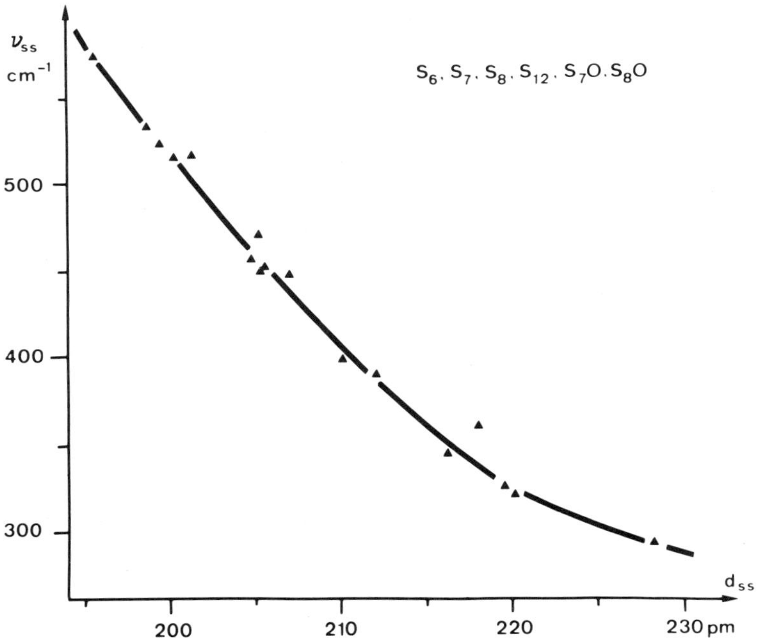

Fig. 4. Relationship between the sulfur-sulfur bond stretching vibration ν_{SS} (in cm^{-1}) of a certain bond in homocyclic sulfur rings and the corresponding bond length d_{SS} (in pm). For equivalent bonds within a molecule (equal d_{SS} values) the ν_{SS} values were averaged to eliminate the splitting by vibrational coupling. The data were taken from the compounds indicated in the Figure

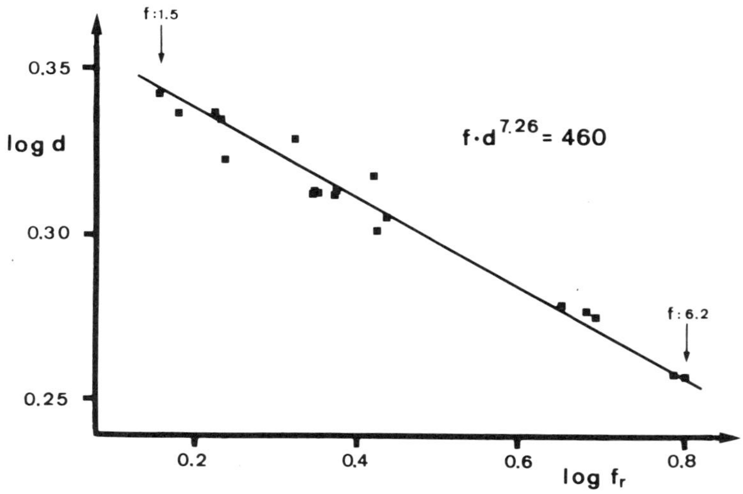

Fig. 5. Relationship between the logarithm of the bond stretching force constant, f_r, of a sulfur-sulfur bond and the logarithm of its length, d_{SS}. The data were taken from S_2 (7 electronic states), S_2^{2-} (in BaS_2), S_6, S_8, S_{12}, S_7 and S_8O

The totally symmetric ring bending vibrations of S_n molecules is a function of the ring size [61] as the following examples show:

$S_6: 266 \quad S_7: 238 \quad S_8: 219 \quad S_9: 188 \quad S_{10}: 178 \quad S_{12}: 127\,cm^{-1}$

Since this vibration gives rise to a very strong Raman line it can easily be identified in the spectra.

The normal vibrations and structural parameters of S_6 [54], S_7 [21], S_8 [57], and S_{12} [57] have been used to calculate several thermodynamic functions of these molecules in the gaseous state. Both the entropy (S°) and the heat capacity (C_p°) are linear functions of the number of atoms in the ring; in this way the corresponding values for S_5, S_9, S_{10} and S_{11} can be estimated by inter- and extrapolation [57]. For a recent review of the thermodynamic properties of elemental sulfur see Ref. [62].

5 Chromatographic Separation and Determination of Sulfur Rings

The unstable rings S_n ($n \neq 8$) spontaneously convert to S_8 and polymeric sulfur at 20 °C or slightly elevated temperatures. The mechanism of this reaction is unknown. A complete and relatively rapid conversion to S_8 can be achieved when a solution of the species S_n is stirred with a little Al_2O_3 powder at 20 °C. Similarly, silica gel catalyzes the conversion both in solution and in the solid state. For this reason attempts to separate mixtures of rings S_n by conventional thin-layer chromatography on SiO_2 failed. Only column chromatography on silica gel at —40 °C using CS_2 as an eluent has been useful in the separation of S_6, S_8 and S_{12}, but partial decomposition of S_6 and S_{12} to S_8 could not be avoided. Mixtures of larger rings (S_x, $\bar{x} = 25$, see above) could not be separated this way [16]. A new chromatographic technique which uses a very inert material as a stationary phase has become known as reversed-phase chromatography. In this case the stationary phase consists of chain-like $C_{18}H_{37}$ radicals fixed to the surface of an SiO_2 support via carbon-oxygen bonds (octadecylsiloxane); polar solvents, e.g. methanol, are used as eluent. With a particle size of 5–10 µm a pressure of 5–40 bar is usually needed to force the solvent with a flow-rate of 1 cm^3/min through a column of 10–30 cm length and 5–10 mm inner diameter (high-pressure liquid chromatography, HPLC). The small particle size results in a very large specific surface yielding a very high separation power (the number of theoretical plates is approximately 5000 per 10 cm). Under these conditions it has been possible to separate the S_n sulfur rings with n = 6–26. The components were detected by their UV absorption at 254 nm which is very intense and allows a trace analysis [15]. Fig. 6 shows the chromatogram of a mixture of S_6, S_7, S_8, S_{10}, S_{12}, α-S_{18}, and S_{20} obtained from the pure components [47]. The excellent resolution allows the detection of even very small amounts of certain rings in the presence of a large excess of others. However, the separation of rings larger than S_{26} has not been achieved yet because of their low solubility in polar solvents.

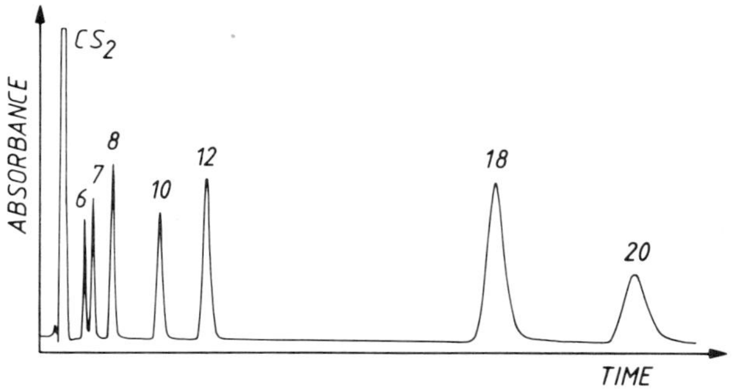

Fig. 6. Chromatogram (HPLC) of a mixture of sulfur rings (S_6, S_7, S_8, S_{10}, S_{12}, α-S_{18}, S_{20}) prepared from the pure compounds (column: Radial-Pak C-18, eluent: methanol, detector: UV absorption at 254 nm)

The retention time t_R is a non-linear function of the ring size n and thus allows the identification of new species of the homologous series S_n by interpolation. Furthermore, the logarithm of the capacity factor $k = (t_R - t_0)/t_0$ (with t_0 = death time of the chromatographic system) is a linear function of the ring size, n, although two such functions are obtained; see Fig. 7 [15].

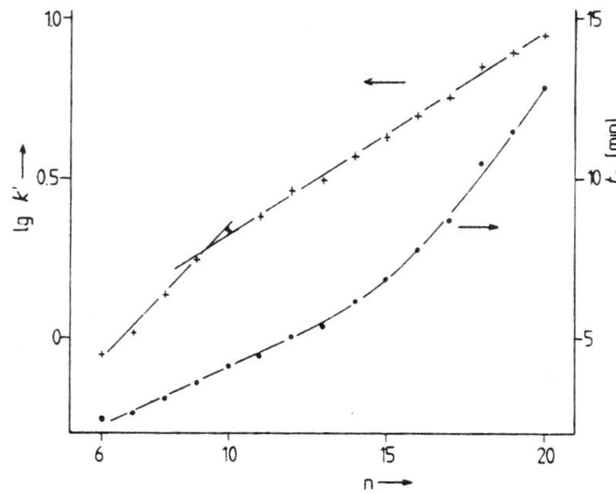

Fig. 7. Gross retention times, t_R, (ordinate on the right side) and logarithm of the capacity factors, $k' = (t_R - t_0)/t_0$, as a function of the number, n, of atoms in the S_n molecules (death time t_0: 1.3 min, column: Nucleosil C-18, eluent: methanol/cyclohexane 80/20)

The area under a chromatographic peak is proportional to the concentration of the corresponding species in the sample analysed. Using empirical calibration functions, a rapid simultaneous qualitative and quantitative analysis of sulfur mixtures by HPLC is now possible. In a similar manner the selenium rings Se_6, Se_7 and Se_8 have recently been separated by HPLC [63].

6 Occurrence of Sulfur Rings in Reaction Mixtures

Gaseous and liquid sulfur

It has been known for some time that the presumably cyclic molecules S_6, S_7 and S_8 are the main constituents of saturated sulfur vapor at temperatures up to about 500 °C. This knowledge comes from mass spectroscopic and vapor pressure measurements [6]. Only recently, however, it was possible to show that liquid sulfur also contains S_6 and S_7 as well as rings larger than S_8. Several of these species, e.g. S_7, S_{12}, α-S_{18} and S_{20}, have been prepared in a pure state from the quenched melt (see above), others were detected by infrared and Raman spectroscopy as well as by HPLC. Chromatograms of a carbon disulfide extract of a quenched sulfur melt show the presence of all rings up to S_{26}, but even larger rings must be present since the average molecular weight of the large ring fraction (termed as S_x, see above) corresponds to 25 atoms per molecule [15–20].

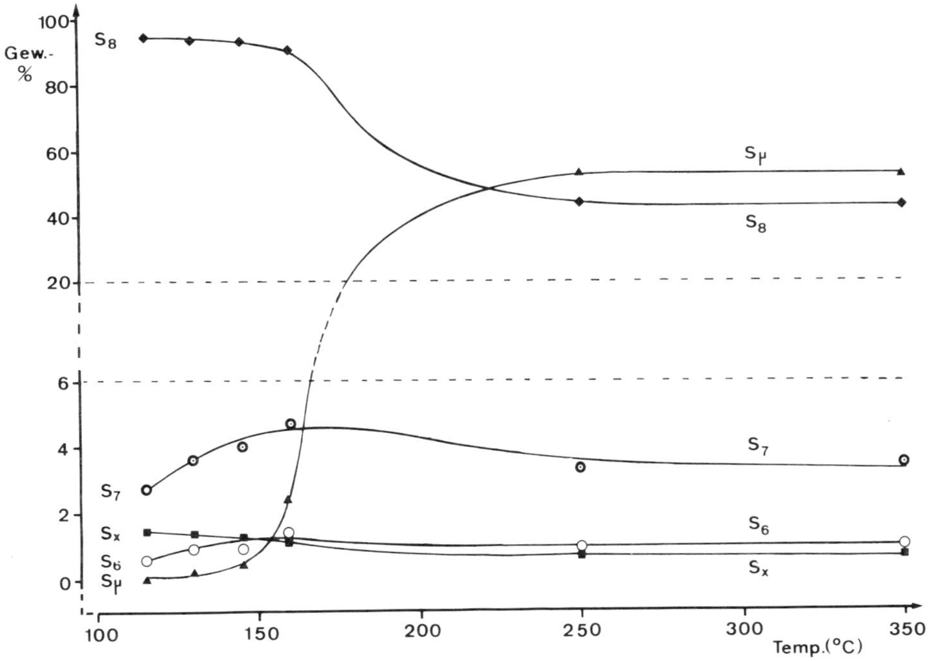

Fig. 8. Equilibrium composition (weight-%) of liquid sulfur in the temperature region 115–350 °C according to spectroscopic investigations after rapid quenching to −196 °C. S_x represents the fraction of rings soluble in CS_2 and with $x > 8$; S_μ is the insoluble, polymeric fraction. Two scales have been used on the ordinate

On heating polycrystalline α-S_8 transforms at 96 °C to monoclinic β-S_8 which melts at 120 °C. Immediately after melting the liquid consists of S_8 molecules only, since the freezing temperature is identical to the melting temperature. However, when the melt is kept at 120 °C for 12 hours or more, equilibrium is reached and the freezing point is now found at 115 °C. The freezing point depression of about 5° according to Raoult's law corresponds to a concentration of 5 mol-% of foreign molecules in the melt (S_n, $n \neq 8$) which in the older literature were termed as π-sulfur (S_π) [64].

Quantitative analyses of the equilibrium melt quenched from temperatures of between 115 and 350 °C led to the following conclusions. At 115 °C the liquid consists of 95.1 % S_8 (by weight), 2.8 % S_7, 0.6 % S_6 and 1.5 % S_x ($x > 8$), while the concentration of polymeric, insoluble sulfur (S_∞ or S_μ) is negligible [20]. The fraction of non-S_8 sulfur amounts to 4.9 % in agreement with the estimates from the freezing point depression metioned above. With increasing temperatures the concentrations of S_7 and S_6 reach a maximum near 160 °C followed by a slight decrease, while the S_x concentration slowly decreases from the beginning. The fraction of S_8 decreases sharply at 160–190 °C at which temperature the amount of S_μ increases dramatically; both concentrations seem to be constant in the region 250–350 °C; see Fig. 8 [20]. These findings now explain most of the unique properties of liquid sulfur but, on the other hand, they do not support the frequently cited polymerization theory of Tobolsky and Eisenberg [65].

Dissolved sulfur

When S_8 or S_7, dissolved in CS_2 or toluene, are heated to 100–150 °C, interconversion reactions take place which finally lead to equilibrium mixtures similar to those found in liquid sulfur. The main equilibrium constituents are S_8, S_7 and S_6 but traces of larger rings are also detectable. However, no polymeric sulfur is formed under these conditions [66]. These reactions, which have been studied kinetically, will be discussed in Section 7.

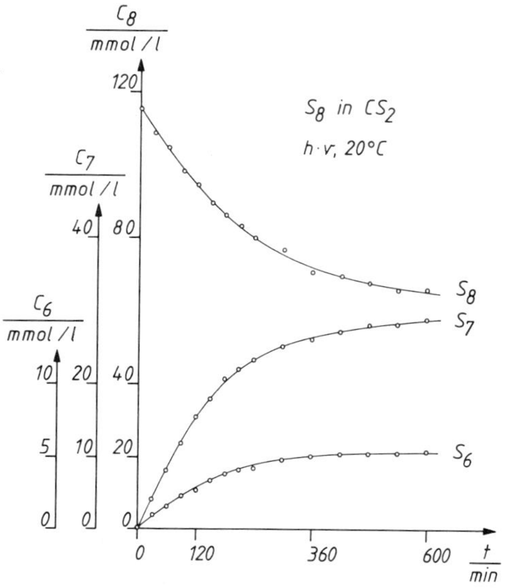

Fig. 9. Formation of S_6 and S_7 from S_8 by UV irradiation of the latter in CS_2 solution at 20 °C for 10 hours. The concentrations were determined by HPLC. Initial S_8 concentration: 117 mmol/l

Irradiation of S_8, dissolved in CS_2, with the UV radiation of a mercury lamp in a quartz apparatus results in the formation of S_6, S_7 and S_{12} together with traces of S_9, S_{10} and some polymeric insoluble material. After 10 hours the concentrations of S_6 and S_7 become approximately constant at levels much higher than in liquid sulfur at 160 °C; see Fig. 9 [67]. Irradiation of either one of S_6, S_7 or S_{10} in CS_2 solution also results in mixtures of S_6, S_7 and S_8 [67].

Acid Decomposition of Aqueous Thiosulfate

Addition of hydrochloric acid to aqueous sodium thiosulfate results in the precipitation of elemental sulfur according to the above equation (Sect. 2). Solvent extraction ($CHCl_3$ or toluene) of this sulfur yields yellow solutions from which S_6 crystallizes on cooling while S_7 and S_8 remain in solution. The molar ratio $S_6:S_7:S_8$ has recently been determined by Raman spectroscopy as approximately 6:2:1, when the $Na_2S_2O_3$ solution is added to cold concentrated hydrochloric acid, and as 2:1.5:1, when the acid is poured into the thiosulfate solution without cooling. In addition, traces of sulfanes, H_2S_n, are formed [13].

According to Davis [68], the reaction mechanism includes the intermediate formation of unstable chain-like sulfane-monosulfonic acids:

$$HS_2O_3^- + HS_2O_3^- \rightleftharpoons HS_3O_3^- + HSO_3^-$$
$$HS_2O_3^- + HS_3O_3^- \rightleftharpoons HS_4O_3^- + HSO_3^-$$

etc., until:

$$HS_7O_3^- \rightleftharpoons S_6 + HSO_3^-$$
$$HS_8O_3^- \rightleftharpoons S_7 + HSO_3^-$$
$$HS_9O_3^- \rightleftharpoons S_8 + HSO_3^-$$

Thermal Decomposition of Di-iodosulfanes

Iodosulfanes (S_nI_2, n = 1, 2, ...) are known to be unstable at 20 °C even in dilute solution decomposing to elemental sulfur and iodine [14]. The iodides are formed on

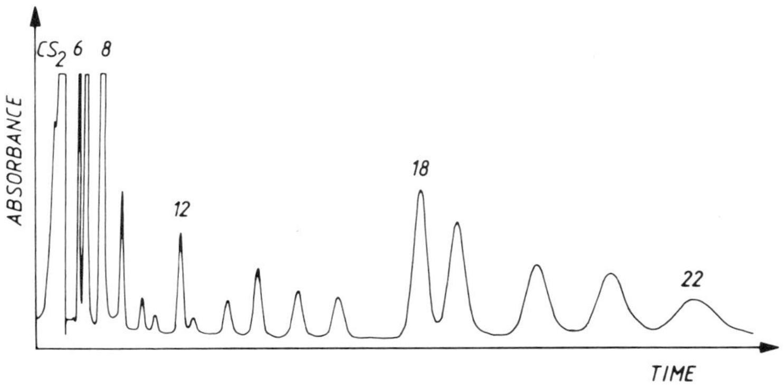

Fig. 10. Chromatogram (HPLC) of the sulfur obtained by reaction of SCl_2 with KI in CS_2 at 20 °C showing the presence of all rings S_n with n = 6–22 (column: Radial-Pak C-18, eluent: methanol, detector: UV absorption at 254 nm)

reaction of the corresponding chlorosulfanes, dissolved in CS_2, with solid or aqueous potassium iodide at 20 °C. The elemental sulfur formed from SI_2 has been found to consist of all S_n rings with n = 6–22 whose concentrations, however, rapidly decrease with increasing ring size; see Fig. 10. From S_2I_2 almost exclusively even-membered rings are formed of which S_6 is the main product which can be isolated in 36% yield based on the equation

$$3S_2Cl_2 \xrightarrow[-6KCl,\,-3I_2]{+6KI} S_6$$

From the remaining sulfur small amounts of S_{12} (1–2%), α-S_{18} (0.4%) and S_{20} (0.4%) could be isolated [14,15].

In similar reactions S_6Cl_2, S_7Cl_2 and S_8Cl_2, respectively, on treatment with KI yield mainly S_6, S_7 and S_8, respectively [67]. It is believed that the reaction mechanism involves the intermediate formation of higher iodosulfanes by intermolecular condensation of the smaller molecules with liberation of iodine:

$$2\,SI_2 \rightarrow S_2I_2 + I_2$$

$$2\,S_2I_2 \rightarrow S_4I_2 + I_2$$

As soon as the sulfur chains are long enough intramolecular formation of molecular iodine and ring closure takes place. Since no S_5 has been observed in these mixtures, the six-membered ring seems to be the smallest possible one under these conditions.

7 Interconversion Reactions of Sulfur Rings

One of the most important reactions of compounds containing cumulated sulfur-sulfur bonds is the interconversion of type

$$m\,S_n^R \rightleftharpoons n\,S_m^R \qquad (n \neq m,\ R = \text{ring})$$

This type of reaction takes place when unstable solid sulfur allotropes decompose, when π-sulfur is formed from S_8 on heating, when liquid sulfur polymerizes and depolymerizes at 160–190 °C, when organic polysulfanes, R_2S_n, decompose and

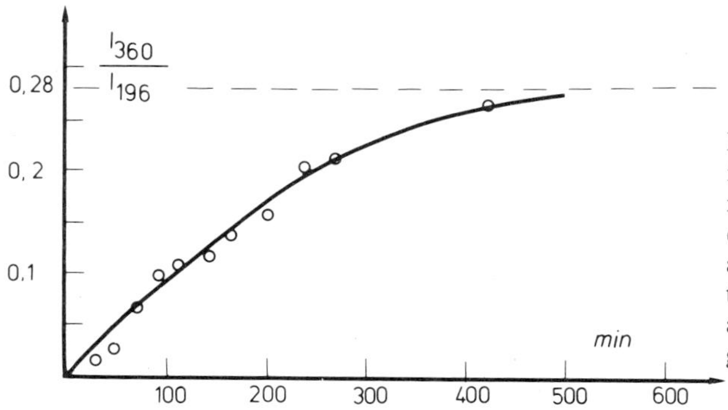

Fig. 11. Formation of S_7 from S_8 in liquid sulfur at 120 °C. The ratio S_7:S_8 was determined after quenching of a sample of the melt at −196 °C. On the ordinate, the ratio of the Raman intensities at 360 (S_7) and 196 cm^{-1} (S_8) is given

rearrange on heating, when rubber is vulcanized by treatment with elemental sulfur, etc. To elucidate the mechanism of such reactions kinetic studies on the thermal interconversions $S_8 \rightarrow S_7, S_6, \dots$ and $S_7 \rightarrow S_8, S_6, \dots$ have recently been made.

The formation of π-sulfur from S_8 at 120 °C is a relatively slow reaction. Figure 11 shows how the S_7 concentration in freshly molten S_8 increases with time. The half-life of this reaction at 120 °C amounts to about 180 min [16].

When S_8 is dissolved in CS_2 and heated to 130–150 °C in a sealed glass ampoule, formation of S_7 and S_6 together with traces of larger rings takes place; see Fig. 12.

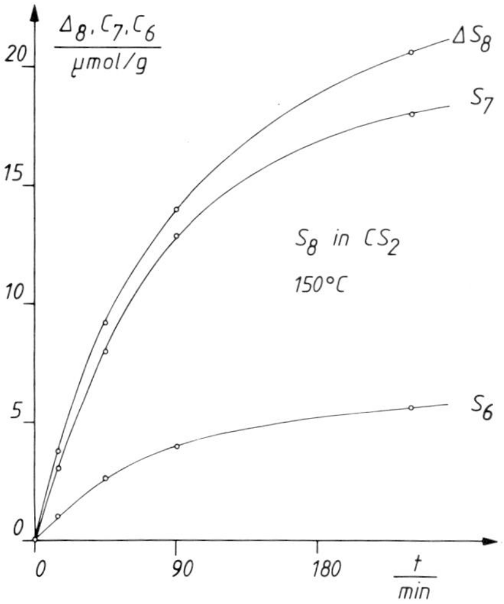

Fig. 12. Formation of S_6 and S_7 from S_8 in CS_2 solution at 150 °C within 4 hours. ΔS_8 is the amount of S_8 consumed. The concentrations were determined by HPLC; initial S_8 concentration: 0.29 mmol/g solution

In solution, the half-life of the S_7 formation from S_8 is approximately independent of the initial S_8 concentration (0.12–0.34 mol · dm^{-3}) but, of course, increases with temperature; see Fig. 13. It decreases from 232 min at 130 °C to 64 min at 150 °C. The apparent activation energy has been calculated as 95 kJ/mol [66].

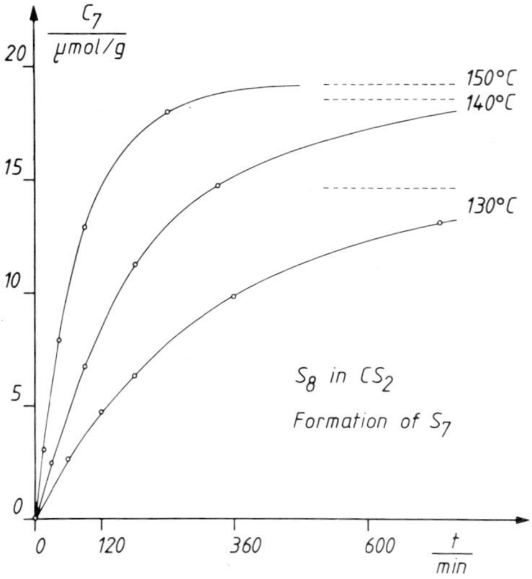

Fig. 13. Effect of temperature on the formation of S_7 from S_8 in CS_2 solution at 130°, 140° and 150 °C. The final equilibrium concentrations of S_7 at these temperatures are indicated by broken lines

From the temperature dependence of the equilibrium concentrations of S_6, S_7 and S_8 the following enthalpy of reaction values have been obtained [66] whose agreement with the best vapor phase data [62] is very satisfactory:

	CS_2 solution	vapor phase (25 °C)
$\frac{3}{4} S_8 \rightleftharpoons S_6$	ΔH° (l): 24	ΔH° (g): 26 kJ/mol
$\frac{7}{8} S_8 \rightleftharpoons S_7$	ΔH° (l): 21	ΔH° (g): 24 kJ/mol

Since $\Delta H^\circ (l) = \Delta H^\circ (g) - \Sigma \Delta H^\circ (v)$ (the last term is the sum of the vaporization enthalpies of products [positive] and reactants [negative]), no perfect agreement between ΔH (l) and ΔH (g) can be expected.

On heating of a CS_2 solution of S_7 to 150 °C, S_6 and S_8 are generated and S_7 disappears with a half-life of 31 min until the same equilibrium concentrations as above are reached. The reaction order has been found near 1.5 [66].

Since all of the above-mentioned interconversion reactions are reversible, any kinetic analysis is difficult. In particular, this holds for the reaction $S_8 \rightarrow S_7$ since the backward reaction $S_7 \rightarrow S_8$ is much faster and, therefore, cannot be neglected even in the early stages of the forward reaction. The observation that the equilibrium is reached by first order kinetics (the half-life is independent of the initial S_8 concentration) does not necessarily indicate that the single steps $S_8 \rightarrow S_7$ and $S_8 \rightarrow S_6$ are first order reactions. In fact, no definite conclusions about the reaction order of these elementary steps are possible at the present time. The reaction order of 1.5 of the S_7 decomposition supports this view. Furthermore, the measured overall activation energy of 95 kJ/mol, obtained with the assumption of first order kinetics, must be a function of the true activation energies of the forward and backward reactions. The value found should therefore be interpreted with caution.

Under these circumstances any discussion of the possible reaction mechanism will be somewhat speculative. For future work, however, it may be helpful to compare the various possible pathways for the interconversion of sulfur rings:

a) Homolytic ring opening ($S_8^{ring} \rightarrow S_8^{chain}$) followed by polymerization ($S_8^{chain} + n\, S_8^{ring} \rightarrow S_\infty^{chain}$) and depolymerization ($S_\infty^{chain} \rightarrow S_6 + S_7 + S_8 + ...$). This free radical mechanism has often been discussed on the basis of the observed first order rate equations found for the formation of π-sulfur [69] and the decomposition of organic polysulfides [70–74] as well as on the basis of the experimentally determined apparent activation energies (50–150 kJ/mol [65, 69–74]). However, this type of mechanism seems rather unlikely at least at moderate temperatures since (a) no free radicals have been detected in liquid sulfur below 170 °C [75, 76], (b) only the highest activation energies observed agree with the energy needed to generate free spins in liquid sulfur (see below), and (c) a reaction of this type should be autocatalytic and should show the features of a chain-reaction which has not been observed. The activation energy of free radical formation in liquid sulfur, i.e. the enthalpy of the reaction

$$—S—S— \rightarrow —S\cdot + \cdot S—$$

has recently be determined by magnetic measurements as 1.6 eV (150 kJ/mol) at 550 °C [75, 77] but it is by no means clear which molecules have been involved.

Obviously, species with weak bonds like S_7 will dissociate more easily than S_8. At lower temperatures the activation energy of radical formation is considerably larger, which has been explained by the influence of impurities, X, which react with the radicals according to

$$-S\cdot + X \rightarrow -S-X$$

thus decreasing the equilibrium spin concentration [75–77]. Sulfur radicals have been found to react rapidly with sulfur-sulfur bonds in a displacement reaction according to

$$-S\cdot + -S-S- \rightarrow -S-S- + \cdot S-$$

(activation energy 13 kJ/mol [76]) and to some extent with the formation of ions according to

$$-\ddot{\underset{\cdot\cdot}{S}}\cdot + -\ddot{\underset{\cdot\cdot}{S}}- \rightarrow -\ddot{\underset{\cdot\cdot}{S}}{:}^{\ominus} + -\dot{\underset{\cdot\cdot}{S}}^{\oplus}-$$

(activation energy 120 kJ/mol [78]) leading to electrical conductivity.

b) Isomerization of S_8 yielding S_7S with a structure analogous to S_7O [41] followed either by sulfur atom transfer to another ring

$$S_8 \rightarrow S_7{=}S \xrightarrow{+S_8} S_7 + S_8{=}S \rightarrow S_7 + S_9$$

or by an attack on neighboring rings with formation of a zwitterion:

$$S_7{=}S + S_8 \rightarrow S_7^{\oplus}{-}S{-}S{-}S{-}S{-}S{-}S{-}S{-}S{-}S^{\ominus}$$

The zwitterion may either grow in chain-length or split off smaller rings or rearrange forming one large ring:

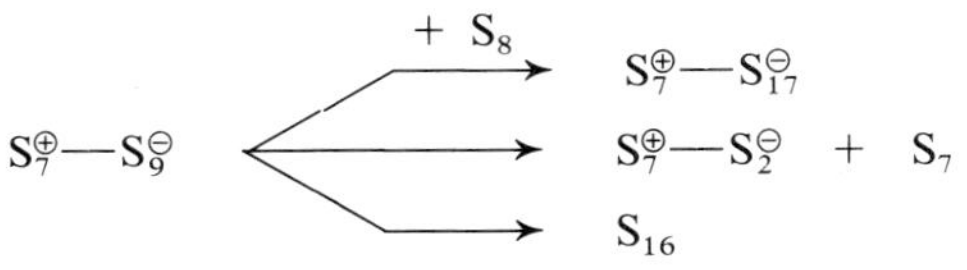

The S_{16} formed by dimerization of S_8 would then rapidly decompose to either $S_7 + S_9$ or $S_6 + S_{10}$ or react with S_8 or other species forming larger rings (see below).

Several compounds of type $X_2S{=}S$ have been prepared (X = F, Cl, Br) [79–81] and the existence of branched sulfur chains at low temperatures has been discussed [82]. The isomerization

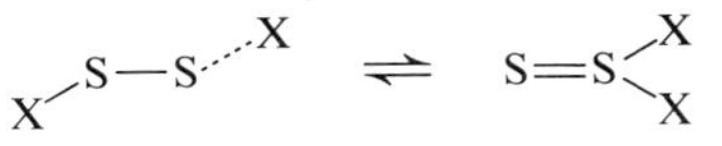

has also been investigated by CNDO/2 calculations of the energy hypersurface, and the interconversion energy barrier (based on the more stable isomer) has been calculated as 1–2 eV for X=F, 1.5 eV for X=Cl, and 3.9 eV for X=H. Only in case of X = F the branched structure (symmetry C_s) is more stable than the chain (symmetry C_2) [83]. Since sulfur resembles chlorine in atomic structure and electronegativity, the barrier of the isomerization $S_8 \rightarrow S_7{=}S$ can be expected to amount to ca. 150 kJ/mol. However, reactions of species $X_2S{=}S$ are not well known and the sulfur atom transfer ($S_7{=}S + S_8 \rightarrow S_7 + S_8{=}S$) postulated above may require some activation energy. Another possibility is the reaction of $S_7{=}S$ with S_8, forming a sulfurane-like spiro compound as an activated complex which could then decompose to S_7 and S_9:

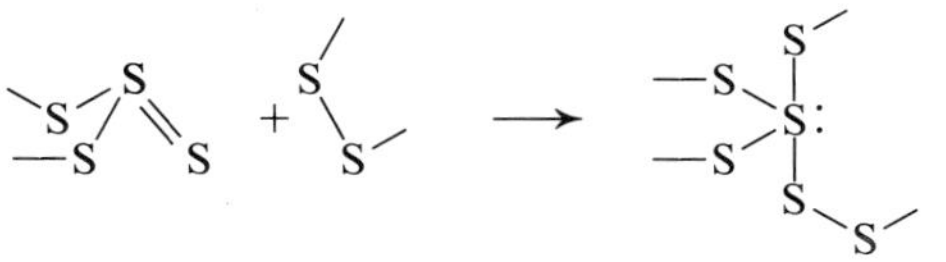

It should be pointed out, however, that the decomposition of S_6 according to this mechanism would require contraction to the rather strained five-membered ring in $S_5{=}S$ which seems an unlikely process.

c) Dimerization of S_8 yielding S_{16} and possibly higher oligomers which then in turn dissociate to smaller species. Since symmetrical four-center reactions like

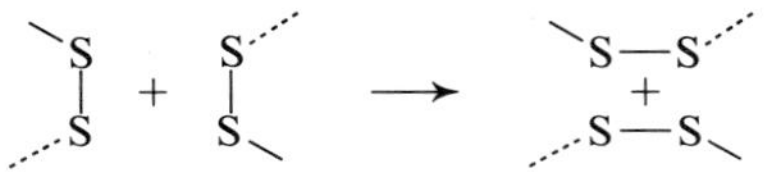

with either a parallel or perpendicular approach of the two bonds to be broken (transition states square-planar and tetrahedral, respectively) seem to be forbidden by orbital symmetry, the first step of the dimerization could involve a sulfurane-like transition state with a four-coordinated sulfur atom. After pseudorotation at the central atom the complex could either rearrange forming one large ring or dissociate to $S_7 + S_9$:

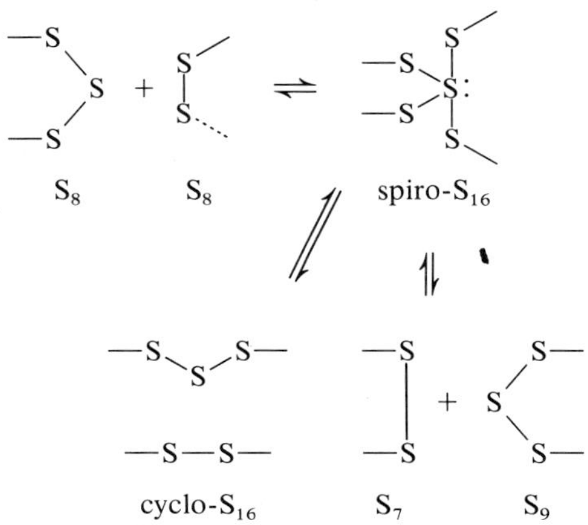

A reaction of this type is known from the reversible addition of chlorine to organic sulfides

$$R_2S + Cl_2 \rightleftharpoons R_2SCl_2$$

which takes place at 20 °C and which is allowed by symmetry. The sulfurane R_2SCl_2 (R = p-Cl-C_6H_4), like SF_4 and similar compounds of organic selenides with chlorine and bromine, is of symmetry C_{2v} at the sulfur atom (pseudo-trigonal-bipyramidal coordination) [84]. Sulfuranes with an S—S bond are known (e.g. FS—SF_3). When S_6, S_7 or S_8 are treated with elemental chlorine at 0 °C in CCl_4 the corresponding chlorosulfanes (S_6Cl_2, S_7Cl_2, S_8Cl_2) are formed [85]. These reactions are likely to proceed in an analogous way as the chlorination of organic sulfides and as the proposed S_8 dimerization:

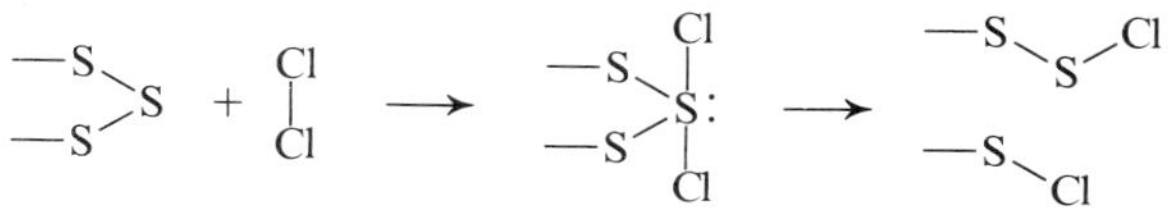

Interestingly, attempts to prepare the unstable S_5 resulted in the formation of S_{10} [28,85] and treatment of S_6O with $SbCl_5$ in CS_2 solution yielded the dimerization product $S_{12}O_2 \cdot 2\,SbCl_5$ [86]. The latter reaction is the first definite dimerization of a homocyclic sulfur molecule.

d) Dissociation of S_8 to S_6 and S_2 followed by either oligomerization of S_2 (→ S_4, S_6, etc.) or attack of rings S_n by S_2 with formation of species S_{n+2}. The first step of the S_2 formation from S_8 could again be an isomerization as discussed above:

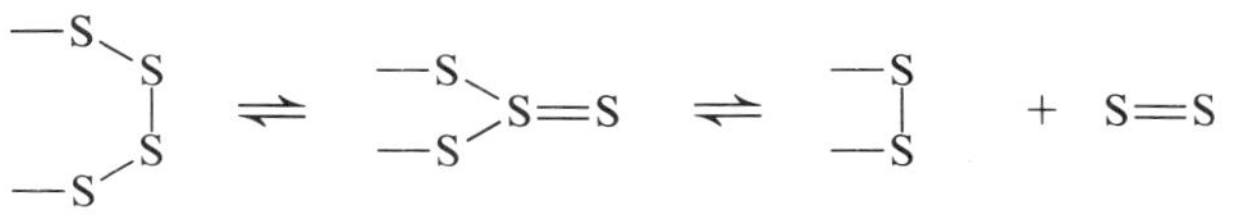

A reaction of this type would resemble the well known thermal extrusion of sulfur monoxide from thiirane oxides [87]:

The thermodynamic data of gaseous S_8, S_6 and S_2 [62,88] suggest that the standard enthalpy of the dissociation reaction should be near 130 kJ/mol when the S_2 molecule is formed in its $^3\Sigma_g^-$ ground state and near 185 kJ/mol when singlet $S_2(^1\Delta g)$ is formed in order to conserve the total spin momentum. The first possibility can almost be ruled out since the principle of spin conservation would require a two-step mechanism with initial homolytic scission of an SS single bond whose dissociation energy amounts to already more than 150 kJ/mol (see above). On the other hand, formation of singlet S_2 is very unfavorable energetically. Furthermore, only even-membered rings could be formed from S_8 by this mechanism unless large rings are assumed as intermediates which then could dissociate to odd-membered fragments.

The enthalpy of the reaction

$$S_8(g) \rightarrow S_5(g) + S_3(g)$$

(150 kJ/mol at 600 K [89]) is identical to the energy needed to break an SS bond homolytically and therefore this reaction is an alternative route for the unimolecular decomposition of S_8. Rings larger than S_8 can easily split off S_3, S_4 or S_5. From the somewhat preliminary thermodynamic data of these species [89] it follows that the reactions

$$S_2(g) \rightarrow \frac{2}{3} S_3(g) \qquad \Delta H^0_{600} \sim -37 \text{ kJ/mol}$$

$$S_2(g) \rightarrow \frac{1}{2} S_4(g) \qquad \Delta H^0_{600} \sim -59 \text{ kJ/mol}$$

$$S_2(g) \sim \frac{2}{5} S_5(g) \qquad \Delta H^0_{600} \sim -80 \text{ kJ/mol}$$

are all strongly exothermic, making the dissociation of, for example, S_9 to $S_6 + S_3$ slightly more favorable than to $S_7 + S_2(^3\Sigma_g^-)$. Since S_3, S_4 and S_5 are likely to exist in singlet ground states, spin conservation is no problem.
Assuming the same mean bond energy in S_9 as in S_6 and S_7 the enthalpy of the reaction $S_9 \rightarrow S_6 + S_3$ can be calculated as 85 kJ/mol. From the structures of S_{10}, S_{12}, S_{18} and S_{20}, it can be estimated that the mean bond energies of these molecules are also slightly lower than in S_8. Under these circumstances the dissociation to smaller rings (S_5 ... S_{10}) should proceed with a reaction enthalpy of less than 80 kJ/mol. It may be for this reason that the equilibrium concentrations of rings larger than S_8 in liquid sulfur and sulfur solutions are so small compared with those of S_6 and S_7. Unimolecular dissociation reactions of S_6 and S_7, e.g.

$$S_6 \rightarrow 2\, S_3 \qquad \Delta H^0_{600} = 166 \text{ kJ/mol}$$

$$S_7 \rightarrow S_3 + S_4 \qquad \Delta H^0_{600} = 157 \text{ kJ/mol}$$

are strongly endothermic and therefore much less favorable than similar reactions of larger rings.

The above discussion shows that several possible pathways for the interconversion of sulfur rings exist. However, none of these alone can explain all the experimental observations. It therefore seems likely that several of them are effective simultaneously. Unimolecular dissociation reactions as discussed under (a) and (d) will dominate at high temperatures due to the increase in entropy. At lower temperatures, however, bimolecular reactions like the dimerization (c) may be most important, at least in case of the small rings (S_6, S_7, S_8) whose unimolecular dissociation is strongly endothermic. Larger rings will probably decompose according to mechanism (d), which in a way is the reversal of the dimerization (c).

The mechanisms (a)–(c) can also be discussed in connection with the thermal decomposition and rearrangement reactions of organic polysulfides. For example, dimethyl tetrasulfide when heated to 80 °C for several hours disproportionates to the corresponding tri-, penta- and hexasulfides. On prolonged heating small amounts of disulfide are formed in addition [70–72].

Dimethyltrisulfide disproportionates at 80 °C to approximately equimolar quantities of di- and tetrasulfide the equilibrium being reached within 500 hours [71,72]:

$$2\,R_2S_3 \rightleftharpoons R_2S_2 + R_2S_4$$

When an equimolar mixture of two trisulfides is heated to 130–150 °C for several hours complete scrambling of the substituents takes place:

$$R_2S_3 + R'_2S_3 \underset{k_2}{\overset{k_1}{\rightleftharpoons}} 2RR'S_3 \qquad R: C_2H_5, \qquad R': n{-}C_3H_7$$

Small amounts of tetrasulfide are formed after longer heating times [74]. This reaction and the tetrasulfide decomposition mentioned above have been claimed to obey first order rate equations and the apparent Arrhenius activation energy has been determined as 121 and 153 kJ/mol, respectively [70–72, 74]. Since all these reactions approach some kind of equilibrium they must be reversible. The authors interpreted their results in terms of a unimolecular homolytic dissociation of an S—S bond as the first and rate determining step, and the following reaction sequences were postulated:

Tetrasulfide decomposition [70–72]

$$R_2S_4 \rightarrow 2\,RS_2^{\cdot}$$
$$RS_2^{\cdot} + R_2S_4 \rightarrow R_2S_3 + RS_3^{\cdot}$$
$$RS_3^{\cdot} + R_2S_4 \rightarrow R_2S_5 + RS_2^{\cdot}$$
$$RS_3^{\cdot} + R_2S_5 \rightarrow R_2S_6 + RS_2^{\cdot}$$

Trisulfide rearrangement [74]

$$R_2S_3 \rightarrow RS^{\cdot} + RS_2^{\cdot}$$
$$RS^{\cdot} + R'_2S_3 \rightarrow RR'S_3 + R'S^{\cdot}$$
$$RS_2^{\cdot} + R'_2S_3 \rightarrow RR'S_3 + R'S_2^{\cdot}$$

In particular Tobolsky et al. [70–72] promoted the idea of a free radical mechanism. However, very special assumptions have to be made to obtain the right products. For example, the $RS^{\cdot}$ radicals formed from R_2S_3 have to react exclusively with the β sulfur atom

$$R{-}\overset{\alpha}{S}{-}\overset{\beta}{S}{-}\overset{\alpha}{S}{-}R$$

of the trisulfide and the $RS_2^{\cdot}$ radicals exclusively with the α atoms since otherwise di- and tetrasulfides would be formed with the same rate as the scrambling of the substituents R and R′ takes place. In the case of the tetrasulfide decomposition, a homolytic scission of the central S—S bond only has to be assumed. Furthermore, it was observed that "the $MeS_2^{\cdot}$ radical is remarkably unreactive" since it reacted neither with triphenylmethane at 130 °C within 130 hours nor with cyclohexene at 80 °C within 110 hours [70–72]. It, therefore, may not have been formed at all.

The reaction temperatures and some of the activation energies cited above seem to be too low to support a radical-chain reaction mechanism. Guryanova [73] found that exchange of radioactive elemental sulfur (^{35}S) with the β sulfur atoms of bis-p-tolyl tetrasulfide proceeds at 80–130 °C with an activation energy of only 50 kJ/mol; in the case of the corresponding trisulfide the activation energy was determined as 60 kJ/mol. These data sharply contrast with the observation that liquid sulfur has to be heated to more than 170 °C to detect free radicals by electron spin resonance spectroscopy [75,76], and the activation energy for homolytic SS bond scission has been determined as 150 kJ/mol (see above).

Under these circumstances, it seems reasonable to look for alternative explanations for the polysulfide reactions and in this connection it is interesting to note that a bimolecular mechanism is in better agreement with the data reported for the trisulfide disproportionation discussed above. This reaction has been followed by ^1H n.m.r. spectroscopy at different temperatures and initial concentrations. Equimolar amounts of diethyl and di-n-propyl trisulfide were used and after the equilibrium had been established the equilibrium constant

$$K = \frac{[EtS_3Pr]^2}{[EtS_3Et]\,[PrS_3Pr]}$$

was found to equal 4 in agreement with the expected statistical distribution of the substituents [74]. The authors observed that the exchange rate was not influenced by either oxygen or nitrobenzene (used as a solvent together with benzene) indicating that neither free radicals nor ionic or zwitterionic species play any role as intermediates. Furthermore, the calculated first order rate constant was found to depend on the initial symmetrical trisulfide concentrations (Fig. 2 in loc. cit. [74]) excluding a first order reaction. Actually, a reaction order of 1.7 ± 0.1 can be calculated from the reported data for the reaction at 140 °C using the fractional-life period method (J. W. Moore and R. G. Pearson, Kinetics and Mechanism, 3rd ed., J. Wiley, New York 1981, p. 60).

Under these circumstances there is no need to postulate unimolecular rate determining reactions. Instead, the bimolecular reaction mechanism (c) discussed above would explain all experimental observations much better. The reaction could proceed via a sulfurane-like molecule which after pseudorotation at the central atom could dissociate to the observed products:

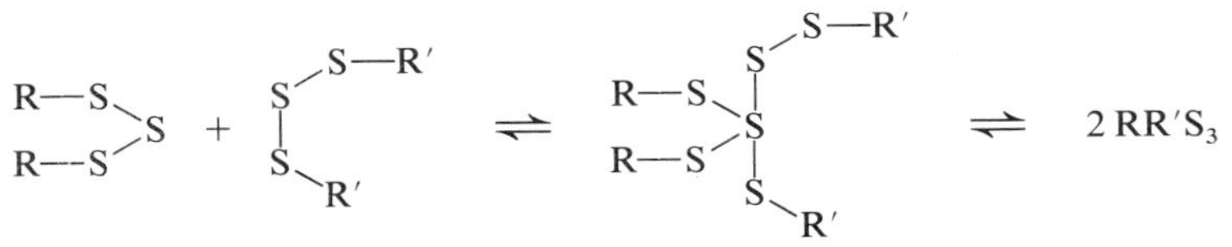

The tetrasulfide decomposition could proceed as follows:

$$2\,R_2S_4 \rightarrow R_2S_3 + R_2S_5$$
$$R_2S_4 + R_2S_5 \rightarrow R_2S_3 + R_2S_6$$
$$2\,R_2S_3 \rightarrow R_2S_2 + R_2S_4$$

In summary, it should be pointed out, however, that the exact mechanism of the thermal rearrangement of sulfur rings and chains is still unknown and that further investigations are necessary in this connection. For any further discussion it may also be interesting to take into account the recent results on the molecular rearrangement reactions of cyclic selenium sulfides [90] and of elemental selenium [63,91], which take place at considerably lower temperatures compared with elemental sulfur and for which — as far as solid selenium is concerned — interesting ionic mechanisms have been proposed [92,93].

8 Other Reactions of Sulfur Rings

Numerous reactions of elemental sulfur with various reagents are known but almost all start with the breakdown of the sulfur ring. It was only in 1974 that the first reaction was found in which the ring was preserved and which thus resulted in a homocyclic derivative [94]:

$$S_8 + CF_3CO_3H \rightarrow S_8O + CF_3CO_2H$$

This reaction takes place in CS_2 or CH_2Cl_2 solution at 0 °C and has also been applied to S_6, S_7, S_9 and S_{10} resulting in the preparation of the corresponding monoxides or, using excess peroxyacid, dioxides, of which only S_7O_2, however, could be isolated as a pure substance [95]. The latter compound is also formed on oxidation of S_8 by excess peroxyacid:

$$S_8 + 4\,CF_3CO_3H \rightarrow S_7O_2 + SO_2 + 4\,CF_3CO_2H$$

This reaction represents the first controlled sulfur ring contraction. The X-ray structural analyses of S_8O [96] and S_7O [41] as well as the vibrational spectra of the other oxides show that the oxygen atoms are always present as sulfoxide groups of type —S—SO—S—. Extensive reviews of this new class of sulfur oxides have been published elsewhere [8].

9 Outlook

The results described above show that despite several decades of extensive research activities the field of homocyclic sulfur molecules is still full of open questions. Basic knowledge concerning the thermodynamic properties of most molecules is lacking, and even the homolytic dissociation energy of the S_8 ring molecule is unknown.

Only a few molecular orbital calculations regarding the structures, conformational changes and bonding of S_n molecules have been published. Fundamental reactions of sulfur rings like interconversions are only poorly understood, and more examples for the formation of homocyclic derivatives directly from the corresponding S_n parent molecules are likely to be found. These few comments show that much more work has to be done before the chemistry of elemental sulfur can be regarded as "well known".

10 Acknowledgements

The author would like to express his sincere thanks to his coworkers whose names appear in the References, for their contributions. Financial support by the Deutsche Forschungsgemeinschaft and Verband der Chemischen Industrie is also gratefully acknowledged.

11 References

1. Gillespie, R. J.: Chem. Soc. Rev. *8*, 315 (1979); Gillespie, R. J., Passmore, J.: Adv. Inorg. Chem. Radiochem. *17*, 49 (1975)
2. Passmore, J. et al.: J. Chem. Soc. Chem. Commun. *1976*, 689
3. Passmore, J., Sutherland, G., White, P. S.: J. Chem. Soc. Chem. Commun. *1979*, 901, and *1980*, 330, Inorg. Chem., in print
4. Burns, R. C., Gillespie, R. J., Sawyer, J. F.: Inorg. Chem. *19*, 1423 (1980)
5. Chivers, T., Drummond, I.: Chem. Soc. Rev. *2*, 233 (1973)
6. Meyer, B.: Chem. Rev. *76*, 367 (1976)
7. Schmidt, M.: Angew. Chem. *85*, 474 (1973)
8. Steudel, R.: In Gmelin Handb. d. Anorg. Chem., 8. Aufl., Schwefel, Ergänzungsband 3, Springer, Berlin 1980; Steudel, R.: Comments Inorg. Chem., in print
9. Kao, J.: Inorg. Chem. *16*, 3347 (1977)
10. Debaerdemaeker, T., Kutoglu, A.: Cryst. Struct. Commun. *3*, 611 (1974)
11. Aten, A. H. W.: Z. Physik. Chem. *88*, 321 (1914)
12. Bartlett, P. D., Roderick, W. R.: Inorg. Synth. *8*, 100 (1966)
13. Steudel, R., Mäusle, H.-J.: Z. Anorg. Allg. Chem. *457*, 165 (1979)
14. Mäusle, H.-J., Steudel, R.: Z. Anorg. Allg. Chem. *463*, 27 (1980)
15. Steudel, R. et al.: Angew. Chem. *93*, 402 (1981), Int. Edit. Engl. *20*, 394 (1981)
16. Steudel, R., Mäusle, H.-J.: Z. Anorg. Allg. Chem. *478*, 156 (1981)
17. Steudel, R., Mäusle, H.-J.: Angew. Chem. *90*, 54 (1978)
18. Steudel, R., Mäusle, H.-J.: Angew. Chem. *91*, 165 (1979)
19. Steudel, R., Mäusle, H.-J.: Chem. uns. Zeit *14*, 73 (1980)
20. Mäusle, H.-J., Steudel, R.: Z. Anorg. Allg. Chem. *478*, 177 (1981)
21. Steudel, R., Schuster, F.: J. Mol. Struct. *44*, 143 (1978)
22. Steudel, R. et al.: Z. Naturforsch. *35b*, 1378 (1980)
23. Fehér, F.: In Handb. d. Präparativen Anorg. Chem. (Brauer, G. (Ed.)), Bd. 1, 3. Aufl., S. 384, F. Enke, Stuttgart 1975
24. Köpf, H., Block, B., Schmidt, M.: Chem. Ber. *101*, 272 (1968)
25. Schmidt, M., Wilhelm, E.: J. Chem. Soc. Chem. Commun. *1970*, 111
26. Schuster, F.: Dissertation Techn. Univ. Berlin 1980
27. Steudel, R., Sandow, T.: unpublished
28. Schmidt, M. et al.: Angew. Chem. *80*, 660 (1968)
29. Steudel, R. et al.: Z. Naturforsch. *33b*, 1198 (1978)
30. Steidel, J., Steudel, R.: unpublished

31. Steidel, J., Pickardt, J., Steudel, R.: Z. Naturforsch. *33b*, 1554 (1978)
32. Coppens, P. et al.: J. Am. Chem. Soc. *99*, 760 (1977)
33. Goldsmith, L. M., Strouse, C. E.: J. Am. Chem. Soc. *99*, 7580 (1977)
34. Watanabe, Y.: Acta Crystallogr. *30B*, 1396 (1974)
35. Reinhardt, R., Steudel, R., Schuster, F.: Angew. Chem. *90*, 55 (1978)
36. Steidel, J., Steudel, R., Kutoglu, A.: Z. Anorg. Allg. Chem. *476*, 171 (1981)
37. Schmidt, M. et al.: Z. Anorg. Allg. Chem. *405*, 153 (1974)
38. Steudel, R.: Angew. Chem. *87*, 683 (1975), Int. Edit. Engl. *14*, 655 (1975)
39. Kao, J., Allinger, N. L.: Inorg. Chem. *16*, 35 (1977)
40. Fehér, F., Lex, J.: Z. Anorg. Allg. Chem. *423*, 103 (1976)
41. Steudel, R., Sandow, T., Reinhardt, R.: Angew. Chem. *89*, 757 (1977)
42. Wiewiorowski, T. K., Touro, F. J.: J. Phys. Chem. *70*, 3528 (1966)
43. Herman, Z. S., Weiss, K.: Inorg. Chem. *14*, 1592 (1975); see also Allinger, N. L., Hickey, M. J., Kao, J.: J. Am. Chem. Soc. *98*, 2741 (1976)
44. Meyer, B., Oommen, T. V., Jensen, D.: J. Phys. Chem. *75*, 912 (1971)
45. Steudel, R., Rebsch, M.: unpublished; Rebsch, M.: Dissertation, Techn. Univ. Berlin 1972
46. Retcofsky, H. L., Friedel, R. A.: J. Am. Chem. Soc. *94*, 6579 (1972)
47. Steudel, R.: unpublished
48. Berkowitz, J., Chupka, W. A.: J. Chem. Phys. *40*, 287 (1964)
49. Berkowitz, J., Lifshitz, C.: J. Chem. Phys. *48*, 4346 (1968)
50. Záhorsky, U.-I.: Angew. Chem. *80*, 661 (1968)
51. Buchler, J.: Angew. Chem. *78*, 1021 (1966)
52. Cocke, D. L., Abend, G., Block, J. H.: J. Phys. Chem. *80*, 524 (1976)
53. Nimon, L. A. et al.: J. Mol. Spectrosc. *22*, 105 (1967)
54. Berkowitz, J. et al.: J. Chem. Phys. *47*, 4320 (1967)
55. Gautier, G., Debeau, M.: Spectrochim. Acta *A30*, 1193 (1974) and *A32*, 1007 (1976)
56. Steudel, R., Rebsch, M.: J. Mol. Spectrosc. *51*, 189 (1974)
57. Steudel, R., Mäusle, H.-J.: Z. Naturforsch. *33a*, 951 (1978)
58. Nimon, L. A., Neff, V. D.: J. Mol. Spectrosc. *26*, 175 (1968)
59. Eysel, H. H.: personal communication
60. Steudel, R.: Z. Naturforsch. *30b*, 281 (1975)
61. Steudel, R.: Spectrochim. Acta *31A*, 1065 (1975)
62. Chao, J.: Hydrocarbon Processing *59*, 217 (1980)
63. Steudel, R., Strauss, E.-M.: Z. Naturforsch. *36b*, 1085 (1981)
64. Review: Steudel, R.: Z. Anorg. Allg. Chem. *478*, 139 (1981)
65. Tobolsky, A. V., Eisenberg, A.: J. Am. Chem. Soc. *81*, 780 (1959) and *82*, 289 (1960); MacKnight, W. J., Tobolsky, A. V.: In Elemental Sulfur, (Meyer, B. (Ed.)), p. 95, Interscience Publ., New York 1965; see also Klement, W.: J. Polymer Sci. *12*, 815 (1974)
66. Steudel, R., Strauss, R.: unpublished; Strauss, R.: Diplomarbeit, Techn. Univ. Berlin 1981
67. Steudel, R., Strauss, E.-M.: unpublished; Woldt, E.-M.: Diplomarbeit, Techn. Univ. Berlin 1981
68. Davis, R. E.: J. Am. Chem. Soc. *80*, 3565 (1958)
69. Wiewiorowski, T. K., Parthasarathy, A., Slaten, B. L.: J. Phys. Chem. *72*, 1890 (1968)
70. Kende, I., Pickering, T. L., Tobolsky, A. V.: J. Am. Chem. Soc. *87*, 5582 (1965)
71. Tobolsky, A. V. (Ed.): The Chem. of Sulfides, Interscience Publ., New York 1968
72. Pickering, T. L., Saunders, K. J., Tobolsky, A. V.: J. Am. Chem. Soc. *89*, 2364 (1967)
73. Guryanova, E. N.: Quart. Rep. Sulfur Chem. *5*, 113 (1970)
74. Trivette, C. D., Coran, A. Y.: J. Org. Chem. *31*, 100 (1959)
75. Koningsberger, D. C., Neef, T.: Chem. Phys. Lett. *4*, 615 (1970) and *14*, 453 (1972)
76. Gardner, D. M., Fraenkel, G. F.: J. Am. Chem. Soc. *78*, 3279 (1956)
77. Radscheit, H., Gardner, J. A.: J. Non-Cryst. Solids *35/36*, 1263 (1980)
78. Edeling, M., Schmutzler, R. W., Hensel, F.: Phil. Mag. B *39*, 547 (1979)
79. Haas, A., Willner, H.: Spectrochim. Acta *35A*, 953 (1979) and references cited therein
80. Feuerhahn, M., Vahl, G.: Chem. Phys. Lett. *65*, 322 (1979)
81. Chadwick, B. M., Grzybowski, J. M., Long, D. A.: J. Mol. Struct. *48*, 139 (1978)
82. Steudel, R.: Z. Naturforsch. *27b*, 469 (1972)
83. Solouki, B., Bock, H.: Inorg. Chem. *16*, 665 (1977)

84. Baenziger, N. C. et al.: J. Am. Chem. Soc. *91*, 5749 (1969)
85. Steudel, R., Mäusle, H.-J.: unpublished; Mäusle, H.-J.: Dissertation, Techn. Univ. Berlin 1980
86. Steudel, R., Steidel, J., Pickardt, J.: Angew. Chem. *92*, 313 (1980)
87. Albersberg, W. G. L., Vollhardt, K. P. C.: J. Am. Chem. Soc. *99*, 2792 (1977)
88. Huber, K. P., Herzberg, G.: Molecular Spectra and Molecular Structure, IV. Constants of Diatomic Molecules, Van Nostrand Rheinhold, New York 1979
89. Detry, D. et al.: Z. Physik. Chem. N.F. (Frankfurt) *55*, 314 (1967)
90. Steudel, R., Laitinen, R.: Top. Curr. Chem. *102*, 177 (1982)
91. Gobrecht, H., Willers, G., Wobig, D.: Z. Physik. Chem. N.F. (Frankfurt) *77*, 197 (1972)
92. Kastner, M., Fritzsche, H.: Phil. Mag. B *37*, 199 (1978)
93. Kastner, M., Adler, D., Fritzsche, H.: Phys. Rev. Lett. *37*, 1504 (1976)
94. Steudel, R., Latte, J.: Angew. Chem. *86*, 644 (1978); Steudel, R., Sandow, T.: Inorg. Synth., in press
95. Steudel, R., Sandow, T.: Angew. Chem. *90*, 644 (1978)
96. Luger, P. et al.: Chem. Ber. *109*, 180 (1976)

Cyclic Selenium Sulfides[1]

Ralf Steudel[1] and Risto Laitinen[2]

[1] Institut für Anorganische und Analytische Chemie, Technische Universität Berlin, D-1000 Berlin 12, FRG
[2] Department of Chemistry, Helsinki University of Technology, SF-02150 Espoo 15, Finland

Table of Contents

---

[1] Discussion lecture delivered at the 3rd IRIS Meeting held in Graz (Austria), August 17–22, 1981.

1 Introduction

There has recently been a marked growth of interest in inorganic ring systems, which is clearly reflected by the multitude of conferences and monographs. It has also led to increased research activities in the field of cyclic selenium sulfides. Compounds of this type have been known for over a hundred years, but even in 1949, when the only review up to that time on the sulfur-selenium binary system [1] appeared, it was not completely clear whether the different crystalline phases containing sulfur and selenium in varying proportions were true chemical compounds with covalent sulfur-selenium bonds or only mixed crystals consisting of discrete eight-membered sulfur and selenium ring molecules. The advent of modern analytical techniques, e.g. X-ray crystallography, mass spectroscopy, Raman spectroscopy, and high-pressure liquid chromatography, has resulted in considerable progress in the understanding of these phases. It has been shown that cyclic sulfur-selenium compounds with the ring sizes of eight and twelve do exist, but in both cases molecules with different selenium content crystallize together forming phases of complex molecular composition with general formulae of Se_nS_{8-n} and Se_nS_{12-n}. The existence of molecules with ring sizes of six and seven has also been indicated. In addition, it has been possible to prepare mixed crystals with discrete S_8 and Se_8 molecules [2,3].

The present review covers the literature from 1950 till July 1981. The emphasis is on the many different preparation procedures and on structural investigations with the most powerful methods. The uses and applications are mentioned but briefly.

2 Preparation

2.1 The Molten Mixtures of Sulfur and Selenium

In the liquid state sulfur and selenium are known to mix in all proportions. The provisional phase diagram [1] shows an eutectic point at 40 mol-% of selenium (m.p. 105 °C). Mixtures with lower selenium content should show freezing points between 105 and 118 °C while those with higher selenium content are expected to have their freezing points at considerably higher temperatures. In practice equilibrium crystallization of the melt is hindered by supercooling and therefore only the melting points can be studied.

Sulfur-selenium phases can be prepared by cooling molten mixtures of the elements either slowly or by quenching followed by extraction with carbon disulfide, carbon tetrachloride or benzene. The crystals are obtained upon evaporation or cooling of the resulting solutions. Their colour deepens from yellow to ruby red with increasing selenium content [1]. In the older literature there has been some confusion whether to consider these phases as mixed crystals of discrete S_8 and Se_8 molecules or as binary compounds containing SeS bonds.

The first convincing evidence that the molten mixtures of elemental sulfur and selenium and the vapour phase above the melt must contain binary compounds was presented by several Russian authors [4-8]. It was shown that:

a) the pressure and the composition of the vapour above the melts show considerable departure from the behaviour expected for an ideal solution of the two elements in each other,

b) the densities of the saturated and unsaturated vapours of sulfur-selenium mixtures are not additive from the vapour densities of the elements,

c) the volatility of selenium from the mixtures is much higher and that of sulfur much lower than what is predicted by Raoult's law for the mixture of the two elements.

These observations can be explained by the formation of compounds with SeS bonds. The composition of the compounds formed and the molecular size was, however, left undiscussed in these papers [4–8].

Fergusson et al. [9] were the first to report the existence of binary compounds with a general formula Se_nS_{8-n} in these melts. They carried out an extensive investigation by X-ray powder diffraction and by absorption spectroscopy in the infrared, visible, and ultraviolet regions over the whole composition range of molten mixtures of sulfur and selenium cooled down to 20 °C. They also examined phases obtained by recrystallization of the cooled melts from carbon disulfide. All phases were isomorphic with one of the allotropes of S_8 and Se_8 indicating that the structures also consist of cyclic eight-membered molecules:

Phase I	0–18 weight-% Se	α-S_8 structure
Phase II	20–49 weight-% Se	γ-S_8 structure
Phase III	50–68 weight-% Se	α-Se_8 structure

The phases with highest selenium content are the least soluble and the least volatile, allowing fractionation by recrystallization and vacuum sublimation, though it was thought possible that rearrangement reactions occur during these processes. The absorption spectra of these phases have been interpreted to indicate the presence of cyclic Se_nS_{8-n} molecules. All phases exhibit an infrared absorption near 470 cm^{-1}, typical for SS bonds, which becomes progressively weaker in phases I, II, and III. However, its presence even in phase III which had a selenium content of 68% indicates random distribution rather than alternation of the atoms in the molecules [9].

Earlier X-ray powder diffraction studies by de Haan and Visser [2] provided similar results. These authors investigated the impact of the selenium content on the structure type of the crystals and deduced the existence of eight-membered ring molecules but were unable to decide whether the crystals consisted of binary compounds of sulfur and selenium or simply of Se_8 and S_8.

In 1963 Hawes [10] claimed the preparation of the first stoichiometrically pure compounds, namely Se_4S_4 (red tabular crystals), Se_2S_6 (orange needles), and SeS_7 (obtained only as an addition compound, 2 $SeS_7 \cdot SnI_4$). Se_4S_4 (m.p. 113 °C) and Se_2S_6 (m.p. 121.5 °C) were obtained from the mixed melts by extraction with boiling benzene and crystallization at 20 °C. Though the observed selenium contents and molecular weights support the formulae, no convincing evidence for the presence of pure stoichiometric compounds was given. Later it was shown [11] that the adduct "2 $SeS_7 \cdot SnI_4$" of Hawes, prepared from the mother liquor of the Se_2S_6

crystallization by adding SnI_4, consists of mixed crystals of type $2\, Se_nS_{8-n} \cdot SnI_4$; the Raman spectrum of this phase revealed the presence of both SeS and SeSe bonds.

The isolation of pure compounds from mixed sulfur-selenium melts was further questioned by mass spectroscopic studies [3,12–16]. The existence of SeS_7, Se_2S_6, Se_3S_5, and of molecules of even higher selenium content was seen in the mass spectra taken from the samples of both quenched melts and of crystals obtained by solvent extraction of them. It must be taken into account, however, that even gentle heating of the samples may initiate disproportionation reactions. Also the volatility of different ring molecules may be quite different. Therefore the composition of the vapour does not necessarily reflect the composition of the solid phase in all details.

Ailwood and Fielding [15] have made an extensive study on the recrystallization products of cooled molten mixtures. Starting from a sulfur-selenium phase with an atomic ratio of 4:1 they obtained 18 fractions by repeated recrystallization from benzene. The selenium content in these fractions varied from 3.2 to 52.1 mol-%. The colour of the crystals ranged from pale yellow to deep red. Obviously any composition between S_8 and Se_4S_4 can be obtained this way. The mass spectra of the samples indicated the presence of all species of the Se_nS_{8-n} series.

It has also been reported that crystals with selenium contents up to 80 weight-% can be grown from solutions [1,16].

It can be concluded that pure stoichiometric selenium sulfides cannot be obtained from molten mixtures of the elements even by repeated recrystallization from organic solvents.

2.2 The Reaction of Selenium Dioxide and Hydrogen Sulfide

The reaction of aqueous SeO_2 (selenous acid) with H_2S at 20 °C yields a yellow precipitate of composition SeS_2 which has earlier been named as "selenium disulfide" [1].

$$2\, H_2S + SeO_2 \rightarrow SeS_2 + 2\, H_2O$$

Fergusson et al. [9] pointed out that the X-ray powder diagram of the product was characteristic of their solid solution phase II indicating γ-sulfur structure and cyclic eight-membered molecules. Recently Weiss [17] confirmed it by determining the crystal structures of three phases with compositions of $Se_{2.9}S_{5.1}$, $Se_{3.3}S_{4.7}$, and $Se_{4.7}S_{3.3}$ which were formed on extraction of "SeS_2" with organic solvent followed by crystallization.

The preparation of "SeS_2" can also be carried out in polyalcohols [18] or using Na_2S [19,20] or $(NH_4)_2S$ [19] instead of H_2S. For example, 40.5 g Na_2S in 150 cm^3 of water yields, after addition of 115 cm^3 of hydrochloric acid and 32.2 g of H_2SeO_3 at 5–7 °C, 33.4 g of "SeS_2" with a selenium content of 57–62 weight-% [20].

2.3 The Reaction of Selenium Dioxide and N-Benzylthiamides

SeS_3 (m.p. 111 °C) is said to be formed in the reaction of various N-benzylthiamides ($R—CS—NH—CH_2—C_6H_5$) and selenium dioxide in ethanol. After repeated recrystallization from carbon disulfide long red hexagonal prisms of SeS_3 are obtained from the crude product [21].

2.4 The Reaction of Sulfur Dioxide and Hydrogen Selenide

The reaction of sulfur dioxide and hydrogen selenide in water yields an orange precipitate of the approximate composition Se_2S [22] the nature of which has not been elucidated. The product is partly soluble in carbon disulfide.

2.5 The Reaction of Chlorosulfanes and Hydrogen Selenide

In 1970 Cooper and Culka [23] described the reaction of hydrogen selenide with S_7Cl_2. Applying the dilution principle the starting materials were reacted in carbon disulfide/diethyl ether mixture at 15 °C to give a yellow needle-like product which after recrystallization from carbon disulfide/benzene mixture showed a composition of $SeS_{7.4}$. The mass spectrum recorded at a sample temperature of 40 °C revealed the parent ions S_8^+ and SeS_7^+ with high intensities. The presence of both S_8 and SeS_7 has also been shown by Raman spectroscopy [24].

The reaction of S_2Cl_2 with H_2Se takes place under similar conditions as described above and yields yellow crystals of composition Se_2S_6 (m.p. 123–124 °C) together with orange crystals of composition Se_3S_5 (m.p. 119–120 °C) [25]. Molecular weight determinations indicate the existence of eight-membered ring molecules, but as no spectra or crystal structure determinations of these samples have been reported, the question whether they contain SeSe bonds remains unsettled. The authors themselves did not think that adjacent selenium atoms were present. However, their observation that Se_2S_6 and Se_3S_5 react more slowly with triphenylphosphine than Se_8 can be explained by the lower number of SeSe bonds in the mixed compounds.

2.6 The Reaction of Chloroselanes and Sulfanes

A mixture of sulfanes (H_2S_n, n = 3, 4, ...; $n_{ave} = 6$) known as "crude sulfane" reacts with both $SeCl_4$ and Se_2Cl_2 in carbon disulfide at low temperatures to form cyclic selenium sulfides of types Se_nS_{8-n} and Se_nS_{12-n} [26]. On extraction and recrystallization from carbon disulfide a mixture of twelve-membered ring molecules was obtained as yellow crystals with a selenium content of up to 29 weight-% corresponding to the formula $Se_{1.7}S_{10.3}$. As this sample was much less soluble in CS_2 than both S_{12} and compounds of type Se_nS_{8-n}, it could easily be separated from the latter. The twelve-membered ring molecules have so far been characterized by X-ray crystallography only (see 3.2).

2.7 The Reaction of Bromoselanes and Hydrogen Sulfide

Powdered amorphous selenium and liquid bromine are shaken together in dry methanol. After 30 minutes a saturated solution of H_2S in methanol is added and a part of the solvent is distilled off. Upon slow cooling yellow to orange crystals of Se_nS_{8-n} appear with n amounting to approximately 1 or 4 depending on the initial ratio of selenium and hydrogen sulfide [27].

2.8 The Reaction of Dichlorodiselane, Chlorosulfanes, and Potassium Iodide

It has been shown recently that at 20 °C the unstable iodosulfanes S_nI_2 (n = 1, 2, ...) decompose to give homocyclic sulfur molecules and iodine [28]. In a similar fashion Se_2I_2 prepared from Se_2Cl_2 and solid KI in carbon disulfide solution produces Se_8 and polymeric selenium [29]. Mixed sulfur-selenium ring molecules can be prepared by the reaction of mixtures of Se_2Cl_2 and chlorosulfanes (S_nCl_2; n = 1, 2, 4, 6–8) with potassium iodide in carbon disulfide solution. The subsequent decomposition of the unstable iodides results in the formation of mixed Se_nS_{8-n} molecules. The general reaction scheme can be presented as follows:

$$\left.\begin{array}{l} S_mCl_2 \xrightarrow{KI} S_mI_2 \\ Se_2Cl_2 \xrightarrow{KI} Se_2I_2 \end{array}\right\} \rightarrow Se_nS_{8-n} + I_2$$

For example, a mixture of SCl_2 and Se_2Cl_2 reacts with KI to give a mixture of Se_nS_{8-n} compounds with n = 1–8. Investigations by high-pressure liquid chromatography have indicated that the reaction mechanism is very complex and probably involves the rupture of both SS and SeSe bonds. It is also possible that ring rearrangement reactions take place in the solutions [29].

2.9 The Reaction of Titanocene Pentasulfide and Dichlorodiselane

Titanocene pentasulfide, $(C_5H_5)_2TiS_5$ (often abbreviated as Cp_2TiS_5), is known to react with S_2Cl_2 to give cycloheptasulfur S_7 and Cp_2TiCl_2 [30]. Se_2Cl_2 reacts similarly with Cp_2TiS_5 in carbon disulfide solution at 0 °C:

$$Cp_2TiS_5 + Se_2Cl_2 \rightarrow Se_2S_5 + Cp_2TiCl_2$$

The unstable seven-membered ring molecule disproportionates rapidly to form eight-membered Se_3S_5 and a few minor components. The Se_3S_5 was characterized by chemical analysis, molecular weight determination, high-pressure liquid chromatography (HPLC), X-ray crystallography, and Raman spectroscopy and was shown to contain mainly the isomer 1,2,3-Se_3S_5 [31] (see also 3.2–3.4).

2.10 Miscellaneous Reactions

Electrolytic reduction of H_2SO_4/H_2SeO_4 mixtures using aluminium electrodes results in the formation of a sulfur-selenium coating at the electrode [32].

Irradiation of a solution of Se_8 in carbon disulfide by sunlight for 1–2 hours at 20 °C results in the formation of various cyclic selenium sulfides [33]. The formation of Se_nS_{8-n} compounds is also initiated by refluxing the above solution [29].

3 Structural, Spectroscopic, and Physical Properties

3.1 Isomerism

Cyclic selenium sulfides constitute a system of considerable structural complexity. In addition to the various stoichiometric compositions of the molecules there are many possibilities for positional and optical isomerism. In the case of the eight-membered ring molecules the following 28 molecules (excluding optical isomers) can exist:

SeS_7	
Se_2S_6	(1,2-, 1,3-, 1,4-, and 1,5-isomers)
Se_3S_5	(1,2,3-, 1,2,4-, 1,2,5-, 1,3,5-, and 1,3,6-isomers)
Se_4S_4	(1,2,3,4-, 1,2,3,5-, 1,2,3,6-, 1,2,4,5-, 1,2,5,6-, 1,2,4,7-, and 1,3,5,7-isomers)
Se_5S_3	(5 isomers like in Se_3S_5)
Se_6S_2	(4 isomers like in Se_3S_5)
Se_7S	

Many of the molecules are chiral. Six of the above-mentioned compounds contain SeSe bonds and no isolated sulfur or selenium atoms (structural units of type —S—Se—S— or —Se—S—Se—), nine contain SeSe bonds and isolated sulfur or selenium atoms, and 13 contain no SeSe bonds. The number of SeS bonds in these molecules can be either 2, 4, 6, or 8.

3.2 Crystal and Molecular Structures

Preliminary X-ray investigations [1,2,9] have been followed by several detailed crystal structure determinations of various sulfur-selenium phases with ring sizes of eight and twelve. The unit cell parameters of the eight-membered species are given in Table 1. The unit cell parameters of monoclinic γ-sulfur [34] and of monoclinic α-selenium [35] are included for comparison.

It can clearly be seen from the crystal data of Table 1 that the phases form two isomorphic series with a break at about 50 mol-% of selenium. The sulfur-rich species crystallize in the monoclinic γ-sulfur lattice and the selenium-rich species in the monoclinic α-selenium lattice (see Fig. 1).

Table 1. The unit cell parameters of cyclic eight-membered selenium sulfides

Approximate Composition	Space Group	a (Å)	b (Å)	c (Å)	β (°)	Ref.
γ-S_8	$P2/c$	8.442(30)	13.025(10)	9.356(50)	124.98(30)	[34]
$Se_{1.1}S_{6.9}$	$P2/c$	8.34(9)	13.11(4)	9.30(6)	123.9	[27]
$Se_2S_6(1)$ [a]	$P2/c$	8.536	13.241	9.343	124.56	[16]
$Se_2S_6(2)$ [a]	$P2/c$	8.580	13.332	9.383	124.35	[16]
$Se_{2.9}S_{5.1}$	$P2/c$	8.578	13.386	9.368	124.3	[17]
$Se_3S_5(1)$ [a]	$P2/c$	8.595	13.397	9.396	124.25	[16]
$Se_3S_5(2)$ [a]	$P2/c$	8.581	13.364	9.385	124.25	[16]
Se_3S_5	$P2/c$	8.55	13.32	9.28	124.0	[36]
Se_3S_5	$P2/c$	8.564(5)	13.354(9)	9.368(7)	124.32(5)	[37]
Se_3S_5	$P2/c$	8.550(5)	13.340(8)	9.336(4)	124.17(5)	[31]
$Se_{3.3}S_{4.7}$	$P2/c$	8.579	13.405	9.354	124.1	[17]
$Se_{3.7}S_{4.3}$	$P2/c$	8.40(8)	13.26(10)	9.37(2)	124.5(8)	[27]
$Se_{4.7}S_{3.3}$	$P2_1/n$	8.737	9.104	11.316	90.8	[17]
$Se_5S_3(1)$ [a]	$P2_1/n$	8.787	9.129	11.389	90.96	[16]
$Se_5S_3(2)$ [a]	$P2_1/n$	8.999	9.109	11.550	90.79	[16]
α-Se_8	$P2_1/n$	9.054(3)	9.083(5)	11.601(6)	90.81(5)	[35]

a The unit cell parameters shown in the table are the mean values obtained from four single crystals. The phases (1) and (2) show the same composition, but have been prepared by starting from different molar ratios of sulfur and selenium: Se_2S_6 (1) S:Se = 7:1, (2) S:Se = 6:2; Se_3S_5 (1) S:Se = 5.5:2.5, (2) S:Se = 5:3; Se_5S_3 (1) S:Se = 1:1, (2) S:Se = 2:6.

Table 2. Details of the structure determinations of cyclic eight-membered selenium sulfides

Phase Designation	Approximate Composition S:Se	Number of Reflections	2θ max (°)	*R*-value	Bond length (Å)	Bond Angle (°)	Ref.
SeS_7	2:6	543[a]	60	0.056	2.070	107.5	[16]
Se_2S_6	2:6	1167[a]	60	0.104	2.173	106.3	[16]
Se_4S_4 [a]	3:5	1113[a]	60	0.085	2.238	105.8	[16]
Se_5S_3	3:5	1075[a]	60	0.082	2.273	105.8	[16]
$Se_{2.9}S_{5.1}$	5.1:2.9	1266[b]	60	0.069	2.173	106.1	[17]
$Se_{3.3}S_{4.7}$	4.7:3.3	1597[b]	60	0.074	2.175	105.9	[17]
$Se_{4.7}S_{3.3}$	3.3:4.7	1729[b]	68	0.114	2.228	106.0	[17]
$Se_{1.1}S_{6.9}$	6.9:1.1	309	—	0.080	2.087	105.6	[27]
$Se_{3.7}S_{4.3}$	4.3:3.7	568	—	0.097	2.127	106.2	[27]
1,2,3-Se_3S_5	5:3	1817[a]	60	0.056	2.185[c]	104.6[c]	[31]
Se_3S_5	5:3	1206	—	0.054	2.166	106.2	[37]

a $I \geqq 3\sigma(I)$; b $I \geqq 2.48\sigma(I)$; c Both sulfur and selenium have been refined independently by constraining the bond lengths loosely to the expected values of 2.05, 2.20 and 2.35 Å for SS, SeS, and SeSe bonds, respectively. The average of the final calculated bond parameters: $r_{SS} = 2.067$ Å, $r_{SeS} = 2.181$ Å, $r_{SeSe} = 2.308$ Å, $\alpha_{SeSeSe} = 96.3°$, $\alpha_{SeSeS} = 100.1°$, $\alpha_{SeSS} = 109.1°$, and $\alpha_{SSS} = 112.9°$.

Detailed structure determinations [16,17,27,31,36,37] have confirmed that in all cases the lattices are built up of eight-membered crown-shaped ring molecules like S_8 and Se_8 as was expected due to the morphology of the crystals. All the structures were found to be disordered with sulfur and selenium distributed statistically

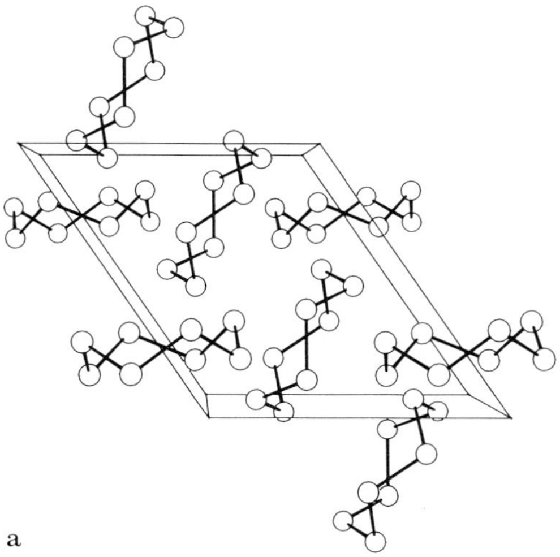

a

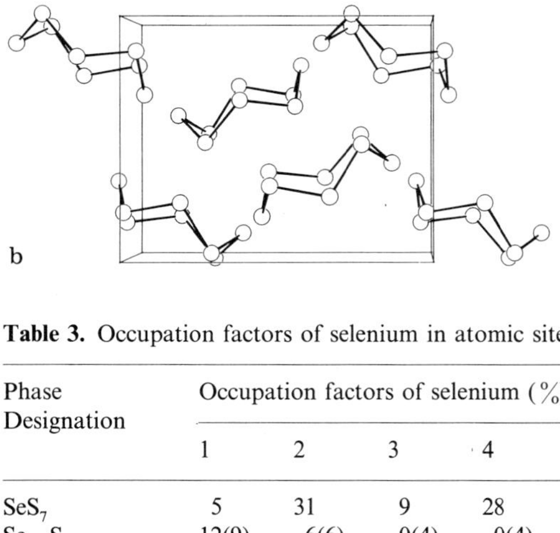

b

Fig. 1a and b. The unit cell contents of (**a**) monoclinic γ-sulfur (a horizontal, c vertical) and (**b**) monoclinic α-selenium (c horizontal, b vertical) as completed to show full molecules. Atomic coordinates of monoclinic γ-sulfur are according to Watanabe [34] and those of monoclinic α-selenium according to Cherin and Unger [35]

Table 3. Occupation factors of selenium in atomic sites of various Se_nS_{8-n} species

Phase Designation	Occupation factors of selenium (%)[a]								Ref.
	1	2	3	4	5	6	7	8	
SeS_7	5	31	9	28	18	0	12	0	16)
$Se_{1.1}S_{6.9}$	12(9)	6(6)	0(4)	0(4)	26(16)	21(13)	20(13)	24(15)	27)
Se_2S_6	18	63	15	25	35	10	30	4	16)
$Se_{2.9}S_{5.1}$	45	38	32	32	35	35	39	38	17)
Se_3S_5	62	42	23	17	34	39	45	40	37)
1,2,3-Se_3S_5	71	41	10	8	30	27	33	37	31)
$Se_{3.3}S_{4.7}$	53	43	36	32	43	43	42	42	17)
$Se_{3.7}S_{4.3}$	86(17)	58(10)	16(7)	16(8)	48(9)	42(9)	48(9)	41(9)	27)
Se_4S_4 [a]	75	50	44	55	37	28	53	57	16)
$Se_{4.7}S_{3.3}$	67	45	47	62	44	27	38	69	17)
Se_5S_3	92	60	56	73	53	40	63	63	16)

a The atoms have been renumbered to correspond each other

over the atomic sites. Details of the structure determinations of the eight-membered species are given in Table 2 and the comparison of occupational factors in Table 3.

Because of the disorder, the bond parameters have no physical significance as they represent only the average values of all the possibilities in the mixed crystal. Therefore it is impossible to deduce the nature of the bonds in the crystal from X-ray diffraction data.

The disorder mainly arises from the presence of several Se_nS_{8-n} molecules in the same lattice. However, the atomic radii of sulfur and selenium are fairly close together and it is possible that any one ring molecule can assume random orientation. This is indicated by the phase prepared from titanocene pentasulfide and dichlorodiselane [31] the main component of which is 1,2,3-Se_3S_5 with only a few minor components. The disorder observed is very similar to all other cases investigated (see Table 3). This renders X-ray crystallography a doubtful means for characterizing even the purely stoichiometric compounds. No superstructure has been observed in the crystals.

Intermolecular distances in these sulfur-selenium phases are comparable to those observed for monoclinic γ-sulfur [34] and monoclinic α-selenium [35]. The shortest distances in both isomorphic series are much shorter than the sum of the van der Waals' radii of the elements. In the isomorphic series of the sulfur-rich species the shortest distance between the molecules gets shorter with increasing selenium content of the phase but remains effectively constant in the selenium-rich series.

Weiss and Bachtler [26] have determined the crystal structure of the twelve-membered selenium sulfide. Recrystallization from benzene yields needle-like orthorhombic crystals with the space group *Pnnm* and unit cell dimensions a = 4.774, b = 9.193, and c = 14.680 Å. These values are comparable to the cell parameters of S_{12} [38]. If the crystallization is carried out in carbon disulfide solution and adduct $Se_nS_{12-n} \cdot CS_2$ is obtained [39]. This behaviour is also observed for S_{12} [38]. Both

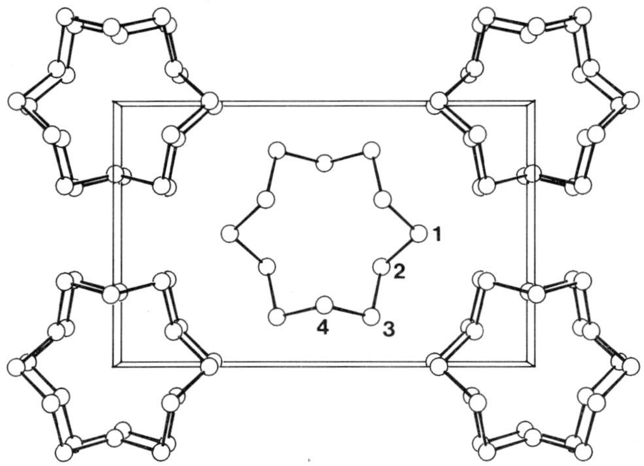

Fig. 2. The unit cell contents of Se_nS_{12-n} as completed to show full molecules (c horizontal, b vertical). Atomic coordinates are according to Weiss and Bachtler [26]. Of the four independent atomic sites two (2 and 4) are occupied by sulfur only while the other two (1 and 3) are occupied by both sulfur and selenium. The occupation factor of selenium in these sites is 25%

$Se_nS_{12-n} \cdot CS_2$ and $S_{12} \cdot CS_2$ crystallize in the trigonal space group $R\bar{3}m$. The molecular geometry ot he twelve-membered selenium sulfide ring molecule resembles that of S_{12}. The structure, however, is also disordered. The distribution of sulfur and selenium over the atomic sites is shown in Fig. 2.

There are no reports on the structures of selenium sulfides with a ring size other than eight or twelve though some indications have been observed for their existence (see 3.4).

3.3 Vibrational Spectra

The vibrational spectra of homocyclic sulfur molecules (S_n, n = 6–10, 12, 18, 20) are extremely characteristic since they reflect not only the size and symmetry of the molecules but also the bond distance pattern [40]. Thus vibrational spectroscopy is a powerful means to detect and identify cyclic S_n molecules even in the mixtures of several members of this homologous series. Force constant calculations have been carried out for many of these molecules [40–42] and also for Se_8 [43].

The fundamental vibrations of 13 cyclic Se_nS_{8-n} molecules have been calculated by using force constants and structural parameters adapted from those of S_8 and Se_8 [44]. The wave numbers obtained allow the identification of single molecules in simple mixtures [31]. The SS stretching modes are found in the region 480–430 cm^{-1}, the SeSe stretching modes at 270–220 cm^{-1}, and the SeS stretching modes at 400–320 cm^{-1}. The ring bending and torsional modes occur below 250 cm^{-1} and therefore interfere only with the SeSe stretching modes [44].

The i.r. absorption intensity of sulfur rings is low and the same can be expected for selenium sulfides. Bands at 471–463 cm^{-1} [9] and 355 cm^{-1} [45], characteristic for SS and SeS stretching modes, have, however, been observed. The latter absorption

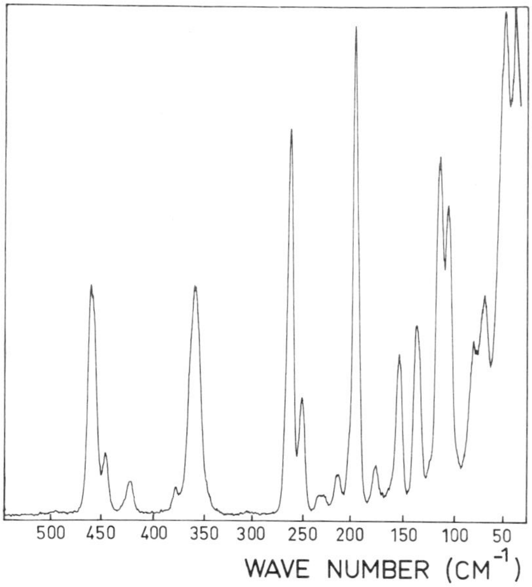

Fig. 3. Raman spectrum of solid 1,2,3-Se_3S_5 recorded at about −100 °C [31]

band has been found in samples with as low a sulfur content as 2.5 mol-% and has been attributed to the Se_3S_5 molecule [45]. The observed broad band indicates, however, that the sample is more likely to contain a mixture of several ring and chain-like molecules.

Raman scattering from crystalline S_8, Se_8, and cyclic selenium sulfides is very intense. The Raman spectrum of 1,2,3-Se_3S_5 [31] is shown in Fig. 3 as a typical example of the spectra of mixed Se_nS_{8-n} species. The first Raman spectra of liquid and solid sulfur-selenium phases were reported in 1968 by Ward [46]. For a mixture of composition $Se_{0.05}S_{0.95}$ he observed nine weak Raman lines not present in the spectrum of S_8 and assigned them to different Se_nS_{8-n} (n = 0–4) molecules by comparison with the spectra of S_8 and Se_8. According to the more recent calculations [44] the assignment given in Table 4 is the most likely one.

Table 4. The nine weak Raman lines of liquid and solid sulfur-selenium phase with composition $Se_{0.05}S_{0.95}$ [46]. The assignment of the lines is given according to the calculations by the present authors [44]. Only lines not arising from S_8 have been included in the Table

Raman lines (cm^{-1})		Polarization	Assignment
Solid (25 °C)	Liquid (120 and 153 °C)		
122	122	depol.	Ring bending modes
129	128	depol.	
138	138	depol.	
180			
201	202	pol.	
347	344	depol.	$\nu_{asym}(SeS)$[a]
363	360	pol.	$\nu(SeS)$[b]
382	380	pol.	$\nu_{sym}(SeS)$[a]
434		pol.	$\nu(SS)$

a In structural units of the type –S–Se–S–
b In structural units of the type –S–Se_n–S– with $n \geqq 2$

In 1977 Datta and Krishnan [47] reported a "compound" with composition Se_4S_4 which was obtained by extracting the cooled equimolar melt of the elements with benzene. The authors stated that "Se_4S_4 does not contain either SeSe or SS bonds and thus belongs to the ring isomer having SeS bonds only" but as Eysel and Sunder [48] have pointed out the spectrum of Se_4S_4 by Datta and Krishnan shows Raman lines typical for covalent SS bonds (near 470 cm^{-1}) and SeSe bonds (263 cm^{-1}). It is also more likely according to the method of the preparation that a mixture of various Se_nS_{8-n} molecules crystallize together rather than that a pure stoichiometric compound is formed (see 2.1).

Eysel and Sunder [49] showed by Raman spectroscopy that all sulfur-selenium phases obtained by recrystallization of commercial "SeS_2" and of an equimolar melt contain SeSe bonds even at very low selenium contents. The composition of the samples studied varied from $Se_{0.05}S_{7.95}$ to $Se_{5.9}S_{2.1}$. These results were confirmed and extended by the present authors [44] who measured the low temperature laser Raman spectra of four sulfur-selenium phases obtained by recrystallization of the quenched melts of the two elements. The composition of the phases varied from Se_2S_6 to Se_5S_3. In all cases SS, SeS, and SeSe stretching frequencies were found. The relative intensities of the SS lines decreased and those of SeSe lines increased with increasing selenium content. S_8 was detected in the phases of composition Se_2S_6 and Se_3S_5. It was concluded that all crystals contain several members of the series Se_nS_{8-n} and that selenium is mostly present in SeSe bonds rather than as "isolated" selenium atoms. This latter conclusion, however, needs revision. In the SeS stretching region of the spectra of the four phases there are up to four lines between 382 and 346 cm^{-1}. Eight-membered ring molecules with cumulated selenium atoms like in 1,2,3-Se_3S_5 [31] give rise to SeS stretching frequencies in the region 366–357 cm^{-1} (both symmetric and asymmetric stretching vibrations of the SeS bonds in the unit —S—Se_n—S— with n = 2–6). Compounds containing isolated selenium atoms in the ring (unit —S—Se—S—) show Raman lines at 384–372 cm^{-1} (ν_{sym}) and 350–340 cm^{-1} (ν_{asym}) [44]. According to these results the various phases obtained from the melts clearly contain compounds with both structural units.

The Raman spectrum of the reaction product of titanocene pentasulfide and dichlorodiselane shows the presence of 1,2,3-Se_3S_5 as the main species [31] (see Fig. 3). The normal coordinate treatment using a modified Urey-Bradley force field with 12 independent force constants resulted in a complete assignment of the spectrum and in a very good agreement between the observed and calculated wave numbers. The composition of the phase as determined from the HPLC data, molecular weight measurement, and selenium analysis agrees with the vibrational analysis.

The most characteristic Raman line of the S_8 molecule is the totally symmetric ring bending mode (a_1) occurring in the solid state (orthorhombic α-sulfur) at 218 cm^{-1} with high intensity. As the wave number of this vibration depends on the ring size and on the degree of substitution of sulfur in the ring it can be used to identify eight-membered sulfur-rich rings in the sulfur-selenium phases:

S_8O	219 cm^{-1} [50]
S_7NH	215 cm^{-1} [51]
SeS_7	215 cm^{-1} [24]
1,2,3-Se_3S_5	198 cm^{-1} [31].

With increasing degree of selenium substitution a new strong line occurs near 112 cm^{-1} which is the wave number of totally symmetric ring bending (a_1) mode of Se_8 [43]. For example, 1,2,3-Se_3S_5 exhibits strong Raman lines at 114 and 106 cm^{-1} (see Fig. 3) [31].

There are no spectra reported for sulfur-selenium phases with the ring sizes other than eight.

3.4 High-Pressure Liquid Chromatography

Chromatographic methods seem to offer the only possibility to provide pure stoichiometric selenium sulfides as fractional crystallization and sublimation result in more or less complicated mixtures. Conventional column chromatography has turned out to be unsuccessful, but recently it has been shown that high-pressure liquid chromatography (HPLC) is a suitable method to separate at least some members of the Se_nS_{8-n} series [29,31]. Columns with octadecylsiloxane (C-18) as a stationary phase and methanol as eluent have been applied. The very intense absorption of sulfur and selenium compounds at 254 nm allows the detection of trace amounts by means of a UV absorbance detector.

A chromatogram of a sample prepared from SCl_2, Se_2Cl_2, and KI is shown in Fig. 4 [29]. The eight peaks represent those members of the series Se_nS_{8-n} (n = 1–8)

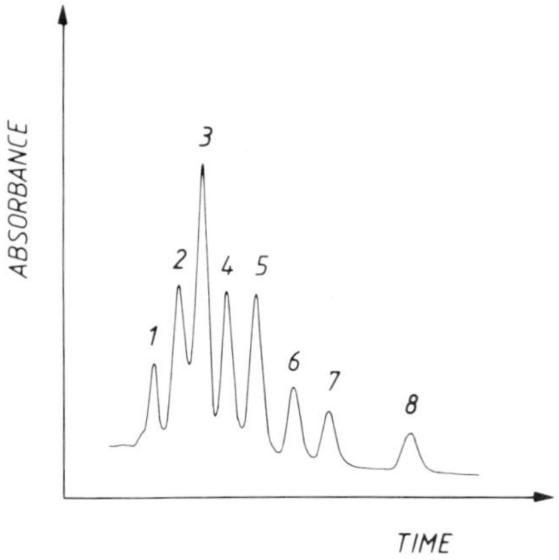

Fig. 4. HPLC-Chromatogram of the selenium sulfide mixture obtained from the reaction of SCl_2 and Se_2Cl_2 with potassium iodide. The figures give the numbers of selenium atoms in the molecule [29]

which contain all selenium atoms in adjacent positions. The retention time increases with the number (n) of selenium atoms in the molecule in a regular manner: the logarithm of the capacity factor is a linear function of *n* (see Fig. 5). The non-polar molecules S_8 and Se_8 do, however, not fit this relationship. This result is in agreement with current theories on the influence of the various molecular parameters on the retention time in reversed-phase chromatography [52]. Under high resolution conditions it has even been possible to separate isomeric compounds [29].

Certain selenium sulfide mixtures obtained by the reactions described in sections 2.8. and 2.9. yielded, on investigation by HPLC, chromatographic peaks at retention times lower and higher than those of the various eight-membered rings.

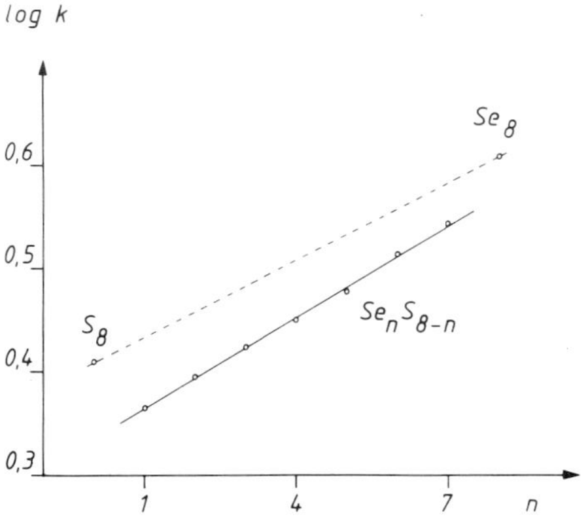

Fig. 5. Dependence of the logarithm of the capacity factor on the number of selenium atoms in the eight-membered ring molecules of type Se_nS_{8-n}; the values of S_8 and Se_8 are given for comparison [29]

These peaks are assigned to six- and seven-membered rings (low retention time) and twelve-membered rings (high retention time) since they occur with similar retention times as has been observed for the corresponding homocyclic molecules S_6, S_7, and S_{12} [29].

3.5 Thermodynamic Properties

Only a few thermodynamic data on cyclic selenium sulfides are available. The appearance potential of gaseous SeS_7^+ (9.3 ± 0.2 eV) is by 0.25 eV lower than that of S_8^+ but 0.7 eV higher than that of Se_8^+ [3]. The enthalpy of mixing of liquid sulfur and selenium is positive throughout the composition range at temperatures of 345 to 460 °C. Obviously the reaction

$$-S-S- + -Se-Se- \rightarrow 2 -Se-S-$$

is slightly endothermic. In an equimolar mixture the enthalpy of mixing amounts to 1.3 kJ/mol mixture [53]. It is interesting to note that also the reaction of the gaseous diatomic molecules

$$S_2 + Se_2 \rightleftarrows 2\,SeS$$

is endothermic ($\Delta H_0^0 = 5.4 \pm 2.9$ kJ/mol S_2) [54].

According to the above evidence, homonuclear sulfur and selenium bonds seem to be slightly more favourable energetically than two heteronuclear sulfur-selenium bonds. The entropy change ΔS_0^0 of the above reactions must, however, be positive due to the loss of symmetry. Therefore the Gibbs free energy which determines the equilibrium position will be negative at moderate or high temperatures, thus shifting the equilibria of the above reactions to the right with increasing temperature.

The polymerization threshold temperature which is 158 °C for pure sulfur is lower for mixtures of sulfur and selenium. It decreases monotonically with increasing selenium content, being 94 °C at 75 mol-% of selenium [55].

No polymorphic changes of solid Se_nS_{8-n} phases have been observed according to a DSC investigation [56]. The melting process was found to be irreversible due to the decomposition of the samples on melting. The reported linear relationships of the enthalpy of melting with the selenium content in the two isomorphic series have no physical significance owing to the mixed crystal nature of the phases.

4 Chemical Properties, Toxicity, and Uses

4.1 Reactions

Very few chemical reactions of cyclic selenium sulfides have been reported and no systematic study is available. The most important reaction seems to be the thermal decomposition which has been reported both in the solid state and in the solution at 20 °C. Umilin et al. [12)] observed that a sulfur-selenium mixture subjected to mass spectroscopy immediately after fusion and cooling contained molecules of higher selenium content with larger abundance than samples which were kept at room temperature for about a month after fusion. It has also been reported [2)] that the X-ray powder diffraction diagram of the cooled equimolar melt changed on storage. The sample of presumably polymeric material was kept at room temperature for a week after which time additional lines due to hexagonal selenium were observed.

Hawes [10)] claimed that the phase of composition Se_2S_6 which was obtained from the melt decomposed in benzene solution within a few weeks at room temperature. With boiling benzene the change was effected much faster. According to Hawes the reaction products are Se_4S_4 and S_8.

Recently it has been observed that Se_8 in dilute carbon disulfide solution dissociates giving rise to an equilibrium with Se_6 and Se_7 [33)]. Similarly, after standing for 30 minutes in carbon disulfide solution, 1,2,3-Se_3S_5 showed additional bands in the chromatogram in the region which is typical for six- and seven-membered sulfur rings [29)]. These bands are not observed in the freshly dissolved sample.

Triphenylphosphine reacts with cyclic selenium sulfides to produce Ph_3PS and Ph_3PSe [25,57)]. The reaction with triphenylarsine proceeds in an analogous manner [57)].

4.2 Toxicity

Commercial selenium disulfide "SeS_2" has been found to be very untoxic (for mice LD 50 per os: 3.7 g/kg). Thus the material is 100 times less toxic than selenites. Daily oral administration of 5, 25, or 125 mg/kg of SeS_2 for seven and a half weeks was tolerated by mice without signs of intoxication [58)]. As "SeS_2" is used as a constituent of shampoos several investigations on the effects of the sulfide on the eyes, skin, and hair have been published [59–68)]. Also the chemotherapeutic activity of commercial selenium disulfide in experimental leptospirosis has been investigated [69)].

4.3 Uses

The uses of selenium sulfides of compositions SeS and SeS_2 in antidandruff shampoos [18,70–75)], in detergent bars [76)], in fireworks [77)], as an inhibitor for polymerization [78)], for the colouration of glass [79)], and — as the electric resistance of selenium-rich sulfur-selenium phases decreases on irradiation with visible light [1)] — in the manufacture of electrophotographic plates [80)] have been proposed. It has also been suggested [19)] that SeS_2 precipitation can be used in the analytical estimation of selenium. The sample containing selenium is oxidized, the resulting selenous acid allowed to react with salt-like sulfide, and the SeS_2 precipitated is filtered, washed, dried, and weighed.

5 Summary and Outlook

Recent developments in the chemistry of cyclic selenium sulfides have revealed that sulfur and selenium form a very complex binary system. Cyclic eight-membered molecules with the general formula Se_nS_{8-n} can be prepared from the molten mixtures of sulfur and selenium and also by a variety of chemical reactions. In all cases mixtures of different selenium sulfides are formed which it has not yet been possible to separate. In the reaction of titanocene pentasulfide with dichlorodiselane, however, the resulting mixture is simple enough to allow the major component to be identified as 1,2,3-Se_3S_5 [31].

The structures and compositions of the sulfur-selenium phases have been studied by X-ray crystallography, mass spectroscopy, Raman spectroscopy, and high-pressure liquid chromatography. With crystallographic investigations it has been possible to prove that the molecules present are either eight- or twelve-membered puckered rings, but as all structures are disordered and are likely to be so even in the case of pure stoichiometric compounds, it will be difficult to obtain information on the bond distances, bond angles and other molecular parameters by this method.

Raman spectroscopy is a powerful means for studying selenium sulfides. The Raman spectra show the presence of SS, SeS, and SeSe bonds in all samples. From the splitting of the Raman lines in the SeS stretching region it can be seen that selenium atoms are present in the mixtures either as cumulated or isolated atoms (structural units —S—Se_n—S—; $n \geqq 2$ or —S—Se—S—, respectively). Also the presence of S_8 and Se_8 in some of the phases can be deduced from the Raman spectra.

High-pressure liquid chromatography has turned out to be a suitable method for at least a partial separation of the sulfur-selenium mixtures into their constituents according to the different sizes of the molecules. Up to nine chromatographic peaks have been observed. In addition, S_8 and Se_8 are also seen in the chromatograms of some phases. HPLC has so far been applied only at analytical scale. Due to the rapidly decreasing solubility of the phases with increasing selenium content investigation on a preparative scale may be difficult to carry out.

Though Raman spectroscopy and HPLC provide important tools for the chemistry of cyclic selenium sulfides it will be essential to design new chemical reactions for the preparation of as pure phases as possible. With identification and characterization of each new molecule a new standard for HPLC has been obtained and more information gained of the nature of sulfur-selenium phases. Synthetic chemistry, Raman spectroscopy, and HPLC may well provide the key for a better understanding of the molecular composition of the cyclic selenium sulfides prepared according to the above-mentioned methods.

6 Acknowledgements

We are grateful to Professor Lauri Niinistö for his interest in this work and to the Deutsche Forschungsgemeinschaft and the Verband der Chemischen Industrie for financial support.

7 References

1. Gmelin Handb. d. Anorg. Chem., 8. Aufl., Selen B, Springer-Verlag, Berlin—Heidelberg—New York 1949, p. 160
2. de Haan, Y. P., Visser, M. P.: Physica *26*, 127 (1960)
3. Berkowitz, J., Chupka, W. A.: J. Chem. Phys. *45*, 4289 (1966)
4. Zhuravleva, M. G., Churfarov, G. I.: Zhur. Prikl. Khim. *24*, 28 (1951)
5. Illarionov, V. V., Lapina, L. M.: Dokl. Akad. Nauk. USSR *114*, 1021 (1957)
6. Lapina, L. M.: Zh. Neorg. Khim. *3*, 1386 (1958)
7. Lapina, L. M., Illarionov, V. V.: Zh. Neorg. Khim. *3*, 1210 (1958)
8. Devyatyk, G. G., Odnesetsev, A. I., Ulmin, V. A.: Russ. J. Inorg. Chem. *7*, 996 (1962)
9. Fergusson, J. E. et al.: J. Inorg. Nucl. Chem. *24*, 157 (1962)
10. Hawes, L. L.: Nature *198*, 1267 (1963)
11. Laitinen, R., Steidel, J., Steudel, R.: Acta Chem. Scand. *A34*, 687 (1980)
12. Umilin, V. A. et al.: Russ. J. Inorg. Chem. *9*, 1345 (1964)
13. Cooper, R., Culka, J. V.: J. Inorg. Nucl. Chem. *27*, 755 (1965)
14. Cooper, R., Culka, J. V.: J. Inorg. Nucl. Chem. *29*, 1217 (1967)
15. Ailwood, C. R., Fielding, P. E.: Austr. J. Chem. *22*, 2301 (1969)
16. Laitinen, R., Niinistö, L., Steudel, R.: Acta Chem. Scand. *A33*, 737 (1979)
17. Weis, J.: Z. Anorg. Allg. Chem. *435*, 113 (1977)
18. Kapral, A. M.: Australian Pat. 223372 (June 1959), Chem. Abstr. *55*, 16920d (1961)
19. Taimni, I. K., Agarwal, R. P.: Anal. Chim. Acta *9*, 121 (1953)
20. Kraft, M. Ya., Borodina, G. M., Rubina, Z. O.: USSR Pat. 125797 (Febr. 1960), Chem. Abstr. *54*, 16764f (1960)
21. Boudet, R.: Ann. Chim. (Paris) *10*, 178 (1955)
22. Rathke, B.: Ann. Chem. Pharm. *152*, 186 (1869)
23. Cooper, R., Culka, J. V.: J. Inorg. Nucl. Chem. *32*, 1857 (1970)
24. Eysel, H. H.: J. Mol. Struct., in press
25. Schmidt, M., Wilhelm, E.: Z. Naturforsch. *25b*, 1348 (1970)
26. Weiss, J., Bachtler, W.: Z. Naturforsch. *28b*, 523 (1973)
27. Boudreau, R. A., Haendler, H. M.: J. Solid State Chem. *36*, 289 (1981)
28. Mäusle, H.-J., Steudel, R.: Z. Anorg. Allg. Chem. *463*, 27 (1980)
29. Steudel, R., Strauss, E.-M., Laitinen, R.: Unpublished results
30. Schmidt, M. et al.: Angew. Chem. *80*, 660 (1968)
31. Laitinen, R. et al.: Z. Anorg. Allg. Chem., in press
32. Mita, I., Hashio, A.: Aruminyumu Hyomen Shori Kenkyu Chosa Hokuku *75*, 50 (1973), Chem. Abstr. *87*, 92466d (1977)
33. Steudel, R., Strauss, E.-M.: Z. Naturforsch. *36b*, 1085 (1981)
34. Watanabe, Y.: Acta Crystallogr. *B30*, 1396 (1974)
35. Cherin, P., Unger, P.: Acta Crystallogr. *B28*, 313 (1972)
36. Kawada, I. et al.: Acta Crystallogr. *A28*, S 61 (1972)
37. Calvo, C. et al.: Acta Crystallogr. *B34*, 911 (1978)
38. Steidel, J., Steudel, R., Kutoglu, A.: Z. Anorg. Allg. Chem. *476*, 171 (1981)
39. Kutoglu, A.: private communication
40. Steudel, R.: Spectrochim. Acta *31A*, 1065 (1975)
41. Steudel, R., Schuster, F.: J. Mol. Struct. *44*, 143 (1978)
42. Steudel, R., Mäusle, H.-J.: Z. Naturforsch. *33a*, 951 (1978)
43. Steudel, R.: Z. Naturforsch. *30a*, 1481 (1975)
44. Laitinen, R., Steudel, R.: J. Mol. Struct. *68*, 19 (1980)
45. Ohsaka, T.: J. Non-Cryst. Solids *17*, 121 (1975)
46. Ward, A. T.: J. Phys. Chem. *72*, 4133 (1968)
47. Datta, A., Krishnan, V.: Indian J. Chem. *16A*, 335 (1978)
48. Eysel, H. H., Sunder, S.: Indian J. Chem. *18A*, 447 (1979)
49. Eysel, H. H., Sunder, S.: Inorg. Chem. *18*, 2626 (1979)
50. Steudel, R., Eggers, D. F.: Spectrochim. Acta *31A*, 871 (1975)
51. Steudel, R.: J. Phys. Chem. *81*, 343 (1977)
52. Möckel, H. J., Freyholdt, T.: Fresenius Z. Anal. Chem. *368*, 401 (1981)

53. Maekawa, T., Yokokawa, T., Niwa, K.: Bull. Chem. Soc. Japan *46*, 761 (1973)
54. Drowart, J., Smoes, S.: J. Chem. Soc., Faraday Trans. II *73*, 1755 (1977)
55. Datta, A., Khrishnan, V.: J. Thermal Anal. *17*, 31 (1979)
56. Laitinen, R., Niinistö, L.: J. Thermal Anal. *13*, 99 (1978)
57. Khrishnan, V., Datta, A., Narayana, S. V. L.: Inorg. Nucl. Chem. Lett. *13*, 517 (1977)
58. Henschler, D., Kirchner, C. A.: Arch. Toxikol. *24*, 341 (1969)
59. Shapiro, E. M., Pomerat, C. A., Mullins, J. F.: J. Invest. Dermatol. *24*, 423 (1955)
60. Robinson, H. M., Yaffe, S. N.: J. Am. Med. Assoc. *162*, 113 (1956)
61. Butcher, E. O.: J. Invest. Dermatol. *29*, 377 (1957)
62. Archer, V. E., Luell, E.: J. Invest. Dermatol. *35*, 65 (1960)
63. Rosenthal, J. W., Adler, H.: Southern Med. J. *55*, 318 (1962)
64. Maguire, H. C., Kligman, A. M.: J. Invest. Dermatol. *39*, 469 (1962)
65. Parran, J. J., Brinkman, R. E.: J. Invest. Dermatol. *45*, 89 (1965)
66. Brotherton, J.: J. Gen. Microbiol. *49*, 393 (1967)
67. Plewig, G., Kligman, A. M.: J. Soc. Cosmet. Chem. *20*, 767 (1969)
68. Cummins, L. M., Kimura, E. T.: Toxicol. Appl. Pharmacol. *20*, 89 (1971)
69. Goble, F. C., Konopka, E. A., Zoganas, H. C.: Antimicrob. Agents Chemother. *1968*, 531, Chem. Abstr. *70*, 10176w (1969)
70. Baldwin, M. M., Young, A. P.: U.S. Pat. 2694669 (Nov. 1954), Chem. Abstr. *49*, 2686b (1955)
71. Spoor, H. J.: Drug & Cosmetic Ind. *77*, 44, 134 (1955), Chem. Abstr. *50*, 2927c (1956)
72. Matson, E. J.: J. Soc. Cosmet. Chem. *7*, 459 (1956)
73. Lehne, R. K.: Ger. Pat. 1301006 (Aug. 1969), Chem. Abstr. *72*, 93290u (1970)
74. Nonogawa, M.: Japan. Kokai 7412041 (Febr. 1974), Chem. Abstr. *81*, 54333b (1974)
75. Nowak, G. A.: Parfüm. Kosmet. *56*, 29 (1975), Chem. Abstr. *83*, 48077w (1975)
76. Anstett, R. M., Wellman, W. S., Andrews, H. W.: U.S. Pat. 3340196 (Sept. 1967), Chem. Abstr. *67*, 109953x (1967)
77. Heiskell, R. H.: U.S. Pat. 2726943 (Dec. 1955), Chem. Abstr. *50*, 5293b (1956)
78. Potts, M. F., Hudson, P. S., Smith, W. L.: U.S. Pat. 2775594 (Dec. 1956), Chem. Abstr. *51*, 4759 (1957)
79. N. V. Philip's Gloeilampenfarbrieken, Neth. Appl. 6605388, Chem. Abstr. *68*, 52924c (1968)
80. Kanazawa, K. K., Street, G. B.: U.S. Pat., 3666554 (May 1972), Chem. Abstr. *77*, 133214j (1972)

Polyhedral Oligosilsesquioxanes and Their Homo Derivatives*

Mikhail G. Voronkov[1] and Vladimir I. Lavrent'yev[2]

[1] Institute of Organic Chemistry, Siberian Division of the USSR Academy of Sciences, 664033 Irkutsk, USSR
[2] Institute of Inorganic Chemistry, Siberian Division of the USSR Academy of Sciences, 630090 Novosibirsk, USSR

Table of Contents

---

* Because of outward circumstances this plenary lecture could not be delivered at the 3rd IRIS Meeting held in Graz (Austria), August 17–22, 1981

[1] Author to whom the correspondence should be addressed

A critical analysis of the present state of synthetic methods and mechanisms of the formation of polyhedral oligosilsesquioxanes and their homo derivatives is given. The data available on the structure, physico-chemical properties, reactivity, and application of the above compounds are summarized and systematized.

1 Introduction

Polyhedral silsesquioxanes represent a rather versatile class of three-dimensional organosilicon oligomers which are of considerable theoretical and practical interest. They are composed of a polyhedral silicon-oxygen skeleton containing intermittent siloxane chains which bear organic or inorganic substituents attached to silicon atoms. The molecules of these compounds have the general formula $(XSiO_{1.5})_n$ where n is an odd number ($n \geqq 4$) and X = H, organyl, halogen, etc. The structure of lower oligosilsesquioxanes is represented by structural formulae I–IV. These compounds may be considered as the products of complete hydrolytic condensation of the corresponding trifunctional monomers, $XSiY_3$ with Y = Hal, OH, OR, OCOR, etc. [1].

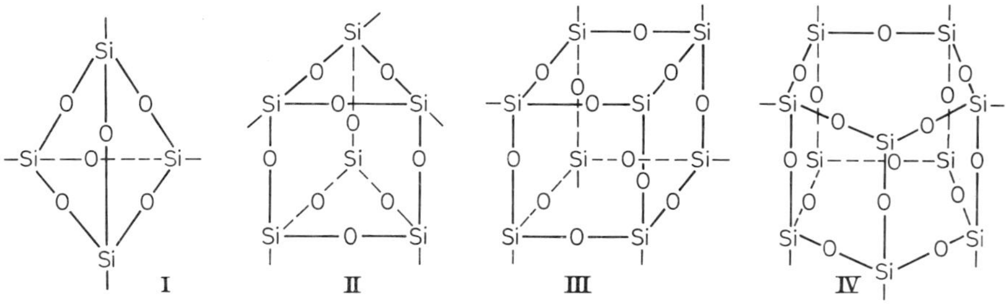

I II III IV

Polyhedral oligomers, called homosilsesquioxanes [2] are structurally rather similar to oligosilsesquioxanes. They differ from regularly built oligosilsesquioxanes in that the Si—O bond of the latter is inserted by a XX'SiO group which is a homologous link in linear and cyclic oligo- and polysiloxanes. Homooligosilsesquioxanes are described by the general formula $(XSiO_{1.5})_nOSiXX'$, and the structure of their lower members is shown by formulae V–VIII. These compounds are the by-products of the synthesis of oligosilsesquioxanes.

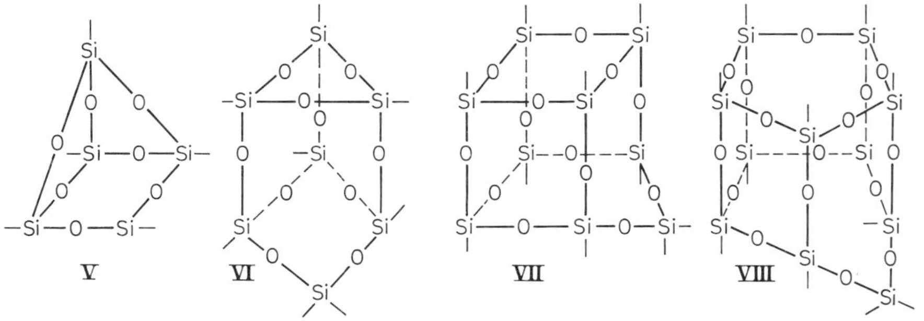

V VI VII VIII

The chemistry of polyhedral silsesquioxanes and their homo derivatives is being studied extensively without imitating nature. Natural compounds of these types have not been found. The first reports on compounds of the general structure $(XSiO_{1.5})_n$ date back to the second half of the nineteenth century. Substances of this type with X = H and C_6H_5 were obtained by hydrolysis of $XSiCl_3$ by

Buff and Wohler [3] and Ladenburg [4] in an attempt to synthesize silicon analogs of carboxylic acids, XSiOOH and were mistaken for anhydrides of the latter, $(XSiO)_2O$. Compounds $(HSiO_{1.5})_n$ were further described by Friedel and Ladenburg [5], Gatterman [6] and Stock [7]. The polymer-like character of the compounds which were assigned the formula $(XSiO)_2O$, was first reported by Meads and Kipping as early as 1914 [8]. In 1930, Palmer and Kipping [9] prepared another compound of the type under consideration, $(C_6H_{11}SiO_{1.5})_n$. Later, Hyde and Daudt patented the process for the manufacture of thermoplastic polymers having the formula $(C_6H_5SiO_{1.5})_n$ [10]. All the above mentioned authors did not suspect that the obtained solids of the general formula $(XSiO_{1.5})_n$ might contain volatile polyhedral oligosilsesquioxanes.

The first oligoorganylsilsesquioxane, $(CH_3SiO_{1.5})_n$ was isolated along with other volatile compounds by Scott in 1946 through thermolysis of the polymeric products of methyltrichlorosilane dimethylchlorosilane co-hydrolysis [11]. However, due to the poor solubility of their crystals in organic solvents used in cryoscopic studies, it was not possible to determine the molecular weight (n value). The present data on the volatility of oligomethylsilsesquioxanes suggest that Scott isolated octa(methylsilsesquioxane). Later, Barry and Gilkey [12] confirmed the existence of such a silicon compound. Upon alkaline thermolysis of the products of the hydrolysis of organyltrichlorosilanes, $XSiCl_3$ with $X = C_2H_5$, C_3H_7, C_4H_9 they isolated subliming crystals identified as the corresponding octa(organylsilsesquioxanes). Somewhat later, oligoalkylsilsesquioxanes, $(XSiO_{1.5})_n$ with $n = 8$ and $X = CH_3$, C_2H_5, C_3H_7, C_4H_9, cyclo-C_6H_{11} and also with $n = 12$ and $X = CH_3$ were described in more detail [13] [1]. These results stimulated a rather extensive development of the chemistry of polyhedral oligosilsesquioxanes.

The first report on homoorganylsilsesquioxanes appeared in 1955 when Sprung and Guenther [14] identified products of the acid hydrolysis of methyltributoxysilane of the structure $(CH_3SiO_{1.5})_8OSiCH_3(OC_4H_9)$.

This article gives a systematic survey of all the data available in the periodic and patent literature (until the middle of 1981) on the synthesis, structure, physical and chemical properties of polyhedral oligosilsesquioxanes and their homo derivatives. Data on the possibility of practical application of oligosilsesquioxanes are also reported.

2 Nomenclature

The nomenclature of oligosilsesquioxanes (as well as of their homo derivatives) has not been elaborated. These compounds are polyhedral oligocyclic systems of general formula $(X_iSiO_{1.5})_n$ with $n = 2m$ $(m \geqq 2)$, i.e., $n = 4, 6, 8, 10, 12$, etc. and X_i is the substituent at the silicon atom. In the simplest case, all the substituents X_i may be the same. On the basis of the existing IUPAC rules [15] for naming polycyclic compounds, oligosilsesquioxanes may be termed in two ways:

1. As polycyclosiloxane systems the skeleton of which is made up of intermittent silicon and oxygen atoms. These atoms are numbered by the method used for analogously built carbocyclic compounds in which the number of carbon atoms

[1] The synthesis of hexa(phenylsilsesquixane) should be considered as reported erroneously.

is equal to the total amount of silicon and oxygen atoms in the corresponding oligo-silsesquioxanes. In this case, the silicon atoms are always odd numbers and the oxygen atoms are even numbers. A compound with n = 4, 6, 8, etc. is given the prefix "tricyclo-", "tetracyclo-", "octacyclo-", etc., respectively. The number of rings corresponds to that of edges in a polyhedral molecule minus 1 and is equal to the number of scissions in silicon-oxygen chains required for the system to be converted to an open-chain compound. The prefix "cyclo-" is followed by square brackets containing figures in diminishing order separated by points, which indicate the total number of silicon and oxygen atoms in the two branches of the main ring, the primary bridge[2] and the secondary bridge. The main ring and the primary bridge form a bicyclic system whose numbering begins from the right upper silicon atom at the bridgehead and goes counterclockwise along the upper right edge and further in the longest of all possible paths to the second bridge-head; the numbering is then continued from this atom via the longer unnumbered path back to the first bridgehead and is completed by the shortest path.

The positions of the other so-called secondary bridges are shown by super-scripts following the figures denoting the number of oxygen atoms in the bridges, which is nearly always equal to 1. Since the system displays similar secondary bridges their numbering starts from the bridgehead having the greatest figure. The bridge position is indicated in increasing order.

The types and position of substituents at the silicon atom are denoted at the beginning of the name. If all the substituents are the same, their position is not necessary to be shown. In this case, the substituent name is preceded by a prefix "per-" or the total number of substituents is indicated; for example, "permethyl-", "hexaethyl-", "octaethyl-", etc.

2. In an overwhelming number of cases, the compounds are more conveniently named using systematic (trivial) nomenclature. According to the latter, the general name of these compounds contains syllables such as "sil-", "sesqui-" and "oxane" which indicate that each silicon atom is connected to 1.5 oxygen atoms, the prefix "oligo-" denoting a small number of such groups. In this way, the systematic name gives the number of silsesquioxane links, $SiO_{1.5}$ in the molecule and the substituents attached to the silicon atom. If the number of similar substituents in the molecule is equal to or less by 1 than that of the silsesquioxane links, i.e. in the case of molecules of the general structure $(XSiO_{1.5})_n$ or $X_{n-1}X'(SiO_{1.5})_n$, no figures are used to denote the position of substituents. For example,

$(HSiO_{1.5})_8$	Octasilsesquioxane or octa(hydrosilsesquioxane)
$(CH_3)_7C_6H_5(SiO_{1.5})_8$	Heptamethylphenyloctasilsesquioxane .

In cases where the $(XSiO_{1.5})_n$ molecule has less than (n — 1) similar substituents, the numbering of silicon atoms agrees with that mentioned above. Thus, it is necessary to use figures for denominating dimethyltetraphenyl(hexasilsesquioxane), for example, as the latter may exist in three isomers with 1,3-, 1,5- and 1,7-positions of the phenyl group. The substituents are numbered according to the IUPAC rules.

[2] The bridge is, as a rule, an oxygen atom linking two opposite silicon atoms. The two nodal silicon atoms connected by the oxygen bridge are "bridgeheads".

Below some examples of both principles used for the numbering of the compounds considered are given

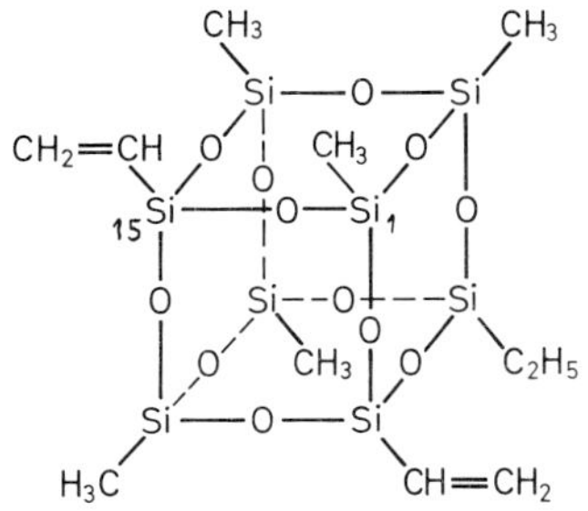

1,3,5,7,13-pentamethyl-9-ethyl-11,15-divinylpentacyclo[9.5.1.$1^{3,9}$.$1^{5,15}$.$1^{7,3}$]octa-siloxane

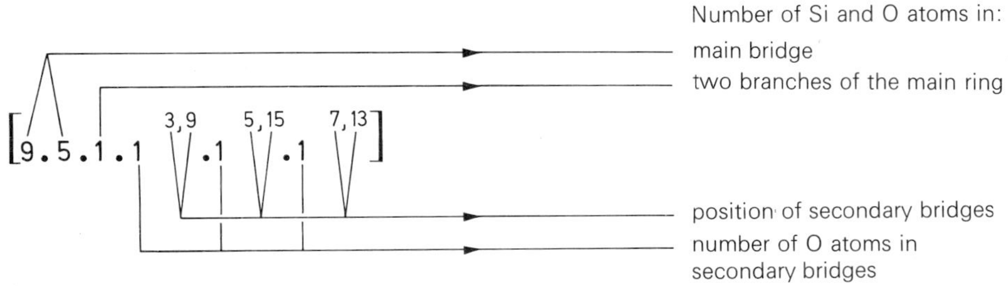

or:
1,3,5,7,13-pentamethyl-9-ethyl-11,15-divinyloctasilsesquioxane

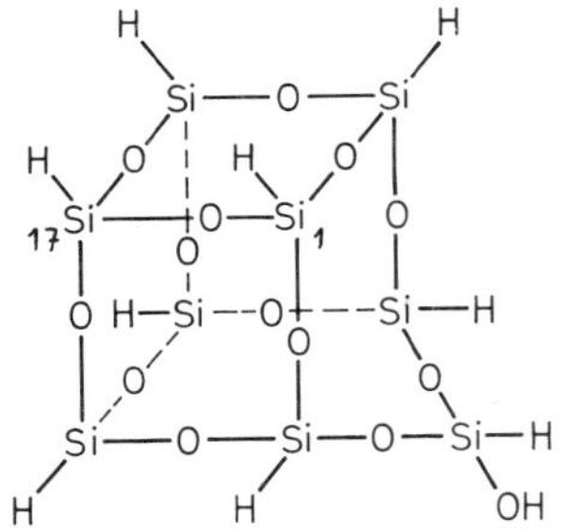

Perhydropentacyclo[11,5.1.$1^{3,9}$.$1^{5,17}$.$1^{7,15}$]nonasiloxanol-11
or:
perhydrohomooctasilsesquioxanol-1 [3]

3 Methods of Synthesis

No industrial methods for the synthesis of oligosilsesquioxanes have so far been described in the literature. However, a great number of reactions leading to the

[3] Position 1 at the silicon atom in the homo group may be assigned if all the other substituents in the compound are the same. Otherwise, numbering is as usual.

formation of polyhedral silsesquioxanes are known. Depending on the nature of the starting materials, all these reactions may be divided into two large groups. The first group includes the reactions giving rise to new Si—O—Si bonds with subsequent formation of the polyhedral framework. These reactions are complex, multistep processes leading to polymers and oligomers which may include oligosilsesquioxanes and their homo derivatives. The above reactions allow polyhedral silsesquioxanes to be synthesized from monomers of the $XSiY_3$ type (with X = a chemically stable substituent, and Y = a highly reactive substituent) or from siloxanes of linear, cyclic or polycyclic structure formed from the above monomers. The second group of reactions covers the processes involving only variations in the structure and composition of substituents at the silicon atom without affecting the silicon-oxygen skeleton of the molecule.

The polyhedral silicon-oxygen skeleton of oligosilsesquioxanes is formed by the following reactions:

1. Hydrolytic condensation of trifunctional monomers, $XSiY_3$.
2. Condensation of Si-functional oligoorganylcyclosiloxanes, $[XYSiO]_m$.
3. Co-condensation of organosilicon monomers and/or oligomers of different structure and composition.
4. Thermolysis of polyorganylsilsesquioxanes.

In some cases, these reactions may be combined in order to obtain certain oligosilsesquioxanes or to increase the yield.

There are only few synthetic routes to oligosilsesquioxanes available with retention of the silicon-oxygen framework. These routes also involve halogenation and nitration.

3.1 Hydrolytic Polycondensation of Trifunctional Monomers, $XSiY_3$

Hydrolytic polycondensation of trifunctional monomers of the type $XSiY_3$ leads to cross-linked three-dimensional as well as network and cis-syndiotactic (ladder-type) polymers, $(XSiO_{1.5})_n$ [16–18]. With increasing amount of solvent, however, the corresponding condensed polycyclosiloxanes, polyhedral oligosiloxanes and their homo derivatives may be formed (Tables 1 and 2). The reaction rate, the degree of oligomerization and the yield of the polyhedral compounds formed strongly depend on the following factors:

1. Concentration of the initial monomer in the solution
2. Nature of solvent
3. Character of substituent X in the initial monomer
4. Nature of founctional groups Y in the initial monomer.
5. Type of catalyst
6. Temperature
7. Addition of Water
8. Solubility of the polyhedral oligomers formed

The influence of all these factors, both individual and together, has been studied only in general, without any quantitative estimation of their effect on the reaction course. This may be explained by the complicated character of the polycondensation process and the strong mutual effect of the above factors. Nevertheless, the hydro-

Table 1. Oligosilsesquioxanes, $(XSiO_{1.5})_n$

X	n	Method of Synthesis	Y [a]	Solvent	Catalyst	Yield (%)	Ref.
H	8	3.1	OCH_3	Cyclohexane	$HCl + CH_3COOH$	13	19, **20**)
H	10, 12, 14, 16	3.1	Cl	Benzene	$H_2SO_4 + SO_3$	15–35	20)
CH_3	6	3.1 + T°	OC_2H_5	Benzene	HCl	—	21)
CH_3	8	3.1, 3.4	Cl	Methanol	HCl	37	13, 14, 21 – **22**, 23 – 26)
CH_3	10, 12	3.4	OC_2H_5	Benzene	KOH	—	13, **23**)
C_2H_5	6	3.1 + T°	OC_2H_5	Benzene	HCl	—	**27**, 28)
C_2H_5	8	3.1, 3.2, 3.4	Cl	Methanol	HCl	37	2, 13, **22**, 24, 27, 28, 29)
C_2H_5	10	3.1, 3.2	Cl	Butanol	HCl	16	2, 29, 30)
C_3H_7	8	3.1 + T°, 3.4	Cl	Methanol	HCl	44	13, **22**)
$CH(CH_3)_2$	4	3.1 + T°	Cl	Ether	HCl	55.5	31)
$CH(CH_3)_2$	8	3.1 + T°	Cl	Methanol	HCl	17	22)
C_4H_9	8	3.1 + T°, 3.4	Cl	Methanol	HCl	38	**22**, 13)
$C(CH_3)_3$	4	3.1 + T°	Cl	Ether	HCl	94,7	**31**, 32)
C_5H_{11}	8	3.1 + T°	OC_2H_5	i-butyl methyl ketone	$(C_2H_5)_4NOH$	20	33)
C_6H_{13}	8	3.1 + T°	Cl	Ether	HCl	—	34 – **36**)
C_6H_{13}	12, 20	3.1 + T°	Cl	Toluene	HCl	—	35)
C_7H_{15}	6	3.1 + T°	Cl	Ether	HCl	—	34, 35)
C_8H_{17}	6	3.1 + T°	Cl	Ether	HCl	—	34 – **36**)
C_8H_{17}	12, 24	3.1 + T°	Cl	Toluene	HCl	—	36)
$i\text{-}C_9H_{19}$	6	3.1 + T°	Cl	Ether	HCl	—	34, **36**)
$i\text{-}C_9H_{19}$	8, 10, 18, 28	3.1 + T°	Cl	Toluene	HCl	—	36)
C_6H_{11}	6	3.1 + T°	Cl	Acetone	HCl	7	37)
C_6H_{11}	8	3.1 + T°	OCH_3	Nitrobenzene	OH	—	13, 37)
$CH{=}CH_2$	8	3.1	OCH_3	Methanol	HCl	20	25, **38**, 39)
$CH{=}CH_2$	10	3.1	OCH_3	Butanol	HCl	—	39)
$CH_2CH{=}CH_2$	8	3.1	OCH_3	Butanol	HCl	—	40)

Table 1. (continued)

X	n	Method of Synthesis	Y [a]	Solvent	Catalyst	Yield (%)	Ref.
C_6H_5	8	3.1 + T°, 3.4	OCH_3	Benzene	$C_6H_5CH_2(CH_3)_3NOH$	88	22, 33, **41**–43)
C_6H_5	10	3.1 + T°	OC_2H_5	Tetrahydrofuran	Me_4NOH	—	**41**, 42)
C_6H_5	12, 22, 24	3.1 + T°	OC_2H_5	Tetrahydrofuran	Me_4NOH	—	41, 43, 44)
2-$CH_3C_6H_4$	c	3.4	Cl	Ether	HCl	—	13)
4-$CH_3C_6H_4$	8	3.1 + T°	Cl	Ethanol	HCl	19	42)
x-$NO_2C_6H_4$	8	3.5	X=C_6H_5	HNO_3	HNO_3	50	42)
1-$C_{10}H_7$	8	3.1 + T°, 3.4	Cl	Methanol	HCl	52	42)
2-C_4H_3S	8	3.4	OCH_3	Methanol	$(C_2H_5)_3N$	13	45)
2-C_4Br_3S	8, 12	3.5	X = 2-C_4H_3S	HBr	Br_2	55	45)
C_6Cl_5	6	3.1	Cl	Ether	HCl	—	**46**, 47)
$C_6H_kCl_{5-k}$ k = 1–4	c	3.1	Cl	Ether	HCl	—	**46**, 47)
$ON(CH_3)_4$	8	3.3	OH	$H_2O + (CH_3)_4NOH$		—	48)
$OSi(CH_3)_3$	8	3.5	X=$ON(CH_3)_4$	2-propanol + H_2O	HCl	75	48)

a Functional group Y in initial monomer $XSiY_3$ or X in $(XSiO_{1.5})_8$
b Maximum yield and synthesis conditions are taken from the reference in bold type
c Non-identified structure
T° Additional thermolysis (reflux)

Table 2. Polyhedral homooligo(organylsilsesquioxanes), $(XSiO_{1.5})_nOSiXX'$

X	X′	n	Method of Synthesis	Y [a]	Solvent	Catalyst	Ref.
CH_3	OCH_3	8	3.4	OC_4H_9	Benzene	KOH	[23]
CH_3	OC_2H_5	8	3.4	OC_2H_5	Benzene	KOH	[23]
CH_3	OC_4H_9	8	3.1; 3.4	OC_4H_9	Benzene	HCl	[14]
C_2H_5	H	4	3.2	$(C_2H_5SiHO)^*_{4;5}$	Acetone, Ethanol	H_2O_2	[49]
C_2H_5	H; OCH_3	6	3.2	$(C_2H_5SiHO)^*_{4;5}$	Ether	KOH	[2]
C_2H_5	OH; OC_4H_9	6	3.1	Cl	Butanol	HCl	[2]
C_2H_5	H; OH; OCH_3	8	3.2	$(C_2H_5SiHO)^*_{4;5}$	Ether	KOH	[2, 50]
C_2H_5	OC_4H_9	8	3.1	Cl	Butanol	HCl	[2, 29]
C_2H_5	H	10	3.2	$(C_2H_5SiHO)^*_{4;5}$	Ether	KOH	[2, 50]
$CH{=}CH_2$	OC_4H_9	6, 8	3.1	Cl	Butanol	HCl	in press
C_6H_5	OH	8	3.1	Cl	Acetone	HCl	[41, 43]
C_6H_5	OH	10	3.1	OC_2H_5	Tetrahydrofuran	$(CH_3)_4NOH$	[41, 43]
CH_3	CH_3	6	3.4	Cl	Butyl acetate	HCl	[51]

a Functional group Y in the initial monomer $XSiY_3$
* Oligoethylhydrocyclosiloxanes as the initial compounds

lytic polycondensation is the most universal and traditional synthetic route to oligosilsesquioxanes and their homo derivatives. Most polyhedral compounds of this type known at present (Tables 1 and 2) have been synthesized by this method, the corresponding organyltrichlorosilanes, less often organyltrialkoxysilanes, as well as trichlorosilane and trimethoxysilane being used as the starting monomer. As a whole, the synthesis of oligosilsesquioxanes is a time-consuming multistep process. At present, however, some procedures by which these compounds are obtained in fairly high yields have been developed [22,36,41,42].

The effects of factors 1–8 on the polycondensation of monomers of the type $XSiY_3$ may be summarized as follows.

3.1.1 Concentration of the Initial Monomer, $XSiY_3$, in Solution

Due to the extremely high reactivity of trifunctional organosilicon monomers of the above type (mainly organyltrichlorosilanes), the synthesis of the most important oligomers is carried out in an organic solvent with the addition of water and in the presence of an appropriate acid or base catalyst. The high concentration of the reagents facilitates the formation of high polymers. When diluted solutions are used, intramolecular cyclization predominates leading to polyhedral oligomers along with other volatile products. The most suitable $XSiY_3$ concentration in the preparation of polyhedral silsesquioxanes depends on the character of substituents X and Y in the initial monomer, the solvent nature, the temperature, the amount of water added and the catalyst concentration. The concentration of alkyltrichlorosilanes having lower alkyl substituents, which is most favorable for the preparation of the corresponding polyhedral octamers, ranges from 0.1 to 0.2 M. In the $XSi(OR)_3$ hydrolytic polycondensation, more concentrated solutions may be used (0.3–0.5 M). The synthesis of oligoalkylsilsesquioxanes bearing higher alkyl substituents requires even more concentrated solutions (2.2 M). It should be taken into consideration that too low concentrations of the initial monomer considerably decrease the rate of polymerization.

3.1.2 Nature of the Solvent

The effect of solvent polarity on the $XSiY_3$ hydrolytic polycondensation has not been studied in detail. Polyhedral silsesquioxanes are formed in both polar and non-polar solvents. The optimum initial monomer concentration depends, as stated above, on the solvent nature. Thus, in alcohol, the most suitable $XSiY_3$ concentration is 0.1–0.2 M whereas in an inert solvent such as benzene, toluene, cyclohexane, ether, this value may be somewhat higher. Non-polar solvents (cyclohexane, benzene, hexamethylsiloxane) favor the formation of oligohydrosilsesquioxanes from $HSiCl_3$. At the same time, the use of acetone or ethyl acetate leads to high molecular weight products. $(HSiO_{1.5})_n$ with $n = 10, 12, 14, 16$ can only be prepared in aromatic hydrocarbons [20]. Acid hydrolysis of organyltrialkoxysilanes used for the preparation of the corresponding oligosilsesquioxanes is most effectively performed in benzene rather than in alcohol [14,28,35]. With solvents having no oxygen atom, the addition of a small amount of ethyl or propyl alcohol to the reaction mixture increases the yield of octa(organylsilsesquioxanes). This restricts the number of silanol units in the intermediates, thus decreasing the degree of intermolecular association via hydrogen bonds and, consequently, preventing the polycondensation processes. The use of alcohols as the hydrolysis medium also influences the degree of oligomerization of the oligoalkylsilsesquioxanes formed and their homo derivatives. Thus, oligomers with n = 4 and 6 could be obtained in only non-polar or weakly polar solvents (Tables 1 and 2) but not in alcohols. The formation of 1-hydroperethylhomotetrasilsesquioxane (Table 2) in ethanol may be explained by the fact that the condensation was carried out with a catalyst not breaking the siloxane bond [49]. Since methanol cleaves the Si—O—Si group in hexaethylsilsesquioxane [27] it is unsuitable for the preparation of the latter.

The synthesis of oligoalkylsilsesquioxanes from alkyltrichlorosilanes in alcohol affords higher yields than that performed in ketones, organic acids, their anhydrides or esters [22]. Some contradictory data on the effect of the alcohol nature on the oligomer yield are associated, first of all, with variations in the concentration of water and the conditions of its addition to the reaction mixture. The use of higher alcohols decreases the rate of polycondensation [22] leading predominantly to oligomers with a low n value [2].

The hydrolytic condensation of higher alkyltrichlorosilanes is more efficiently carried out in ethers and ketones with a high boiling point [31, 33–36] than in toluene [36], ethanol [33] or methanol [21]. The use of alcohols increases the degree of siloxane chain solvation, thus preventing intramolecular condensation of the intermediates.

Octa(phenylsilsesquioxane) is more readily formed in benzene, nitrobenzene, benzyl alcohol, pyridine, or ethylene glycol dimethyl ether and the corresponding dodecamer in tetrahydrofuran [41]. The hydrolysis of $C_6H_5SiCl_3$ in acetonitrile, diglyme, acetone, and methyl isobutyl ketone gives high molecular weight polyphenylsilsesquioxanes. Depending on the solvent, the total yield of hetero substituted methylphenyloctasilsesquioxanes decreases in the following order:

$$C_2H_5OH + CH_3COOH > C_2H_5OH > CH_3OH$$ [54].

3.1.3 Character of Substituent X in the Initial Monomer, $XSiY_3$

The rate of the hydrolytic polycondensation of alkyltrichlorosilane markedly depends on the length and branching of the alkyl group at the silicon atom and decreases with X varying in the following order:

$$CH_3 \gg C_2H_5 > C_3H_7 > C_4H_9 > C_5H_{11}$$ [22, 33]

$$CH_3 \gg CH_3CH_2 \gg (CH_3)_2CH \gg (CH_3)_3C$$ [22].

It was not possible to prepare in alcohol the corresponding polyhedral octa(alkylsilsesquioxanes) from alkyltrichlorosilanes containing a tributyl or n-alkyl group with five or six carbon atoms [21]. At the same time, the corresponding tetra(alkylsilsesquioxanes) were obtained in ether from $XSiCl_3$ with $X = (CH_3)_2CH$ and $(CH_3)C$ [33]. Attempts to synthesize analogous oligomers with $X = CH_3$ and C_2H_5 under the same conditions were unsuccessful.

Phenyltrichlorosilane undergoes hydrolytic polycondensation more readily than ethyltrichlorosilane [21] in spite of the steric difference of phenyl and ethyl groups. However, oligophenylsilsesquioxanes are not formed as the end products of this reaction performed in an aqueousacetone medium [44]. Therefore, most of these compounds have been prepared by additional thermolysis of hydrolytic polycondensation products, $C_6H_5SiCl_3$ [35,42,44], for example in the presence of trimethylbenzylammonium [42]. Introduction of a halogen atom or a methyl substituent into the phenyl group of ($C_6H_5SiCl_3$) makes the formation of the corresponding oligoarylsilsesquioxanes more difficult. Thus, octasilsesquioxanes with X = 3-ClC_6H_4, 4-BrC_6H_4 and 4-$CH_3OC_6H_4$ have not been prepared [43]. The structure of the products of the hydrolysis of phenyltrichlorosilane derivatives, $C_6H_kCl_{5-k}SiCl_3$ (with k = 1–4), has not been exactly established [46].

The formation of oligo(vinylsilsesquioxanes) proceeds somewhat more difficult than that of the corresponding methyl- or ethylsubstituted oligomers [25,38].

The formation of polyhedral oligo(organylsilsesquioxanes) by hydrolytic polycondensation of monomers $XSiY_3$ strongly depends on steric and inductive effects of the substituent X. With substituents capable of undergoing p_π–d_π interactions with the silicon atom, this dependence is of a specific character. For acid-catalyzed reactions, the inductive effect of substituent X is weaker and the steric effect is stronger than for analogous processes catalyzed with bases.

The synthesis of oligosilsesquioxanes having reactive substituents X (X = H, OH, OR, Hal, etc.) at the silicon atom is especially difficult.

The $HSiCl_3$ hydrolysis results mainly in polyhydrosilsesquioxanes of various structure [3,5–7,57–59]. However, when this reaction was carried out in 80% sulfuric acid and in the presence of hexamethyldisiloxane, octa(hydrosilsesquioxane) was isolated in 0.2% yield [19]. Later, $(HSiO_{1.5})_n$ with n = 8, 10, 12, 14, 16 was prepared by hydrolysis of $HSiCl_3$ involving the addition of a benzene solution of $HSiCl_3$ to a mixture of benzene and SO_3-enriched sulfuric acid [20]. Due to the susceptibility to sulfuric acid, $(HSiO_{1.5})_8$ is formed under these conditions only in traces. The hydrolysis of trimethoxysilane carried out in cyclohexane-acetic acid in the presence of concentrated hydrochloric acid leads to the octamer in considerably lower yield (13%) [20].

No synthesis of oligosilsesquioxanes with X = OH, OR, Hal has been reported. The preparation of such compounds seems possible from the corresponding oligomers by replacement reactions or halogenation.

$[(CH_3)_4NOSiO_{1.5}]_8$ is presumably formed by the replacement of hydroxy groups by $(CH_3)_4NO$ in polycyclosiloxanol (or $[(HO)SiO_{1.5}]_8$) called silicic acid by the authors [48].

3.1.4 The Nature of Functional Groups Y in the Initial Monomers, $XSiY_3$

Polyhedral oligosilsesquioxanes and their homo derivatives have been prepared from monomers $XSiY_3$ with Y = OH, Cl, OR. The monomers with X = Br, I, OCOR, OM (M = Na, K), NR_2, etc. may also be used. The reactivity of the functional groups decreases in the following order:

$$Cl > OH > OCOR > OR .$$

The susceptibility of organyltrialkoxysilanes towards hydrolysis decreases with increasing length and branching of the alkyl moiety in the alkoxy group [14,23]. The presence of silanol groups in the hydrolysis intermediates favors gel formation [32,33].

The hydrolysis of organyltrialkoxysilanes in the synthesis of oligosilsesquioxanes is most frequently performed in the presence of acid catalysts (Table 1). Nevertheless, the corresponding polyhedral oligomers can be obtained in fairly high yield by thermolysis of the organyltrialkoxysilane alkaline hydrolysis products [23,37,41]. The presence of a π-electron substituent, ($CH_2{=}CH$, C_6H_5) at the silicon atom slightly accelerates the hydrolytic polycondensation of organyltrialkoxysilanes. An increase in the +I-effect of the alkoxy group should produce an analogous effect [60] which is, however, counteracted by a concomitant steric shielding of the reaction center. The steric effect decreases with increasing distance between bulky substituents and the Si—O—Si group. Thus, the logarithm of the equilibrium constant for the $(CH_3)_3Si$-OCOR alcoholysis reaction where the substituent R is separated from the silicon oxygen by the carbonyl group linearly depends on the σ_R^* values even for $R = C(CH_3)_3$ [61].

The co-hydrolysis of alkyltrichloro- and triethoxysilanes in benzene increases slightly the yield of octamers [21,27].

The above principles also apply to the hydrolysis of organyltrichlorosilanes in an alcohol medium since, on the reaction with the alcohol, these compounds are converted to the corresponding alkoxy derivatives (see Sect. 4.1).

The hydrolytic polycondensation of pseudofunctional monomers, $XSiHY_2$, may be considered as a particular case of the preparation of oligoorganylsilsesquioxanes from $XSiY_3$. This method has been used for the synthesis of oligoethylsilsesquioxanes involving simply polycondensation of $C_2H_5SiHCl_2$ in benzene or preferably in butanol [29,2]. Due to the relative stability of the Si—H bond, the hydrolysis of organyldichlorosilanes first affords the corresponding oligoorganylhydrocyclosiloxanes. Then, the HCl-catalyzed hydrocondensation of the latter leads to the formation of oligoorganylsilsesquioxanes along with other products.

3.1.5 Type of Catalyst

The formation of polyhedral silsesquioxanes from linear, cyclic and polycyclic products of the hydrolytic polycondensation of trifunctional monomers, $XSiY_3$, occurs only in the presence of either acid or base catalysts. Only the hydrolysis of lower alkyltrichlorosilanes does require no special catalysts. In this case, the process is an autocatalytic one and the products are fairly reactive. For the preparation of octa(alkylsilsesquioxanes), HCl, $ZnCl_2$, $AlCl_3$, $HClO_4$ and CH_3COOH have been used as catalysts, HCl being most efficient [22]. Low pH values of the medium favor cyclization whereas high pH values facilitate polymerization [17,27]. This seems to account for the fact that all the oligoorganylsilsesquioxanes with a low degree of oligomerization (n = 4, 6) and their homo derivatives (Tables 1, 2) have been prepared using an acid catalyst (HCl). It is surprising that sulfuric acid has proved to be an efficient catalyst for the preparation of oligohydrosilsesquioxanes with n = 10, 12, 14, 16 [20]. The high HCl concentration accelerates the synthesis of octa-(alkylsilsesquioxanes) [22]. It is desirable, however, to carry out the synthesis of $(CH_3SiO_{1.5})_8$ from CH_3SiCl_3 at lower HCl concentrations. Octamers with $X = C_2H_5$, $CH_2{=}CH$ are formed in fairly high yield (~30%) in alcohols without addition of a catalyst [2,38,53]. On the other hand, the hydrolysis of higher alkyltrichlorosilanes proceeds more readily when concentrated HCl is employed [36]. In contrast, the yield of $(C_6H_5SiO_{1.5})_8$ increases with decreasing HCl concentration [22,42]. The maximum yield is obtained with a base catalysts [41].

In acid catalysis (HCl), the variation in pH of the medium affects the yield of oligoorganylsilsesquioxanes to a lesser extent than in the case of alkaline catalysis. Base catalysts are used in the synthesis of oligosilsesquioxanes mainly in the thermolysis of the primary products of $XSiY_3$ hydrolysis (Table 1). The most efficient catalysts, in this case, are those which are readily decomposed upon heating to give inactive products [62]. For example, the use of tetraethylammonium [31,32,45] trimethylbenzyl ammonium [23,37,41] and triethylamine hydroxides [42,45] has proved successful.

3.1.6 Effect of Temperature

The first step of the synthesis of oligosilsesquioxanes involving hydrolytic polycondensation of organyltrichlorosilanes is usually carried out at low temperature. The use of higher alcohols as the solvent prevents cooling of the reaction mixture [2,29,52]. Oligomers, $(XSiO_{1.5})_n$ with the lowest n value are prepared at reduced temperatures, the compound with $n = 6$ and $X = CH_3$, C_2H_5 being obtained at 20 °C [2,21,27,28] and those with $n = 4$ and $X = (CH_3)_2CH$, $(CH_3)_3C$ below 0 °C [31]. For the preparation of oligoalkylsilsesquioxanes from hydrolytically more stable higher alkyltrichlorosilanes, additional heating of the hydrolysis mixture is advantageous. Thus, the synthesis of octa(amylsilsesquioxane) has been carried out in boiling isobutyl methyl ketone (115.6 °C) [33]. The yield of $(C_6H_5SiO_{1.5})_n$ with $n = 8, 10, 12$ upon alkaline (KOH) condensation of the products of the $C_6H_5SiCl_3$ hydrolysis increases when the system is heated to 160 °C [41]. Additional heating (sometimes in vacuum) of the primary products of the $XSiY_3$ hydrolysis is often applied in the preparation of oligoorganylsilsesquioxanes in considerably higher yields (Table 1). In most cases, the alkaline thermolysis of polysilsesquioxane gels affords polyhedral oligomers with $n > 8$ [23,41]. Variation in temperature may control, to a certain extent, the solubility of hydrolytic polycondensation products, thus facilitating the isolation of oligoorganylsilsesquioxanes, and their homo derivatives from the solution [42–44].

3.1.7 Addition of Water

The formation of oligoorganylsilsesquioxanes from organyltrichlorosilanes may occur without the addition of water if methanol or ethanol are used as the solvent [1,22]. In higher alcohols, however, the reaction ceases at the formation of the corresponding alkoxysilanes and lower oligomers of linear and, to a lesser degree, cyclic structures [2,29,52]. Nevertheless, water is involved in the above reactions since it is generated via the interaction of alcohol with hydrogen chloride. The use of aqueous alcohol usually accelerates the reaction and increases the yield of oligoorganylsilsesquioxane. The molar ratio $XSiCl_3 : H_2O$ should be 1:1 because this promotes heterofunctional condensation (2) rather than slower anhydrocondensation (1):

$$\text{—Si—OH} + \text{HO—Si} \rightarrow \text{—Si—O—Si—} + H_2O \qquad \text{Scheme (1)}$$

$$\text{—Si—OH} + \text{Cl} \rightarrow \text{—Si—O—Si—} + \text{HCl} \qquad \text{Scheme (2)}$$

A large excess of water decreases the yield of $(XSiO_{1.5})_n$ formed by hydrolysis of $XSiCl_3$. The maximum yield of $(XSiO_{1.5})_8$ with $X = CH_3$, C_2H_5, C_3H_7 is obtained by use of a 2–6 fold excess of water as compared with the theoretical value ($XSiCl_3:H_2O = 1:1.5$). With a 12-fold excess of water or without water the yield decreases considerably [22]. The hydrolysis of ethyltrichlorosilane in butanol may be carried out with as much as a 20-fold excess of water (except for the initial step of the reaction) under appropriate conditions of water addition [52].

The optimum $XSiY_3:H_2O$ ratio in the hydrolysis of organyltrialkoxysilanes is considered to be 1:3 [27].

3.1.8 Solubility of the Oligosilsesquioxanes Formed

The yield of oligosilsesquioxanes and their homo derivatives greatly depends on the solubility of these compounds in the reaction medium because they are usually isolated as crystals. This dependence is mainly observed with $(XSiO_{1.5})_8$ with X = CH_3, C_5H_{11} or Ar.

The solubility of oligosilsesquioxanes is determined, first of all, by the solvent nature, the character of organic substituents at the silicon atom, the degree of oligomerization, n, and the temperature. In general, $(XSiO_{1.5})_n$ are of limited solubility and insoluble in water. Octasilsesquioxanes are considerably less soluble than their homologs with n larger or smaller than 8. The lowest solubility in organic solvents is displayed by $(CH_3SiO_{1.5})_8$ and $(C_6H_5SiO_{1.5})_8$. Solubility is increased with rising alkyl chain length in oligoalkylsilsesquioxanes and with the introduction of substituents into a phenyl group in oligoarylsilsesquioxanes (Table 4). Heterosubstituted and homooligosilsesquioxanes are usually more soluble than the corresponding oligomers of regular structure.

3.2 Condensation of Si-Functional Oligoorganylcyclosiloxanes, $(XYSiO)_m$

The synthesis of oligoorganylsilsesquioxanes from Si-functional oligoorganylcyclosiloxanes has been first carried out by dehydrocondensation of 1,3,5,7-tetramethylcyclotetrasiloxane with 1,2,5,7-tetraphenylcyclotetra-1,3,5,7-siloxanol in ether in the presence of a base catalyst [55]. The yield of the sym-tetramethyltetraphenyloctasilsesquioxane formed greatly depends on the concentration of the initial siloxanol; a maximum yield (95%) being obtained when the concentration of siloxanol is low (2.5%). The synthesis of $(XSiO_{1.5})_8$ with X = CH_3 and C_2H_5 and n = 4 and 6 in organic solvents and in the presence of methanolic KOH as the catalyst has been also reported [2,30,63]. In this case, either polyhedral oligosilsesquioxanes or high molecular weight ladder polysilsesquioxanes are formed, depending on the reaction conditions. For the synthesis of oligosilsesquioxanes, $(XYSiO)_m$ with Y = Cl have been also used as the initial compounds [62]. No preparation of polyhedral oligomers from oligoorganylalkoxycyclosiloxanes has been reported until now.

3.3 Co-condensation of Organosilicon Trifunctional Monomers and/or Oligomers of Different Structure and Composition

The first heterosubstituted oligosilsesquioxane (with different substituents at the silicon atoms has been prepared by co-condensation of 1,3,5,7,9,11,14-heptacyclo-

Table 3. Heterosubstituted octasilsesquioxanes $X_mX'_{8-m}(SiO_{1.5})_8$

X	X′	m	Method of Synthesis	Initial compounds	Solvent	Catalyst	Yield	Ref.
C_6H_{11}	H	7	3.3	$HSiCl_3$; $(PhSiO_{1.5})_4(OSiPhOH)_3$	Benzene	HCl; C_5H_5N	—	[37]
CH_3	C_2H_5	1–7	3.3	$MeSiCl_3$; Ethylpolycyclosiloxanes	Butanol	HCl	49	[52]
C_2H_5	$CH=CH_2$	1–7	3.3	$EtSiCl_3$; $VinSiCl_3$	Butanol	HCl	25	[53]
CH_3	C_6H_5	7	3.4	$MeSi(OEt)_3$; $PhSi(OEt)_3$	Ethanol	KOH	8	[54]
CH_3	C_6H_5	6	3.4	$MeSi(OEt)_3$; $PhSi(OEt)_3$	Ethanol	KOH	31	[54]
CH_3	C_6H_5	4	3.3	$[PhSi(OH)O]_4$; $(MeSiHO)_4$	Ether	Base	95	[55]
t-C_4H_9	C_6H_5	6	3.3	t-$BuSi(OEt)_3$; $PhSi(OEt)_3$	Ethanol	KOH	—	[56]
$CH=CH_2$	CH_2CH_2Br	7	3.5	$(CH=CH_2SiO_{1.5})_8$	Butanol	Reagent Br_2		in press

Table 4. Solubility of oligosilsesquioxanes, $(XSiO_{1.5})_n$ in organic solvents

X	n	Solvent							
		CCl_4	Chloroform	Dichloromethane	Pentane	Heptane	Cyclohexane	Benzene	Toluene
CH_3	8	N	P	P	N	N	N	N	N
CH_3	10, 12	S					S		S
C_2H_5	8	S	S		S			S	
C_2H_5	6	S	S		S	S	S	S	
C_3H_7	8	S	S		S	S	S	S	
$CH(CH_3)_2$	8	S	S		P			P	
$CH(CH_3)_2$	4			N				N	
C_4H_9	8	S	S		S	S	S	S	
$C(CH_3)_3$	4			N				N	
$CH{=}CH_2$	8	S	S					S	
C_6H_5	8	P		0.017**				P	
$CH_3C_6H_4$	8			0.078					
$O_2NC_6H_4$	8	N	S	S				S	
2-C_4H_3S	8		N	0.007				N	
$ON(CH_3)_4$	8	S						S	

N — non soluble; P — poorly soluble; S — soluble; * — authors' data, ** — g/min

hexyltricyclo[7,3,3.1^{5,14}]heptasiloxane-3,7,11-triol with trichlorosilane in benzene in the presence of pyridine [37]. Until now, most of the heterosubstituted oligosilsesquioxanes known (Table 3) have been obtained by co-hydrolysis of different $XSiY_3$ monomers [53,54,56], Si-functional cyclotetrasiloxanes [55] and $XSiY_3$ monomers with polycyclosiloxanes [37,52].

Most regularities of the hydrolytic polycondensation of $XSiY_3$ monomers described in Section 3.1 are also characteristic of co-condensation reactions of the latter with $X'SiY_3$ monomers.

The co-hydrolysis of a mixture of trifunctional $XSiY_3$ and $X'SiY_3$ monomers usually gives a mixture of heterosubstituted oligosilsesquioxanes with all possible combinations of substituents X and X', the distribution of the substituents being of statistical character [52,53]. When used for the preparation of heterosubstituted octasilsesquioxanes, for example, this reaction leads to nine compounds of the general formula $XX'_{8-n}(SiO_{1.5})_8$ with n = 0–8. Compounds of this type with n = 2–6 may, in turn, represent mixtures differing not only in the composition, but also by the substituent position. The relative yield of heterosubstituted oligosilsesquioxanes with various n values depends mainly on the molar ratio and reactivity of the initial monomer. In the case of an equimolar ratio of both monomers and similar rates of hydrolysis, the compound with an equal number of X and X', $X_4X'_4(SiO_{1.5})_8$, is formed in highest yield. However variations in the molar ratio of the initial monomers give rise to different products and yields. Thus, the $XSiY_3$-$X'SiY_3$ co-hydrolysis in the molar ratio 1:7 leads to the formation of an octamer, $XX'_7(SiO_{1.5})_8$ in maximum yield. The yield of heterosubstituted oligosilsesquioxanes are significantly affected by differences in the reactivity of the initial monomers. Maximum amounts of

Me-thanol	Ethanol	2-Pro-panol	Butanol	Acetone	Ether	Dioxane	CS_2	Tetra-line	Tetra-hydro-furan	Ref.
				N				P		21,23,22)
				S			P			23)
P	P		0.08	S	P		S			27,29,*
P	S	S		S			S			27)
S	S	S		S			S			22)
P	P			P			P			22)
					S				S	31)
S	S			S			S			22)
P	P			S	S	S			S	31)
P	P	P	P	S						*
	P		N	N	N			P		31,22,*
P				P	P					
	P			P	N	S	N			42,*
				N						45)
				P						48)

$C_6H_5(CH_3)_7(SiO_{1.5})_8$ and $(C_6H_5)_2(CH_3)_6(SiO_{1.5})_8$ in the mixture of the heterosubstituted octasilsesquioxanes formed [55] are presumably due to different reactivities of $CH_3Si(OC_2H_5)_3$ and $C_6H_5Si(OC_2H_5)_3$ toward hydrolysis. Accordingly, the synthesis of heterosubstituted oligosilsesquioxane of a definite composition presupposes that the monomer, which is more susceptible towards hydrolysis is added to the reaction mixture only after the less reactive monomer has already been hydrolyzed [55].

3.4 Thermolysis of Polyorganylsilsesquioxanes

Polyhedral ologoorganylsilsesquioxanes have been first prepared by thermal depolymerization of the organyltrichloro- and organyltrialkoxysilane hydrolysis products. Although a great number of polyhedral siloxanes were obtained by the application of this method (Tables 1–3), it has however not widely been used. The thermolysis of polymeric products of the hydrolysis of $XSiY_3$ is performed by heating at 200–400 °C in the presence of a base catalyst (sometimes, in a vacuum) [13, 14, 23, 42, 45, 54]. Alkali hydroxides and, in some cases, triethylamine are most frequently used as the catalysts [42, 45]. In the thermal decomposition of polyphenylsilsesquioxane, KOH is a more efficient catalyst than NaOH or $(CH_3)_4OH$ [41]. An advantage of the thermal method is the possibility of preparing oligosilsesquioxanes such as octa(2-thienylsilsesquioxane) [45] as well as oligomers with $n > 8$ which are not formed through hydrolysis of trifunctional monomers.

3.5 Reactions with Retention of the Silicon-Oxygen Skeleton

3.5.1 Nitration

The nitration of $(C_6H_5SiO_{1.5})_8$ by fuming nitric acid leads to octa(nitrophenylsilsesquioxane) [42]. The position of the nitro group on the phenyl ring has not been established. It is not possible to introduce a second nitro group into the aromatic ring by this method. However, the nitration of octa(4-tolylsilsesquioxanes) yields dinitro compounds $[CH_3(NO_2)_2C_6H_2SiO_{1.5}]_8$ [45].

3.5.2 Halogenation

Octa(phenylsilsesquioxanes) can be brominated with Br_2 without cleavage of the Si—O bond [42]. The action of a solution of bromine in hydrobromic acid on octa- and dodeca(thienylsilsesquioxanes) leads to the corresponding perbromothienylsilsesquioxanes [45]. Bromination of octa(vinylsilsesquioxane) in carbon tetrachloride occurs only with difficulty. The main reaction products are heptavinyl(bromoethyl)-octasilsesquioxane and hexavinyl(bromoethyl)bromooctasilsesquioxane. Attempts to achieve a more extensive bromination of $(CH_2{=}CHSiO_{1.5})_8$ under the conditions studied failed.

3.5.3 Silylation

The silylation of $[(CH_3)_4NOSiO_{1.5}]_8$ by hexamethyldisiloxane in an aqueous propanol solution in the presence of hydrochloric acid affords octa(trimethylsiloxysilsesquioxane) in good yield (75%) [48].

4 Mechanisms of Formation

4.1 Hydrolytic Polycondensation of $XSiY_3$ Monomers

The formation of oligosilsesquioxanes in the course of hydrolytic polycondensation of $XSiY_3$ monomers in diluted solvents, which follows a seemingly simple scheme

$$nXSiY_3 + 1.5n\ H_2O \rightarrow (XSiO_{1.5})_n + 3n\ HY\ , \qquad \text{Scheme (3)}$$

is a multistep and rather complicated process. At present, there are two hypotheses explaining the mechanism of this process. They are both mainly based on the nature of the identified intermediates formed in the course of hydrolytic polycondensation of the initial monomers. These intermediates have been isolated and identified by fractionation, GLC, chromatography-mass spectrometry, NMR, IR, and UV spectroscopy. Trimethylsilylation with Me_3SiCl for the isolation and identification of the intermediates [25,26,64] falsifies the composition of the reaction mixture to a considerably greater degree than simple fractionation does since trimethylchlorosilane participates in the cleavage of the siloxane bonds of the intermediates [1]. On the whole, among the intermediates of $XSiY_3$ polycondensation, linear oligosiloxanes containing two to four silicon atoms (most frequently, in blocked alkoxy groups), [21, 25–29, 43, 52, 65], oligocyclosiloxanes, $(XYSiO)_m$ with m = 3–7 and Y = OH, OR [14, 21, 25–29, 33, 34, 43, 52, 65] as well as condensed polycyclosiloxanes have been isolated[4]:

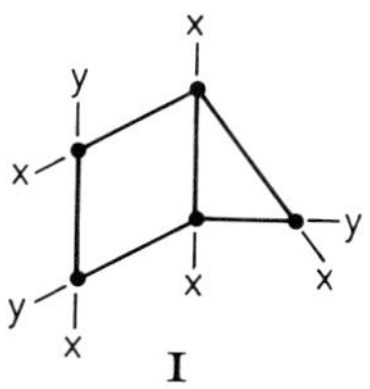

I

14, 26–30, 37, 52, 64)

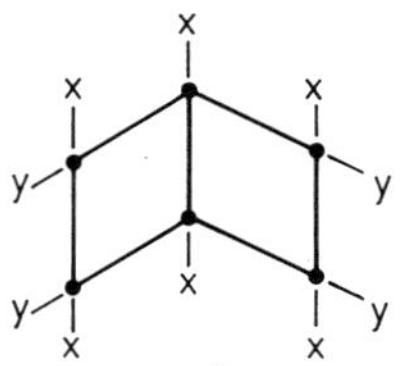

II[5]

21, 25–30, 34, 52, 64)

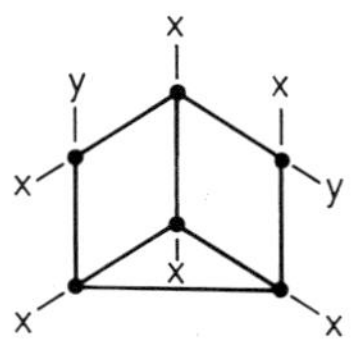

III

28, 30, 34, 37, 52)

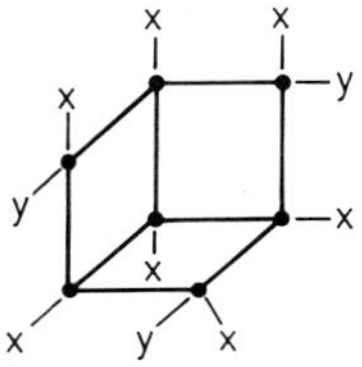

IV

28, 37, 52)

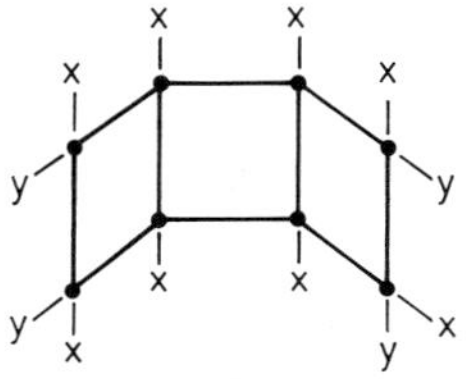

V

25, 26, 30, 34, 43)

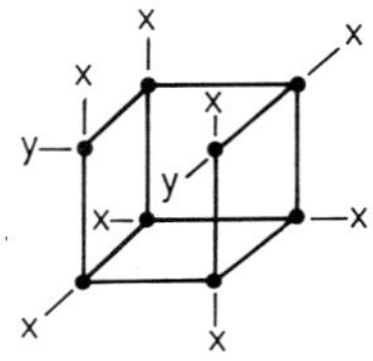

28, 30, 33, 34, 43, 52)

where Y = OH or OR (X see in Refs., Table 1–3). The structure and yield of the intermediates greatly depend on the reaction conditions (Sect. 3.1) and the reactivity of the monomers. The structures of some intermediates or of their analogs have been determined by X-ray diffraction analysis [66, 67]. In the hydrolytic polycondensation of $XSiY_3$ in an organic solvent yielding oligosilsesquioxanes, no intermediate alkylsilanetriols have been found nor cyclic, polycyclic and linear siloxanes containing Si—Cl bonds (excluding disiloxanes [28, 29]).

Sprung and Guenther [27] were the first to postulate that the hydrolysis of organyltrifunctional monomers $XSiY_3$ involves a consecutive formation of linear, cyclic, polycyclic and, finally, polyhedral siloxanes. They assumed the chain growth to be of a random character. This hypothesis was further developed by Brown and Vogt who studied the hydrolysis of cyclohexyl- [37] and phenyltrichlorosilane [43]. They suggested the formation of polyhedral silsesquioxanes and their homo

4 In this case and, in addition, the silicon atoms in complex structures are conventionally denoted by solid circles, the connecting lines being siloxane oxygen atoms.

5 The lack of structural-analytical data has led to the conclusion that this compound might also exhibit a cyclolinear structure

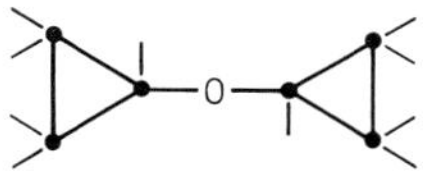

It is unlikely, however, that such a compound be formed in appreciable amounts under the reaction conditions employed.

derivatives as resulting from consecutive stepwise polycondensation of cyclic macromolecules, for example:

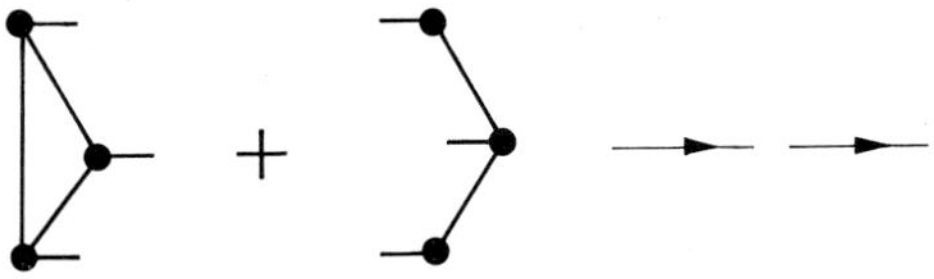

Scheme (4)

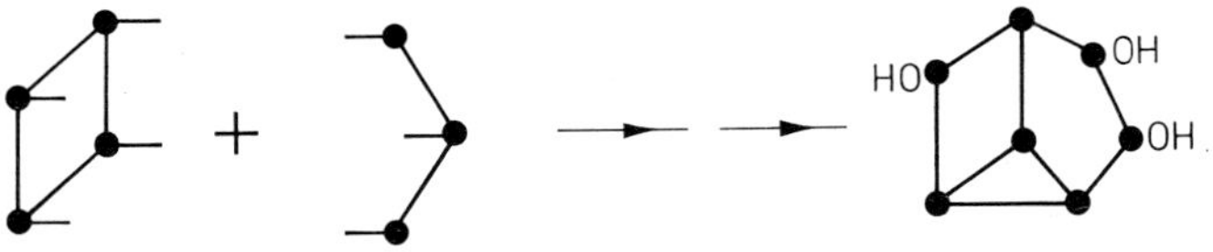

Scheme (5)

On the basis of the polycondensation intermediates found, cyclization at the initial step of the reaction and simultaneous formation of linear and cyclic oligosiloxanes were assumed to occur. Co-condensation of the latter was believed to lead to polycyclosiloxanes [43)]

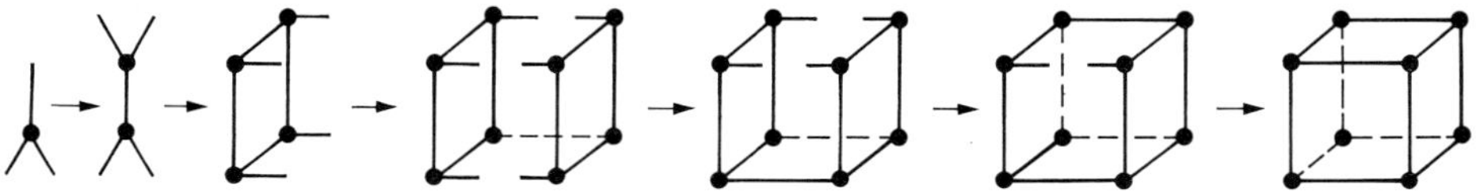

Scheme (6)

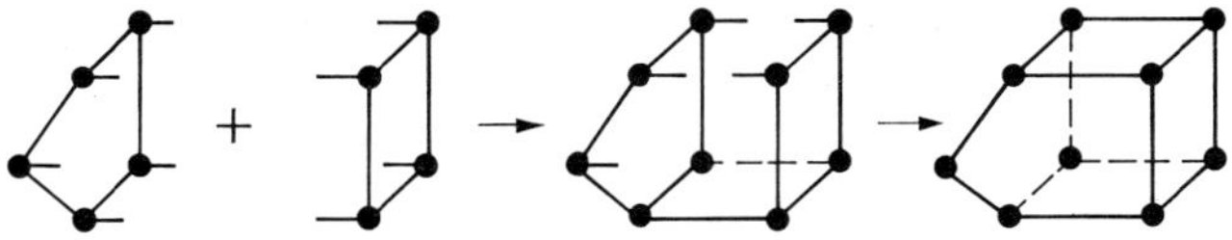

Scheme (7)

which are inert to further intermolecular polycondensation. The authors emphasize that the statistical theory based on an equal reactivity of all the functional groups is not applicable to polycondensations of complex polysiloxanes of type III–VI and their isomers. This is explained, first of all, by structural differences in the functional groups. For the silanol groups in phenylpolycyclosiloxanes, the difference in the reactivity is due to two factors, steric accessibility and ability for hydrogen bonding. These investigations have led to the important conclusion that in the course of $C_6H_{11}SiCl_3$ and $C_6H_5SiCl_3$ hydrolysis the resulting oligosilsesquioxanes are formed by different mechanisms. This conclusion is based on the fact that the end products of the polycondensation of $C_6H_{11}SiCl_3$ are mainly hexa(cyclohexylsilsesquioxane) and tricyclosiloxanetriol of type IV whereas the hydrolysis of $C_6H_5SiCl_3$ affords, under the same conditions, oligo- and polysiloxanes of higher molecular weights.

In most cases, the mechanism of the polycondensation of trifunctional monomers $XSiY_3$ bearing a different substituent X is also different which is evident from the type and yield of linear, cyclic and polycyclic siloxanes formed as intermediates in the course of the corresponding polycondensation.

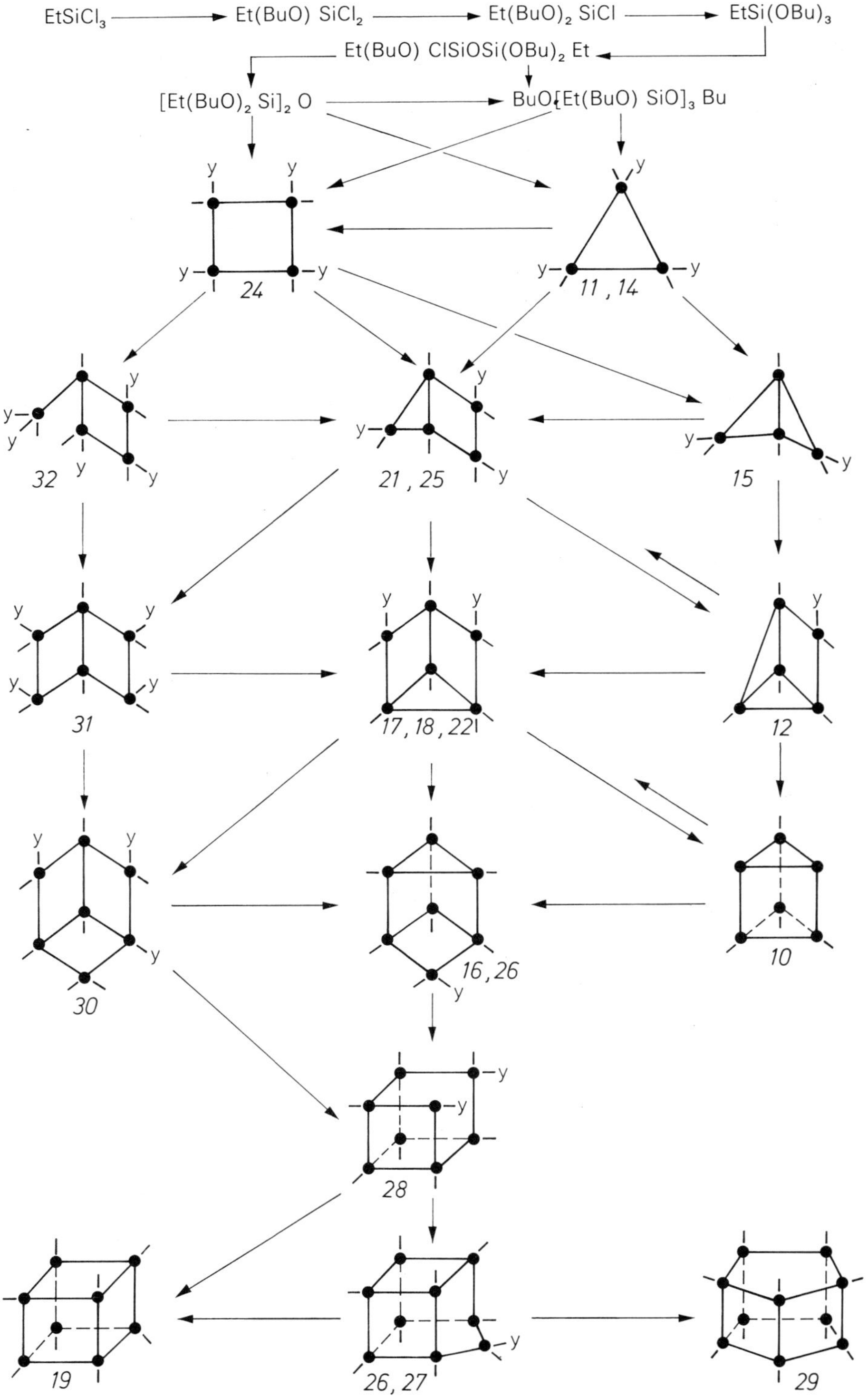
EtSiCl3
Et(BuO) SiCl2
Et(BuO)2 SiCl
EtSi(OBu)3
Et(BuO) ClSiOSi(OBu)2 Et
[Et(BuO)2 Si]2 O
BuO[Et(BuO) SiO]3 Bu
24
11, 14
32
21, 25
15
31
17, 18, 22
12
30
16, 26
10
28
19
26, 27
29

Scheme (8)

The formation of oligo(alkylsilsesquioxanes) by hydrolysis of higher alkyltrichlorosilanes is assumed [34–36] to result from consecutive, stepwise polycondensation according to Scheme (4).

The mechanism by which oligo(ethylsilsesquioxanes) are formed in the hydrolysis of ethyltrichlorosilane [28, 52] and ethyldichlorosilane [29] has been established by chromatography-mass spectrometry. By the application of this method about thirty kinetic reaction products including those formed in negligible amounts (less than 1%) and having short lifetimes (about 1 min) have been identified. By controlling the formation and variation of concentration of the intermediates during the hydrolytic condensation by means of GC/MS it is possible to follow interconversions of these products. From the results of these studies a second mechanism for the formation of oligo(alkylsilsesquioxanes) has been suggested according to which the whole process may be subdivided into two steps. In the first step, rapid consumption of $XSiY_3$ monomers occurs with the formation of XSiYY′Y″ as well as linear, cyclic and polycyclic siloxanes. No polycyclic systems are formed in the first step of the hydrolysis of $C_2H_5SiHCl_2$. The second step involves slow formation of the polyhedral silicon-oxygen framework, due to the interaction of condensed polycyclosiloxanes with monomers and dimers of the type XSiYY′Y″ and $(XSiYY')_2O$ which may contain substituents of both similar and different reactivity (Y = Cl, OH, OR, H) and regenerate due to cleavage and rebuilding of complex and frequently strained polysiloxane molecules. This mechanism does not exclude a possible formation of polyhedral compounds via stepwise polycondensation of cyclic macromolecules (Schemes (4) and (5)). The formation of oligo(ethylsilsesquioxanes) through hydrolysis of $C_2H_5SiCl_3$ in aqueous butanol is shown in Scheme (8) [28], all the products formed being identified by means of GC/MS. This scheme has been confirmed by a study of the polycondensation of methyltrichlorosilane with an equilibrium mixture of polyethylcyclosiloxanes formed in high yield by hydrolysis of ethyltrichlorosilane [52]. The heterosubstituted octasilsesquioxanes, $(CH_3)_m(C_2H_5)_{8-m}(SiO_{1.5})_8$, with m = 1–7 isolated under these conditions imply that the formation of the polyhedral silsesquioxane skeleton is associated with a consecutive addition of trifunctional CH_3SiCl_3 derivatives to polycyclosiloxanes.

The formation of Si—O—Si units in the course of the process leading to oligosilsesquioxanes may be due to homofunctional condensation (anhydrocondensation) according to Schemes (1) and (9)

$$\gt SiOR + RO{-}Si\lt \rightarrow \gt Si{-}O{-}Si\lt + ROR \qquad \text{Scheme (9)}$$

and heterofunctional condensation according to Schemes (2), (10), (11)

$$-SiCl + RO{-}Si{-} \rightarrow -Si{-}O{-}Si{-} + RCl \qquad \text{Scheme (10)}$$

$$-SiOH + RO{-}Si{-} \rightarrow -Si{-}O{-}Si{-} + ROH \qquad \text{Scheme (11)}$$

Reactions (2) and (10) take place predominantly in the first step of the process when the reaction mixture contains a considerable amount of partly hydrolyzed XSiYY′Y″ monomers with Y, Y′, Y″ = Cl, OR, OH. In the second step, reactions (1), (9) and (11) predominate.

Along with the highly reactive substituents OH and OR, cyclo- and polycyclosiloxanes contain Si—O—Si groups which can be readily cleaved [1]. The reactivity of the groups involved in the formation of the polyhedral oligosilsesquioxane framework decreases in the following order [52]:,

$$\text{Si—OH} > (=\text{SiO})_3 > \text{Si—OC}_4\text{H}_9$$

Cleavage of the disiloxane group occurs chiefly in strained cyclotrisiloxane rings $(=\text{SiO})_3$ of condensed polycyclosiloxane systems. The reactivity of $(=\text{SiO})_3$ rings in different polycyclosiloxanes is governed by structural dissimilarity. Thus, the possibility of Si—O—Si group cleavage decreases with increasing structural stress of the molecule:

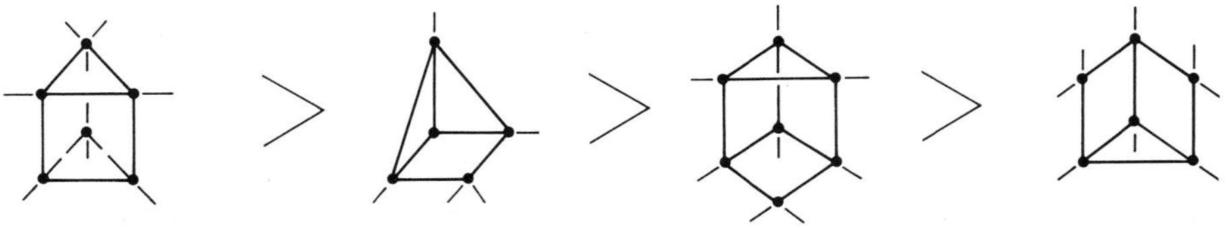

Cleavage of the siloxane bond and rearrangement of polysiloxanes may be accompanied by either elimination of XSiYY′Y″ monomers and, less frequently, $(\text{XSiYY}')_2\text{O}$ dimers.

It is reasonable to assume that the formation of other oligoalkylsilsesquioxanes containing short-chain alkyl substituents also follows Scheme (8).

4.2 Condensation of Oligoorganylcyclosiloxanes, $(\text{XYSiO})_m$

The mechanism of the formation of oligosilsesquioxane produced by hydrolytic condensation of oligoorganylcyclosiloxanes, $(\text{XYSiO})_m$, has only been studied for $(\text{C}_2\text{H}_5\text{SiHO})_{4.5}$ [2,30]. This is very probable the second step of the hydrolytic polycondensation of ethyltrichlorosilane (Sect. 4.1) involving cleavage of Si—H, Si—OH, Si—OR bonds, and Si—O—Si groups. However, the higher yield of polyhedral oligo(ethylsesquioxanes) and their homo derivatives as well as that of the intermediate tricyclosiloxane of type V suggests the Brown-Vogt mechanism (Schemes (4) and (5)).

4.3 Thermolysis of Polyorganylsilsesquioxanes

The mechanism of oligo(organylsilsesquioxane) formation through thermal decomposition of the polymeric products of trifunctional XSiY_3 was first reported by Sprung and Guenther [33]. The mechanism of the formation of polyhedral siloxanes from polyphenylsilsesquioxanes is suggested [33,42] to be based on nucleophilic interaction of terminal silanol groups with neighboring siloxane bonds. There are two possible polymer structures, ladder-type and linear block structures which give rise to the corresponding oligo(phenylsilsesquioxanes):

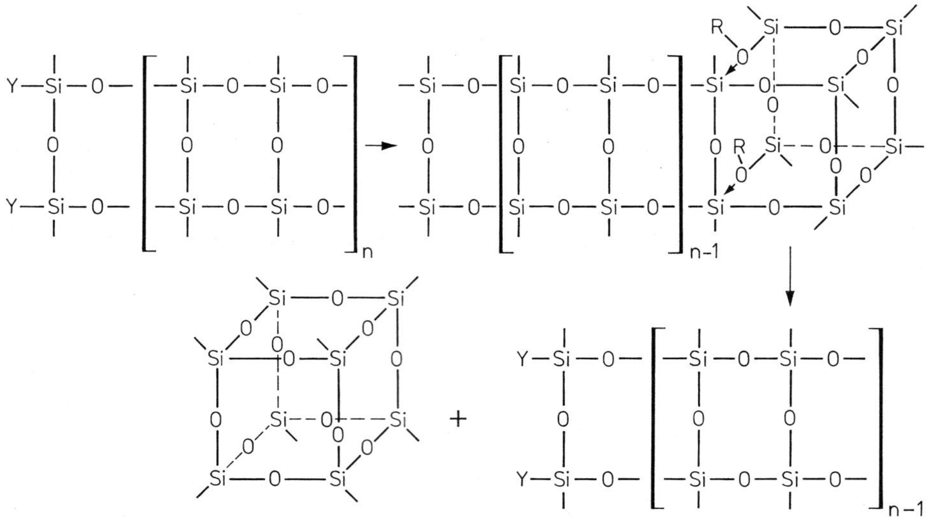

Scheme (12)

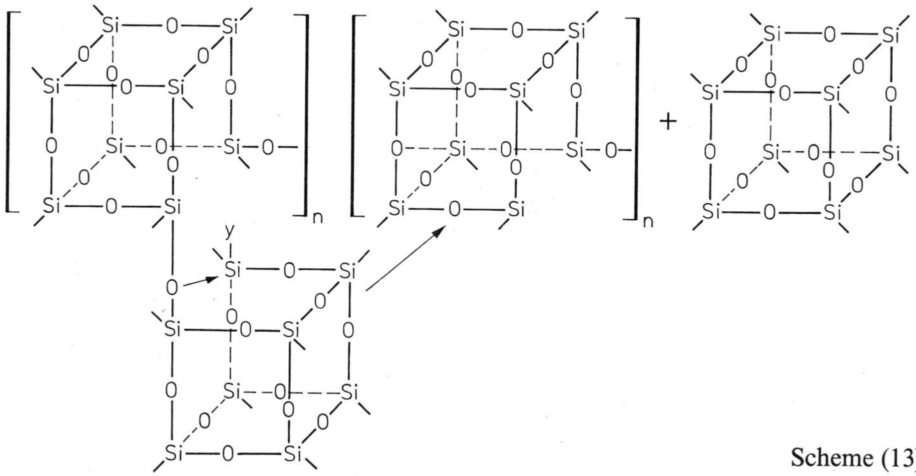

Scheme (13)

Oligomers with n = 6, 10, 12, etc. can be formed in an analogous manner. However, among the products of the thermal decomposition of polyphenylsilsesquioxane in the presence of alcoholic alkali, a great number of incompletely condensed low molecular weight oligosiloxanes containing one and more alkoxy groups have been found [41]. This shows that thermal degradation under drastic conditions involves redistribution of the siloxane skeleton. Polyhedral silsesquioxanes, however, do not result from a one-step decomposition of the polymer as evidenced from a chromatography-mass spectrometrical study of the thermolysis of branched permethylpolycyclosiloxanes [51]. Polyorganylsilsesquioxanes are formed according to Scheme (12) when the cyclotetrasiloxane rings of the polymer are cis-syn-cis bonded. If the initial polymer exhibits a cis-anti-cis structure, it is convert to even a higher molecular weight polymer with a double-chain cis-syndiotactic (ladder-type) structure.

On the whole, the mechanism of the formation of oligo(organylsilsesquioxanes) through pyrolysis of polymers has not been proved experimentally.

5 Structure and Physical Properties

5.1 General Characteristics

Most oligosilsesquioxanes known are colorless crystalline substances. The only compounds of this series, those with long-chain alkyl substituents (starting from C_6H_{13}), are transparent and highly viscous liquids.

Octa(hydrosilsesquioxane) is more volatile and has a lower melting point than octa(methylsilsesquioxane). It is readily soluble in most organic solvents and the only oligosilsesquioxane which is poorly soluble in water. With growing chain length of the alkyl group in octa(alkylsilsesquioxanes), the melting points and densities of

Table 5. Physical properties of oligosilsesquioxanes, $(XSiO_{1.5})_n$

X	n	M_p (°C)	$T_{subl.}$ (°C/P, mm)	d^{20} (g/cm³)	Ref.
H	8	250	—	1.88	20,68)
CH_3	6	209–210	86–120/0.7–0.9	—	21)
CH_3	8	415 dec.	200–210/1	1.49	14,22)
CH_3	10	332.6–334.0	—	—	23)
CH_3	12	270	—	—	23)
C_2H_5	6	59–60	—	—	27)
C_2H_5	8	282–285	160–180/1	1.31	22,27)
C_3H_7	8	219–220; 175	—	1.08	22,68)
$CH(CH_3)_2$	4	80–100	80/v	—	31)
$CH(CH_3)_2$	8	295.5; 296.5	—	1.18	22)
C_4H_9	8	190–195; 73–76	—	1.10	13,22)
$C(CH_3)_3$	4	225	60/v	1.11	31)
C_5H_{11}	8	77–82	No sublimate	—	33)
C_6H_{13}	8	281–285 (2.5)[c]	—	1.035	34,36)
C_6H_{13}	12	—	—	1.040	36)
C_6H_{13}	20	—	—	1.046	36)
C_7H_{15}	6	292–296(3)[c]	—	1.012	34)
C_8H_{17}	6	319–323(1)[c]	—	1.010	34,36)
C_8H_{17}	12	—	—	1.002	36)
C_8H_{17}	24	—	—	1.003	36)
i-C_9H_{19}	6	298–302(2)[c]	—	0.9860	34,36)
i-C_9H_{19}	8	—	—	0.9865	36)
i-C_9H_{19}	10	—	—	0.9872	36)
i-C_9H_{19}	18	—	—	0.9890	36)
i-C_9H_{19}	28	—	—	0.9897	36)
C_6H_{11}	6	265–267	—	—	37)

Table 5. (continued)

X	n	M_p (°C)	$T_{subl.}$ (°C/P, mm)	d^{20} (g/cm³)	Ref.
C_6H_{11}	8	400	—	1.174	13,37)
$CH{=}CH_2$	8	—	150–220/0.1–0.001	1.374	38,39)
$CH_2CH{=}CH_2$	8	—	—	1.28	40)
C_6H_5	8	500 dec.	250–300/v	1.34; 1.3	22,68)
C_6H_5	10	415–418	—	—	41)
C_6H_5	12	385	300/v	—	41)
4-$CH_3C_6H_4$	8	400 dec.	—	1.25	42,68)
x—$NO_2C_6H_4$	8	325	100/v	1.2	42,68)
1-$C_{10}H_7$	8	335–343	—	1.24	42,68)
2-C_4H_3S	8, 12	300	—	—	45)
2-C_4Br_3S	8, 12	360	—	—	45)
C_6Cl_5	6	188–193	—	—	47)

v Vacuum. No residual pressure is reported.
c Boiling point (mm).
dec. Decomposition

these compounds decrease whereas the volatility and solubility in organic solvents simultaneously increase (Table 5). The volatility of these compounds is extremely high. Among the oligomers displaying a similar nature of their substituent X, the octamers have the highest melting points and are the least volatile products.

The saturated vapor pressure of $(CH_3SiO_{1.5})_8$ is described by the following equation:

$$P_{tor} = -5770/T + 11.9 \pm (-8/T + 0.015) \qquad (120\text{–}290\ °C)$$ [69]

From this, $\Delta H_{subl.} = 24.4 \pm 0.1$ kcal/deg × mol and $\Delta S_{subl.} = 41.7 \pm 0.2$ kcal/deg × mol were defined.

5.2 Molecular Structure

X-ray data are available only for the readily accessible octasilsesquioxanes and deca(methylsilsesquioxane) (Table 6). The crystal lattices of octa(alkylsilsesqui-

Table 6. Average geometric parameters of oligosilsesquioxanes, $(XSiO_{1.5})_n$

X	n	αSi-O-Si (deg.)	αO-Si-O (deg.)	Δ_{max} (Å)	dSi-O (Å)	dSi-X (Å)	Molecular volume (Å³)	Ref.
H	8	145.0		—	1.60	1.48	370	70)
CH_3	8	145.3	111.1	0.34	1.61	1.895	589.6	71)
CH_3	10	155.0; 149	109.0	0.30	1.604	1.854	774	72)
$CH{=}CH_2$	8	150.2	108.5	0.30	1.60	2.028	762	73)
$CH_2CH{=}CH_2$	8	150.7	108.2	0.31	1.63	1.83	962	40)
C_6H_5	8	149.2	109.0	0.30	1.614	1.828	1389.7	74,75)

oxanes) with X = CH_3, C_2H_5, $(CH_3)_2CH$, C_3H_7, C_4H_9 are isomorphous [68]. The alkyl groups (two groups of eight) exhibit either a random orientation of rotate (the X-ray diagrams reveal diffuse reflections). The alkyl group orientation of one of three equivalent positions on the triple axis favors their intramolecular interaction with neighboring oxygen atoms. The intermolecular distances in $(HSiO_{1.5})_8$ and $(CH_3SiO_{1.5})_8$ indicate possible intermolecular interactions between the oxygen atoms and the corresponding substituents of neighboring molecules. The strong intermolecular attractions in octa(methylsilsesquioxane) crystals is likely to be higher than in octa(hydrosilsesquioxane), due to O ... O and H ... H "repulsion" of neighboring molecules in the latter crystals[6]. The intermolecular distances in the octa(phenylsilsesquioxane) lattice also suggest possible interactions between a hydrogen atom of the phenyl group and the oxygen atom of the neighboring molecule. At the same time, the lattice packing of $(C_6H_5SiO_{1.5})_8$ allows the formation of impurity compounds (pyridine [74], acetone [75], for example, etc.). Octa(phenylsilsesquioxane) may exist in three crystalline modifications, namely monoclinic, triclinic [68] and rhombohedral [41]. The existence of two melting points for $(XSiO_{1.5})_8$ with X = C_3H_7, $(CH_3)_2CH$ and C_4H_9 (Table 5) seems to be also associated with polymorphism [68].

The molecular structure of all the octamers is a combination of six condensed tetrasiloxane rings (Fig. 1). The silicon atoms in these rings lie, within the limit of error, in one plane whereas the oxygen atoms are located in another plane, parallel to the first one. The largest distance between the planes, Δ_{max}, is practically independent of the type of substituent X. The structural parameters of the silicon-oxygen skeleton of all the $(XSiO_{1.5})_8$ compounds studied are nearly similar. The average Si—O—Si valence angles (145–155°) in the compounds investigated are beyond the upper limit of the valence angle values for oxygen atoms in the corresponding tetra- and pentacyclosiloxanes [66,67]. This is consistent with the

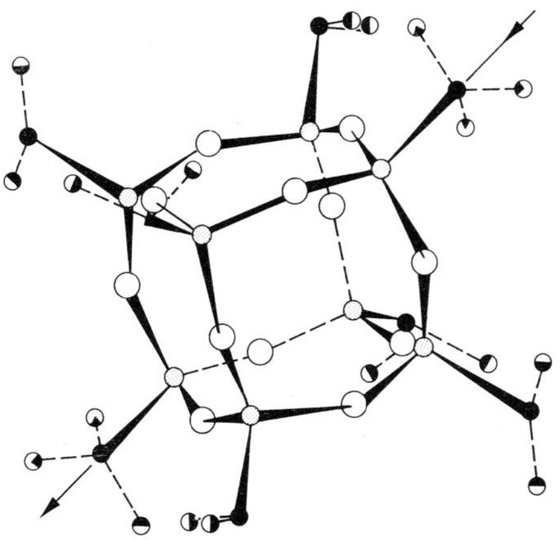

Fig. 1. X-Ray Structure of Octa(methylsilsesquioxanes)

[6] In the crystal lattice of silsesquiovanes with X = C_2H_5, C_3H_7, etc., intermolecular interactions are hindered due to the oxygen atoms shielded by the alkyl groups. Besides, the alkyl group position in such molecules is disordered, the atoms approaching the terminal hydrocarbon chain being more heat-dependent.

concept of p_π—d_π interactions between the silicon and oxygen atoms which are stronger in oligosilsesquioxanes than in cyclosiloxanes; as a result, the Si—O bonds in the former compounds are by 0.04 Å shorter. In $(XSiO_{1.5})_8$ molecules with X = CH_2=CH, CH_2=CH—CH_2 and C_6H_5, the Si—O—Si angle is approximately 150°, i.e. by 5° larger than in isostructural compounds with X = H and CH_3. This points to a strengthening of the silsesquioxane π-system in compounds of the first type. The stability of the silicon-oxygen skeleton in octasilsesquioxanes to ionic reagents and thermolysis is determined by the nature of the substituent X; it decreases in the following order:

$$H > CH_3 > C_6H_5$$

The avergae Si ... Si diagonal length in the silicon-oxygen skeleton is 5.4 Å. The average distances of the Si—C bond lengths are in the range 1.83–2.03 Å. An increase in the Si—C bond length (2.03 Å) in $(CH_2{=}CHSiO_{1.5})_8$ has some importance. The average C=C distances in vinyl groups agree with the limits typical of the C=C double bond. The possibility of vinyl group "rotation" about the Si—O bond, which unfortunately coincides with the triple axis, has not been established experimentally.

The deca(methylsilsesquioxane) molecule [72] consists of five tetrasiloxane and two plicate pentasiloxane rings (Fig. 2). The silicon and oxygen atoms lie in two nearly parallel planes separated by approximately 0.3 Å. The silicon atoms occupy the tops of the pentagonal prism and the oxygen atoms are located at the edges.

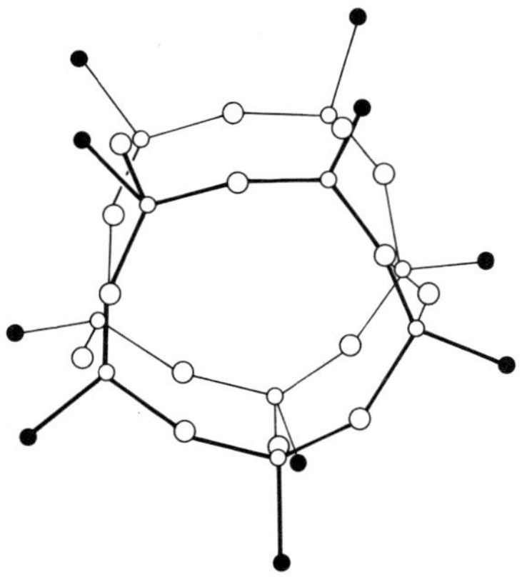

Fig. 2. X-Ray Structure of Deca(methylsilsesquioxanes)

5.3 NMR Spectra

The ^{29}Si-NMR spectra of octa- and deca(hydrosilsesquioxanes) display a signal at ^{29}Si = 5.79 and 5.73 ppm, respectively [20]. This shows the presence of a strain-free structure of these oligomers. In the spectrum of dodeca(hydrosilsesquioxane), two singlets (5.67 and 5.75 ppm) with the intensity ratio of 3:2:1 are observed whereas a dodecamer with structure A (Fig. 3) should display singlets with the intensity ratio of 2:1. This implies that the compounds in question are composed of a mixture of two structural isomers A and B, the signal of isomer being more intense than that of isomer B. For $(HSiO_{1.5})_n$ with n = 14, 16 a larger number of strain-free isomer structures are possible; this is manifested by a broadening of

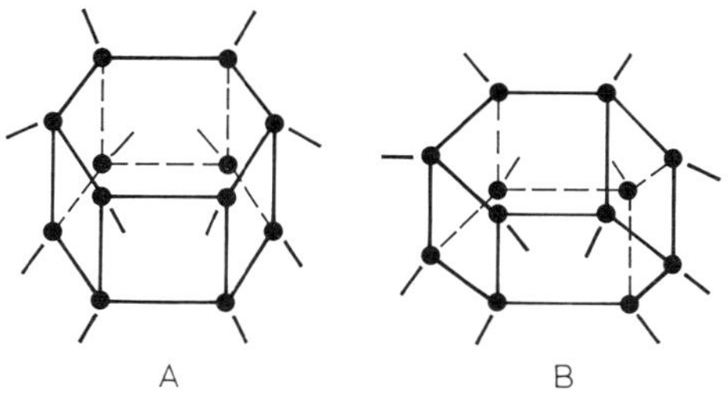

Fig. 3. Stereoisomers of Dodeca(hydrosilsesquioxanes)

the peaks which may contain more than one line. The absolute chemical shifts of oligoorganylcyclosiloxanes are rather high, i.e. the values are more shifted upfield as compared to oligosilsesquioxanes. The ^{1}H-NMR spectra give no information on the structure of oligosilsesquioxanes. Thus, the ^{1}H-NMR spectra of all oligo(methylsilsesquioxanes) with n = 6, 8, 10, 12, 14 show a singlet [23]. ^{1}H-NMR spectroscopy may be useful, however, when the structure of heterosubstituted oligosilsesquioxanes and their homo derivatives has to be determined. Thus, the ratio of the intensity lines of the protons of methyl and phenyl [54] as well as of ethyl and vinyl [53] groups has provided evidence for the structure of the corresponding methylphenyl- and ethylvinyloctasilsesquioxanes.

5.4 IR Spectra

The IR spectra of oligosilsesquioxanes and their homo derivatives (Table 7) show a strong band of a Si—O—Si asymmetric stretching vibration (ν_{as} = 1100–1140 cm^{-1}) the position of which does not markedly depend on the nature of substituent X. In the spectra of hexa(organylsilsesquioxanes) containing a strained ring, this band is displaced to a lower frequency region (ν_{as} = 1057–1085 cm^{-1}). The spectra of octa- and decasilsesquioxanes are characterized by three or four intense lines in the 360–600 cm^{-1} region arising from symmetric deformational vibrations of the silicon-oxygen framework. The spectra of hexasilsesquioxanes contain six to eight well-defined maxima in the same region [76]. The assignment of Si—C bond vibrations and, especially, ν_s(Si—O—Si) in the spectra of oligosilsesquioxanes is difficult, due to an equidistant location of some satellite bands [77]. Besides, when assigning the corresponding absorption bands one should not neglect the effect of Si—O and Si—C interaction. Nevertheless, the assignment of the absorption bands in the IR spectra of oligosilsesquioxanes, corresponding to bond vibrations of the substituents has proved successful: Si—H: 2285–2260, 910,870 cm^{-1} [20]; Si—CH_3: 2920, 1403, 1277, 774 cm^{-1} [22,23,54,76]; Si—CH_2R: 1238–1230, 1220–1180 cm^{-1} [36]; Si—C_6H_{11}: 1150–950 cm^{-1} [37]; Si—CH=CH_2: 2980, 3080, 3035, 1610, 296 cm^{-1} [38,39,40]; Si—C_6H_5: 2990, 1600, 1572, 1434, 1031, 999, 696, 747 cm^{-1} [22,33,41,54]. The degree of oligomerization of polyhedral silsesquioxanes with n = 8, 10, 12 and X = H [20], CH_3 [23], C_6H_5 [41], and alkyl [22,33,36] is very difficult to define from the IR spectra,

Table 7. Si—O absorption bands in the IR spectra of oligosilsesquioxanes, $(XSiO_{1.5})_n$

X	n	ν_{as} (cm^{-1})	ν_s (cm^{-1})			Ref.
H	8	1140	570	465	400	20)
H	10	1140	540 560		400	20)
H	12	1140	565		400	20)
H	14, 16	1140	570		400	20)
CH_3	6	1057				23)
CH_3	8	1122–1128	525	465	390	22, 23, 76)
CH_3	10, 12	1127	—	—	—	23)
C_2H_5	8	1117; 1106	575 540	460	400	22)
C_3H_7	8	1117; 1106	—	—	—	22)
C_4H_9	8	1117; 1106	—	—	—	22)
C_6H_{13}	6	1075	—	—	—	36)
i-C_9H_{19}	8	1125	—	—	—	36)
C_6H_{11}	6	1085	—	—	—	37)
C_6H_{11}	8	1125	—	—	—	37)
$CH{=}CH_2$	8	1115	585 520	475	400	38, 39)
C_6H_5	8	1119–1121	—	—	—	22, 41)
C_6H_5	10, 12	1129	—	—	—	41)
C_6H_5	22, 24	1135–1150	—	—	—	41)
CH_3; C_6H_5	8	1130–1135	—	—	—	54)
C_2H_5, $CH{=}CH_2$	8	1116	585–540			38)

due to similarity of the these spectra. The IR spectra of homooligosilsesquioxanes differ from those of the corresponding $(XSiO_{1.5})_n$ by extra lines characterizing the Si—X groups [23, 41, 43].

5.5 UV Spectra

Oligoorganylsilsesquioxanes display no absorption in the spectral region beyond 180 nm. In the UV spectra of octa- and deca(vinylsilsesquioxane) a vinyl group absorption maximum is observed at 212 nm [39]. The phenyl group absorption in the spectra of octa- and deca(phenylsilsesquioxanes) corresponds to a maximum at 264 nm, the spectrum of $(C_6H_5SiO_{1.5})_{12}$ displaying a slight batochromic shift [41, 78]. The latter is explained by a larger average angle between the C_6H_5 axes in the dodecamer as compared to that in the octa- and decamer.

5.6 Mass Spectra

Mass spectrometry has proved to be one of the most informative and time-saving physical methods for the investigation of oligosilsesquioxanes and their homologs. The mass spectra of these compounds furnish useful information not only on the molecular mass but also on the structure of the silicon-oxygen skeleton and the nature of the substituent attached to the silicon atom. At present, mass spectrometry is used for the investigation of oligomers $(XSiO_{1.5})_n$ with X = H [20], CH_3 [23,79], C_2H_5 [79], $CH_2{=}CH$ [79], $CH_2{=}CHCH_2$ [80], C_6H_5 [80], $OSi(CH_3)_3$ [48], and heterosubstituted silsesquioxanes $X_mX'_{8-m}(SiO_{1.5})_8$ with X = CH_3 and X' = C_2H_5 [52], X' = C_2H_5 and X = $CH_2{=}CH$ [53], X = $CH_2{=}CH$ and X' = C_2H_4Br [80] and homooligosilsesquioxanes $(XSiO_{1.5})_nOSiXX'$ with X = CH_3 and X' = OH, OCH_3, OC_2H_5 [23,51]; X = C_2H_5 and X' = H, OH, OCH_3, OC_4H_9 [2,50]. The basic lines in the spectra of most of the above compounds arise from $(M{-}X)^{+\cdot}$ ions. The lines from doubly charged $(M{-}2X)^{2+}$ ions have an intensity ranging from 20 to 45%, depending on the structure of silsesquioxanes and the nature of substituents attached to silicon. These values are highest for octamers with substituents C_2H_5 and CH_3. Molecular ions in the spectra of oligosilsesquioxanes and their homo derivatives (excluding those of vinyl- and phenyl-substituted derivatives) either exhibit a low intensity or are absent. Upon the action of electron impact on heterosubstituted octa(organylsilsesquioxanes), the substituents containing a longer hydrocarbon chain are more readily split. [49] The ethyl group is more readily expulsed from the polyhedral framework than the vinyl group, possibly due to a higher Si—C= bond order. A more ready loss of C_2H_4Br from heptavinylbromoethyloctasilsesquioxane as compared with the vinyl group may be explained analogously.

Further $(M{-}X)^+$ and $(M{-}2X)^{2+}$ ion fragmentation for all $(XSiO_{1.5})_n$ compounds studied proceeds mainly via peripheric disintegration. In this case, the substituents are eliminated as neutral (X + H) and (X — H) fragments which is displayed in the mass spectra by the appearance of line groups arising from singly and respectively doubly charged ions of gradually decreasing intensity. Sometimes, these series of lines end in ion peaks corresponding to a "bare" silsesquioxane framework. The number of line groups in the series of singly charged ions corresponds to that of the silicon atoms in the oligosilsesquioxane molecule studied in the series of doubly charged ions, this number being by 1 smaller. All the ion types listed above are characterized by fairly intensive isotope ion lines. The appearance of the latter provides information on the elementary composition of at least those ions whose peaks exhibit no multiple overlaps.

The silicon-oxygen framework of polyhedral oligosilsesquioxanes is stable to the electron impact of 500 eV. A slight scission of Si—O bonds was only observed in the case of oligo(vinylsilsesquioxanes). The mass spectra of homooligosilsesquioxanes show in the low mass region characteristic lines of silyl ions which are due to the cleavage of the endocyclic siloxane bond at the homo group.

The mechanism of electron-impact elimination of the substituents of oligo(organylsilsesquioxanes) and their homo derivatives is discussed in [50,81].

6 Chemical Properties

Polyhedral oligo(organylsilsesquioxanes) are highly stable to thermolysis and action of nucleophic and electrophilic agents as compared to analogously Si-substituted linear, cyclic and polycyclic oligosiloxanes. This is due to the rigid silicon-oxygen framework and shorter Si—O bond distances in oligo(organylsilsesquioxanes). In polyhedral oligomers with a strained structure, e.g. in tetra- and hexasilsesquioxanes and their homo derivatives, the Si—O—Si units are cleaved much more readily. A comparatively high stability of silsesquioxanes with different n values has been established in studying the composition of a $(CH_3SiO_{1.5})_n$ mixture formed by heating of a toluene $(CH_3SiO_{1.5})_{10}$ solution in the presence of KOH [23], the ratio of oligomers with n = 6, 8, 10, 12, 14 being 0:1:2:1:0, respectively. A similar oligomer ratio has been found in the disproportionation of octa(phenylsilsesquioxane). However, this regularity by no means indicates a poor stability of higher oligosilsesquioxanes with n > 12. It is merely indicative of a certain statistical possibility of the formation of these compounds. The relative stability of silsesquioxanes with n = 6, 8, 10, 12, 14 is presumed to be mainly determined by the degree of distortion of the Si—O—Si angle and the nature of the intramolecular interaction of the atoms forming the angles concerned. X-ray diffraction data reveal that the O—Si—O angles in oligosilsesquioxanes as well as in other tetravalence silicon compounds are tetrahedral (109.5°). Simple trigonometric calculations show that the strain-free Si—O—Si angles in the oligomers with n = 6, 8, 10, 12, 14 should be 129.5°, 148.5°, 154°, and 151.5°, respectively. This is consistent with the X-ray diffraction data on octa- and decasilsesquioxanes (Table 6). Thus, fairly high angular strains may occur in hexasilsesquioxane molecules, which accounts for the lower stability of the silicon-oxygen skeleton in the series of compounds under consideration.

IR spectrometric data reveal that the thermolysis of octa(methylsilsesquioxane) is a heterogeneous reaction involving the formation of a non-volatile solid phase similar to SiO_2 [82]. In a helium atmosphere at 150–300 °C $(CH_3SiO_{1.5})_8$ almost completely evaporates without marked decomposition. When heated in the air, however, this compound begins to decompose slowly at 270 °C, and at 450 °C an exo-effect attributed to the oxidation by air oxygen is observed [82]. When the temperature is raised to 500 °C, 66% of $(CH_3SiO_{1.5})_8$ are oxidized to SiO_2.

Octa(ethylsilsesquioxane) decomposes slowly already at its melting point (285 °C). Octa(propylsilsesquioxane) decomposes very slowly in the air at 150 °C whereas octa(isopropylsilsesquioxane) does not change under the same conditions [22].

Octa(vinylsilsesquioxane) begins to react with air oxygen at 170 °C with release of heat and at 550 °C it is completely oxidized to SiO_2 [38]. The activation energy of the thermooxidative decomposition of this oligomer at 170–290 °C is 39.7 kcal/mol, the pre-exponential facor being 1.45×10^{16} [83]. The isothermal oxidation of $(CH_2{=}CHSiO_{1.5})_8$ at 90–160 °C is characterized by an increasing weight of the initial sample in which carbonyl and hydroxy-groups are formed as IR data reveal [84]. This may be accompanied by simultaneous elimination of CO, H_2O and HCHO. The silicon content of the substance does not change until 240 °C, the amount of carbon decreasing by a factor of two. All these findings indicate that the oxidation of $(CH_2{=}CHSiO_{1.5})_8$ occurs at vinyl groups and seems to follow a chain mechanism with degenerative branching. In vacuum or an inert atmosphere, octa(vinylsilsesquioxane) decomposes at 290–300 °C to form a solid, non-volatile polymer.

$(C_6H_5SiO_{1.5})_8$ is the most heat-resistant oligosilsesquioxane known. It does not change when heated in the air to its melting point (500 °C). In their fused state (above 415 and 385 °C, respectively) deca- and dodeca(phenylsilsesquioxanes) slowly polymerize [41].

The Si—O—Si group in oligosilsesquioxanes is more stable toward sulfuric acid than in cyclosiloxanes. Oligo(hydrosilsesquioxanes) are susceptible to attack by sulfuric acid, $(HSiO_{1.5})_8$ being more reactive than $(HSiO_{1.5})_{10}$ [20]. Octa(ethylsilsesquioxane) is decomposed by sulfuric acid only on heating [22].

Octa(hydro- and octa(organylsilsesquioxanes) are stable to concentrated hydrochloric and acetic acid [20,22,43]. Octa(ethylsilsesquioxane) does not react with fuming nitric acid, hot aqueous perchloric acid solution, phosphorus pentachloride, 60% aqueous KCl and bromine [22].

Octa(phenylsilsesquioxane) polymerizes under the action of sulfuric acid [85]. In fuming nitric acid this compound is converted to $(O_2NC_6H_4SiO_{1.5})_8$ [42]. The hexadecanitro derivative of octa(4-tolylsilsesquioxane), $[(NO_2)_2C_7H_5SiO_{1.5}]_8$, has been prepared in a similar manner. Nitration of octa(1-naphthylsilsesquioxane) has not given the expected crystalline product. Attempts to reduce the nitro derivatives obtained as well as attempts to oxidize the methyl groups in octa(4-tolylsilsesquioxane) to carboxy groups failed [42]. Fluorine, chlorine and fluorinating agents of the type ClF_3, BrF_3, IF_3, SF_4, XeF_2 split the silicon-oxygen skeleton of $(C_6H_5SiO_{1.5})_8$ [85]. Arylsilsesquioxanes reluctantly react with bromine and iodine [42]. After bromination, each aromatic ring may contain more than one bromine atom, the Si—C bond remaining intact. This bond is not affected either in octa(arylsilsesquioxanes), nor upon attack by other electrophilic agents. Octa- and deca(2-thienylsilsesquioxanes) are brominated to the corresponding tribromo-2-thienylsilsesquioxanes [45].

The siloxane bonds in oligo(organylsilsesquioxanes) and their homo derivatives are split when heated with alkali hydroxides [2,22,23]. Fusion of oligosilsesquioxanes with KOH or NaOH destroys completely the silicon-oxygen skeleton. The above reaction is utilized for the determination of silicon in oligo- and polyorganylsilsesquioxanes. Preliminary treatment of the compounds concerned with alkoxides accelerates cleavage of the siloxane bonds and reduces the analytical time [86].

Heating of oligosilsesquioxanes with a catalytic amount of alkali at 200–250 °C leads to the formation of ladder polymers [41,44,87–93] according to a general scheme:

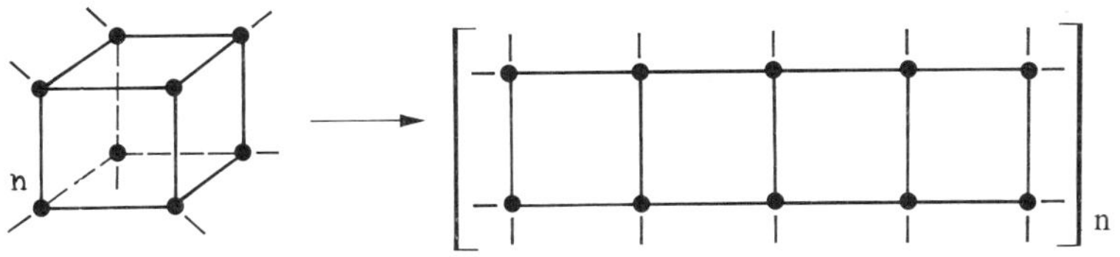

Scheme (14)

The polymers thus formed are very heat-resistant. Polyphenylsilsesquioxane, for example, starts to decompose above 600 °C. At 900 °C only loss of phenyl groups occurs, the silicon-oxygen framework of the macromolecule remaining intact [89]. This sharply differs the above polymers from linear and branched polyphenylsilsesquioxanes.

Polymerization of a mixture of oligo(3-tolylsilsesquioxanes) involves consecutive addition of the oligomer to an active chain end of the polymer molecule. This is a reversible process, the equilibrium depending greatly on temperature. Thus, above 300 °C, oligo(3-tolylsilsesquioxanes) are mainly formed [92]. An electron-microscopic examination of the polymer structure has revealed the formation of a specifically ordered structure of poly-3-tolylsilsesquioxane in the course of synthesis. This structure is characterized by the tendency of elongated macromolecules to form mesomorphous lamellar aggregates.

Upon electron impact of different energy, oligo(organylsilsesquioxanes) may polymerize with either abstraction of hydrocarbon radicals (C—C bond rupture) and formation of ladder-type molecules cross-linked by alkylene bridges or cleavage of siloxane bonds and formation of polymers of irregular structure [94].

The main differences in the chemical properties between homooligosilsesquioxanes and the corresponding oligomers exhibiting a regular structure are due to the presence in the former compounds of an endocyclic homo group, OSiXX'. Nucleophilic or electrophilic cleavage of the siloxane bond in homo(octa- and decasilsesquioxane) occurs predominantly at the silicon homo atom [2]. The reactive group (X') in these compounds may be replaced by a less reactive one or be involved in the condensation process to form the corresponding dimers. In this way, perphenylbishomooctasilsesquioxane was prepared from perphenylhomooctasilsesquioxanol-1 [43]:

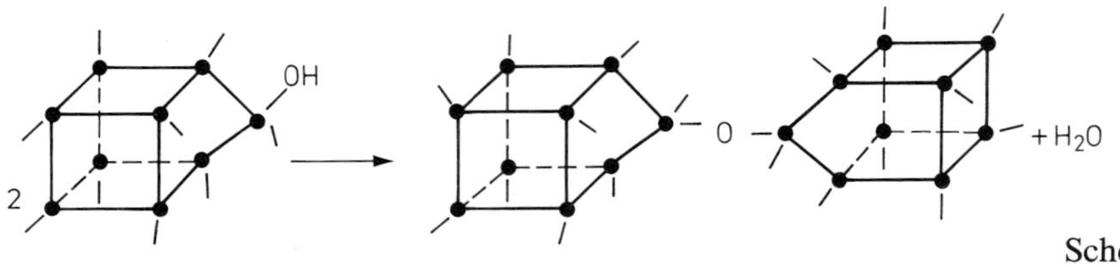

Scheme (15)

7 Applications

Oligo(organylsilsesquioxanes) and their homo derivatives can be used for the preparation of ladder polymers displaying a very high heat resistance and thermo-oxidative resistance [16, 87, 89].

The use of oligosilsesquioxanes as resistors for the electronlithographic manufacture of semiconductor microreliefs is most promising [38, 95, 96]. This enables electron lithography to be performed as a "dry" process [38]. The susceptibility of these compounds to an electron beam is fairly high and depends on the nature of substituent X. For octa(vinylsilsesquioxane), this value is 1×10^{-5} c/cm^2.

Oligosilsesquioxanes with long-chain alkyl substituents have been offered as water-repellents and adhesives for pearlite heat-insulating plates and materials [87, 98] and as damping fluids and structural plasticizers for polymeric ceramic materials [36] which allow to control the properties of the latter in the manufacture of the items concerned.

8 References

1. Voronkov, M. G., Mileshkevich, V. P., Yuzhelevskii, Yu. A.: The Siloxane Bond. London: Consultant Bureau, N.Y. 1978
2. Lavrent'ev, V. I., Voronkov, M. G., Kovrigin, V. M.: Zh. Obshch. khim. *50*, 382 (1980)
3. Buff, H., Wohler, F.: Ann. *104*, 94 (1857)
4. Ladenburg, A.: Ber. *6*, 379 (1873)
5. Friedel, C., Ladenburg, A.: Ann. Chem. Phys. *23*, 430 (1871)
6. Gatterman, L.: Ber. *22*, 186 (1889)
7. Stock, A., Zeidler, F.: Ber. *56*, 986 (1923)
8. Meads, J. A., Kipping, F. S.: J. Am. Chem. Soc. *105*, 679 (1914)
9. Palmer, K. W., Kipping, F. S.: J. Chem. Soc. *1930*, 1020
10. Hyde, J. A., Daudt, W. H.: US Patent 2482276 (1949)
11. Scott, D. W.: J. Am. Chem. Soc. *68*, 356 (1946)
12. Barry, A. J., Gilkey, J. W.: US Pat. 2465188 (1949); C. A. *43*, 6221 (1949)
13. Barry, A. J. et al.: J. Am. Chem. Soc. *77*, 4248 (1955)
14. Sprung, M. M., Guenther, F. O.: J. Am. Chem. Soc. *77*, 6045 (1955)
15. Nomenclature of Organic Chemistry IUPAC, Vol. 2, London: Butterworths 1969
16. Korshak, V. V., Andrianov, K. A.: Uspekhi Khimii *44*, 468 (1975)
17. Andrianov, K. A.: Metody elementoorg. khimii. Kremnii. Moskva: Nauka 1968
18. Brown, I. F.: J. New Sci. *17*, 304 (1963)
19. Müller, R., Khole, R., Sliwinski, S.: J. Pract. Chem. *9*, 71 (1959)
20. Frye, C. L., Collins, W. T.: J. Am. Chem. Soc. *92*, 5586 (1970)
21. Sprung, M. M., Guenther, F. O.: J. Am. Chem. Soc. *77*, 3990 (1955)
22. Olsson, K.: Arkiv Kemi *13*, 367 (1958)
23. Vogt, L. H., Brown, J. F.: Inorg. Chem. *2*, 189 (1963)
24. Wiberg, E., Simmler, W.: Z. Anorg. Allg. Chem. *283*, 401 (1956)
25. Andrianov, K. A. et al.: Zh. Obshch, Khim. *49*, 2692 (1978)
26. Andrianov, K. A., Vasil'eva, T. V., Petrovnina, N. M.: Stroenie i reakts. sposobn. kremnii-organich. soed., p. 278, Irkutsk: Tezisy dokl. I Vses. simp. 1977
27. Sprung, M. M., Guenther, F. O.: J. Am. Chem. Soc. *77*, 3996 (1955)
28. Lavrent'yev, V. I.: Tezisy dokl. III Respubl. konferents. molodykh uchenykh khimikov, p. 87, Tallin 1979
29. Lavrent'yev, V. I., Kostrovskii, V. G.: Izvest. Sib. Otd. AN SSSR, Ser. Khim. *12*, 14 (1979)
30. Voronkov, M. G., Lavrent'yev, V. I., Kovrigin, V. M.: J. Organometal. Chem., *220*, 285 (1981)
31. Wiberg, E., Simler, W.: Z. Anorg. Allg. Chem. *282*, 330 (1955)
32. Schwab, G. M., Grabmaier, J., Simmler, W.: Z. Physik. Chem. (Frankfurt), *6*, 376 (1956)
33. Sprung, M. M., Guenther, F. O.: J. Polym. Sci. *28* (116), 17 (1958)
34. Andrianov, K. A., Izmailov: Zh. Obshch. Khim. *36*, 341 (1966)
35. Andrianov, K. A., Izmailov, B. A.: J. Organometal. Chem. *3*, 435 (1967)
36. Andrianov, K. A., Izmailov, B. A.: Zh. Obshch. Khim. *46*, 329 (1976)
37. Brown, J. F., Vogt, L. H.: J. Am. Chem. Soc. *87*, 4313 (1965)
38. Mshenskaya, T. A. et al.: Elektron. tekhnika. ser. materialy, vyp. *7*, 69 (1979)
39. Voronkov, M. G. et al.: Zh. Obshch. Khim. *49*, 1522 (1979)
40. Podberezskaya, N. V. et al.: Zh. Strukt. Khim. *49*, 000 (1981)
41. Brown, J. F., Vogt, L. H., Prescott, P. I.: J. Am. Chem. Soc. *86*, 1120, (1964)
42. Olsson, K., Gronwall, C.: Arkiv Kemi *17*, 529 (1961)
43. Brown, J. F.: J. Am. Chem. Soc. *87*, 4317 (1965)
44. Brown, J. F.: J. Polym. Sci. *1c*, 83 (1963)
45. Olsson, K., Axen. C.: Arkiv Kemi *22*, 237 (1964)
46. Andrianov, K. A., Odinets, V. A.: Izv. AN SSR, Otd. Khim. Nauk. *1959*, 460
47. Andrianov, K. A., Zhdanov, A. A.: Plastmassy *7*, 24 (1962)
48. Hoebel, D., Wieker, W.: Z. Anorg. Allg. Chem. *384*, 43 (1971)
49. Kovrigin, V. M., Lavrent'yev, V. I., Kostrovskii, V. G.: Tezisy dokl. V Vses. konf. po khimii i primeneniyu kremniiorganich. soed., p. 160, Tbilisi—Moskva 1980
50. Lavrent'yev, V. I., Kostrovskii, V. G.: Zh. Obshch. Khim. *49*, 2013 (1979)

51. Garzo, G. et al.: Acta Chem. Acad. Sci. Hung. *69*, 273 (1971)
52. Lavrent'yev, V. I., Kovrigin, V. M., Treer, G. G.: Zh. Obshch. Khim. *51*, 123 (1981)
53. Voronkov, M. G. et al.: Dokl. AN SSSR *58*, 642 (1981)
54. Andrianov, K. A., Makarova, N. N.: Izv. AN SSSR, Ser. Khim. *1967*, 1381
55. Andrianov, K. A., Tikhonov, V. S., Makhneva, G. P.: Izv. AN SSSR, Ser. Khim. *1973*, 956
56. Makarova, N. N.: Referat kand. diss., Moskva, INEOS 1972
57. West, R.: J. Am. Chem. Soc. *75*, 1002 (1953)
58. Wiberg, E., Simmler, W.: Angew. Chem. *67*, 723 (1955)
59. Poverennyi, V. V. et al.: Otkr. izobr., prom.obr., tov.zn. *46*, 539043 (1976)
60. Csakwary, B., Fabry, L., Gomory, P., Ujszaszy, K.: Acta Chim. Acad. Sci. Hung. *90* (3), 213 (1976)
61. Mileshkevich, V. P., Novikova, N. F.: Usp. Khim. *50*, 85 (1981)
62. Gilbert, A., Kantor, S.: J. Polym. Sci. *40*, 35 (1959)
63. Martynova, T. N.: Ref. 26, p. 242
64. Makarova, N. N., Dubovik, I. I.: Ref. 49, p. 195
65. Sprung, M. M.: Fortschr. Hochpolym.-Forsch. *2*, 442 (1961)
66. Shklover, V. E., Struchkov, Yu. T.: Usp. Khim., *49*, 518 (1980)
67. Shklover, V. E., Palyulin, V. A., Struchkov, Yu. T.: Ref. 49, p. 481
68. Larsson, K.: Arkiv Kemi *16*, 209 (1960)
69. Titov, V. A., Chusov, T. P., Kokovin, G. A.: Izv. Sib. otd. AN SSSR, Ser. Khim. *3* (7), 3 (1975)
70. Larsson, K.: Arkiv Kemi *16*, 215 (1960)
71. Larsson: ibid. *16*, 203 (1960)
72. Baidina, N. A. et al.: Zh. Strukt. Khim. *21*, 125 (1980)
73. Baidina, N. A. et al.: Ibid. *20*, 648 (1979)
74. Shklover, V. E. et al.: Ibid. *19*, 1107 (1978)
75. Hossain, M. A., Hursthouse, M. B., Malik, K. M. A.: Acta Crystallorg. *B35*, 2258 (1979)
76. Smith, A. L.: Spectrochim. Acta *19*, 849 (1963)
77. Lazarev, A. N.: Kolebat. spektry i stroyenie silikatov, L., 300 pp., Moscow: Nauka 1968
78. Brown, J. F., Scott, P. I.: J. Am. Chem. Soc. *86*, 1402 (1964)
79. Voronkov, M. G. et al.: Dokl. AN SSSR *249*, 106 (1979)
80. Kovrigin, V. M., Lavrent'yev, V. I., Voronkov, M. G.: 3rd International Symposium on Inorganic Ring Systems, Abstracts of Papers, Graz 1981, p. 86
81. Lavrent'yev, V. I., Kovrigin, V. M., Treer, G. G.: Ref. 49, p. 548
82. Kanev, A. N., Prokhorova, S. A., Kokovin, G. A.: Ref. 26, p. 243
83. Fedorov, V. E. et al.: Zh. Prikl. Khim. *53*, 1604 (1980)
84. Gimelshtein, F. Ya. et al.: Ref. 49, p. 569
85. Martynova, T. N., Nikonorov, Yu. N.: Ref. 26, p. 241
86. Wetters, J. H., Smith, R. C.: Anal. Chem. *41*, 379 (1969)
87. Katchmann, A.: US Patent 3162614 (1964), Ref. Zh. Khim. *1c*, 738, (1967)
88. Andrianov, K. A. et al.: Vysokomol. Soed. *7*, 1477 (1965)
89. Andrianov, K. A. et al.: Dokl. AN SSSR, *166* (4), 855 (1966)
90. Andrianov, K. A., Makarova, N. N.: Vysikomol. Soed. *12A*, 663 (1970)
91. Papkov, V. S. et al.: ibid. *17A*, 2700 (1975)
92. Papkov, V. S. et al.: ibid. *19A*, 2551 (1977)
93. Papkov, V. S. et al.: ibid. *22A*, 117 (1980)
94. Korchkov, V. P., Martynova, T. N., Semyannikov, P. P.: Ref. 21, p. 293
95. Basikhin, Yu. V.: Ref. 26, p. 302
96. Gribov, B. G. et al.: Elektron. promyshl., vyp. *9* (93), 113 (1980)
97. Yakovlev, D. A., Faintsimmer, R. Z. et al.: USSR Patent 295738, Bull. izobr., No 8 (1971)
98. Andrianov, K. A. et al.: Khim. Prom. *3*, 75 (1973)

Author Index Volumes 101–102

Contents of Vols. 50–100 see Vol. 100
Author and Subject Index Vols. 26–50 see Vol. 50

The volume numbers are printed in italics

A. F. Williams

A Theoretical Approach to Inorganic Chemistry

1979. 144 figures, 17 tables. XII, 316 pages
ISBN 3-540-09073-8

Contents: Quantum Mechanics and Atomic Theory. – Simple Molecular Orbital Theory. – Structural Applications of Molecular Orbital Theory. – Electronic Spectra and Magnetic Properties of Inorganic Compounds. – Alternative Methods and Concepts. – Mechanism and Reactivity. – Descriptive Chemistry. –Physical and Spectroscopic Methods. – Appendices. – Subject Index.

This book outlines the application of simple quantum mechanics to the study of inorganic chemistry, and shows its potential for systematizing and understanding the structure, physical properties, and reactivities of inorganic compounds. The considerable strides made in inorganic chemistry in recent years necessitate the establishment of a theoretical framework if the student is to acquire a sound knowledge of the subject. A wide range of topics is covered, and the reader is encouraged to look for further extensions of the theories discussed. The book emphasizes the importance of the critical application of theory and, although it is chiefly concerned with molecular orbital theory, other approaches are discussed. This text is intended for students in the latter half of their undergraduate studies. (235 references)

Springer-Verlag
Berlin
Heidelberg
New York